Electron and Photon Interactions with Bio(Related) Molecules

Electron and Photon Interactions with Bio(Related) Molecules

Editor

Filipe Ferreira da Silva

MDPI • Basel • Beijing • Wuhan • Barcelona • Belgrade • Manchester • Tokyo • Cluj • Tianjin

Editor
Filipe Ferreira da Silva
Physics
NOVA University of Lisbon
Lisbon
Portugal

Editorial Office
MDPI
St. Alban-Anlage 66
4052 Basel, Switzerland

This is a reprint of articles from the Special Issue published online in the open access journal *International Journal of Molecular Sciences* (ISSN 1422-0067) (available at: www.mdpi.com/journal/ijms/special_issues/Electron_Photon_Interactions).

For citation purposes, cite each article independently as indicated on the article page online and as indicated below:

LastName, A.A.; LastName, B.B.; LastName, C.C. Article Title. *Journal Name* **Year**, *Volume Number*, Page Range.

ISBN 978-3-0365-6579-8 (Hbk)
ISBN 978-3-0365-6578-1 (PDF)

Contents

About the Editor

Filipe Ferreira da Silva

Filipe Ferreira da Silva is Assistant professor with Habilitation at the NOVA University of Lisbon. In 2009, he was awarded a PhD degree in Physics from the Universitat Innsbruck. In 2022, he was awarded Habilitation in Atomic and Molecular Physics from the NOVA University of Lisbon. Currently, he is an integrated research member at the CEFITEC - Centre of Physics and Technological Research. His research group is focused on electron and photon interactions with isolated molecules in the gas phase.

Preface to "Electron and Photon Interactions with Bio(Related) Molecules"

Electron and photon interactions are of vital importance in biological processes. Medical applications, namely diagnostic and therapeutic processes, are also based on electron (low- and high-energy) and photon interactions with biological tissues. Apart from these applications, chemical and pharmaceutical applications centered on molecular manipulation and drug design have attracted special attention from scientists in terms of using electrons and photons as tools in chemical processing.

The aim of this Special Issue is to combine the most recent and outstanding studies on electron and photon interactions with bio(-related) molecules. Research on the association of gas, liquid, and condensed phases with quantum chemical and dynamical studies aims to present new insights and open new research pathways, bringing together science complementarities.

Filipe Ferreira da Silva
Editor

International Journal of *Molecular Sciences*

MDPI

Editorial

Electron and Photon Interactions with Bio (Related) Molecules

Filipe Ferreira da Silva

Centre of Physics and Technological Research (CEFITEC), Department of Physics, NOVA School of Science and Technology, Universidade NOVA de Lisboa, 2829-516 Caparica, Portugal; f.ferreiradasilva@fct.unl.pt

Part of the energy deposited in biological tissue by high-energy radiation is converted to secondary electrons. The knowledge at the molecular level on radiation interaction with biological species has increased due to the contributions of many different scientists working on radiation physics and radiation chemistry. The intention of this Special Issue on electron and photon interactions with (bio) molecules is to bring together different areas of knowledge that focus on radiation interactions with matter.

Photon interactions play an important role, since biological tissue is consistently exposed to photon radiation. It is known that high-energy radiation interactions can produce 4E5 electrons per MeV of incident radiation [1]. Moreover, the produced electrons have energy distribution values up to 100 eV, peaking at 20 eV [2]. As an example, it has been shown that 90% of the formed secondary electrons ejected from water upon high-energy radiation (20 keV electrons) have electron energy distribution values of below 20 eV [3].

In this Special Issue, emphasis is given to electron interactions with biological molecules and their related molecules, focusing on low-energy regimes for scattering and dissociation, as well as photon interactions with isolated molecules and enzymatic activity influenced by radiation.

Electron scattering cross-sections, both elastic and inelastic, are an important type of input information in Monte Carlo simulations for radiotherapy dose planning. Different target molecules are used to obtain electron scattering information. Elastic, integral and momentum electron scattering from the simplest alkane, methane, have been evaluated in the energy range of 50 to 300 eV [4]. The authors show that, at very low angles (<20°), molecular absorption effects play an important role in the cross-sections. Sevoflurane is a fluorinated alkane widely used as an anesthetic drug and contains fluor and oxygen in its formulation, in addition to carbon and hydrogen. As a simple molecule, it can be used as a model for radiation dose planning in radiotherapy protocols. It has been reported that absolute differential elastic electron scattering cross-sections have been experimentally and theoretically obtained for the first time, in the energy range of 50 to 300 eV and for 25° to 125° scattered angles [5].

Benzene is the most stable aromatic ring. Due to its pi-conjugated system, together with its simplicity, composed only of carbon and hydrogen (C_6H_6), benzene is used as a prototype molecule for many relevant biological molecules, such as DNA and RNA nucleobases. Double and triple differential electron cross-sections have been experimentally obtained through electron ionization of benzene parent cation formations at 90 eV electron energy [6]. Furthermore, in the energy range of 10 to 1000 eV, induced cationic fragmentation, together with electron detachment cross-sections in superoxide anion collisions, have been detailed for benzene [7].

High-energy radiation is widely used in medicine in diagnosis and therapy, as well as for other applications, such as sterilization. As a high number of secondary electrons are formed and these electrons will further interact with the environmental compounds, a vast number of reaction pathways may be available. Dissociative electron attachment is one of the possible reaction pathways, which is suitable for electron energy below the ionization

Citation: Ferreira da Silva, F. Electron and Photon Interactions with Bio (Related) Molecules. *Int. J. Mol. Sci.* **2022**, *23*, 15491. https://doi.org/10.3390/ijms232415491

Received: 30 November 2022
Accepted: 6 December 2022
Published: 7 December 2022

energy. Therefore, dissociative electron attachment plays an important role, acting as a secondary mechanism upon high-energy radiation interaction [8,9].

Dissociative electron attachment processes of fluorinated thiophenol derivatives are described in this Special Issue, showing the influence of fluorination on the dissociative electron attachment processes. Moreover, the formation of neutral HF molecules is proposed upon low-energy electron interaction with 2-fluorothiphenol and perfluorothiophenol molecules [10]. Oxaloacetic and citric acids are metabolites of aerobic organisms, produced in the Krebs cycle. These compounds have two carboxyl and one carbonyl groups, in addition to one hydroxyl and three carboxyl functional groups, respectively. The presence of oxygen enhances the electron-accepting properties of these metabolites. Dissociative electron attachment processes have been shown to act via attachment of the extra electron to a valence molecular orbital in the case of oxaloacetic acid and via electron capture in the dipole-bound state in the case of citric acid [11]. Radiosensitizers have been extensively experimentally and theoretically studied using electron interactions. Dissociative electron attachment of tirapazamide [12] and cis-$Pt(CO)_2Br_2$ [13], as well as electron ionization in the latter case, have been described. Tirapazamide dissociates via H_2O loss, leading to the loss of stability of the tirazine ring, followed by a molecular degradation process. The platinum compound shows preferential dissociation of the CO radical groups, in contrast with the dissociative electron attachment to cis-$Pt(NH_3)_2Cl_2$ [14]. The efficiency of radiosensitization of bromodeoxyuridine has been demonstrated under hypoxia conditions through the cell response to ionizing radiation [15].

UV spectroscopy has been used to fully characterize the electronic excited states of bromopyrimidine isomers by absolute photoabsorption cross-section determination [16]. The effect of the pigment pheomelanin on UVB radiation was shown in the oxidation and nitration of tyrosine processes [17]. Moreover, enzyme activity can be controlled by treatment with coherent humid air, leading to heat inactivation. This process is activated by IR radiation of humid air with a 1270 nm wavelength light [18].

Electronic circular dichroism spectroscopy was used to confirm the formation of the ribonucleoprotein complex and to determine its individual constituents [19]. The results show that the RNP complex faces some structural rearrangements in the Cas9 protein. Brillouin spectroscopy was employed to determine the elastic properties of taurine crystals [20]. Identification and temperature dependence from 175 °C to −170 °C of the longitudinal and transverse acoustic modes have been shown. This study describes for the first time the temperature dependences of the sound velocities, acoustic absorption coefficients and elastic constants for taurine crystals.

The present Special Issue offers complementary studies on radiation induced by photons and electrons in biological (and related) target molecules. The present collection provides important inputs on radiation physics and chemistry and on the advance of knowledge in these areas.

Funding: This research received no external funding.

Conflicts of Interest: The author declares no conflict of interest.

References

1. Paul, H.; Berger, M. *Atomic and Molecular Data for Radiotherapy and Radiation Research;* IAEA-TECDOC-799; International Atomic Energy Agency: Vienna, Austria, 1995; pp. 415–545.
2. Pimblott, S.M.; LaVerne, J.A.; Mozumder, A.; Green, N.J.B. Structure of electron tracks in water. 1. Distribution of energy deposition events. *J. Chem. Phys.* **1990**, *1*, 488. [CrossRef]
3. Cobut, V.; Frongillo, Y.; Patau, J.P.; Goulet, T.; Fraser, M.J.; Jay-Gerin, J.P. Monte Carlo simulation of fast electron and proton tracks in liquid water—I. Physical and physicochemical aspects. *Radiat. Phys. Chem.* **1998**, *51*, 229–243. [CrossRef]
4. Vukalović, J.; Maljković, J.B.; Tőkési, K.; Predojević, B.; Marinković, B.P. Elastic electron scattering from methane molecule in the energy range from 50–300 ev. *Int. J. Mol. Sci.* **2021**, *22*, 647. [CrossRef] [PubMed]
5. Vukalović, J.; Maljković, J.B.; Blanco, F.; García, G.; Predojević, B.; Marinković, B.P. Absolute differential cross-sections for elastic electron scattering from sevoflurane molecule in the energy range from 50–300 ev. *Int. J. Mol. Sci.* **2022**, *23*, 10021. [CrossRef] [PubMed]

6. Lozano, A.I.; Costa, F.; Ren, X.; Dorn, A.; Álvarez, L.; Blanco, F.; Limão-Vieira, P.; García, G. Double and triple differential cross sections for single ionization of benzene by electron impact. *Int. J. Mol. Sci.* **2021**, *22*, 4601. [CrossRef] [PubMed]
7. Guerra, C.; Kumar, S.; Aguilar-Galindo, F.; Díaz-Tendero, S.; Lozano, A.I.; Mendes, M.; Oller, J.C.; Limão-Vieira, P.; García, G. Total Electron Detachment and Induced Cationic Fragmentation Cross Sections for Superoxide Anion (O_2^-) Collisions with Benzene (C_6H_6) Molecules. *Int. J. Mol. Sci.* **2022**, *23*, 1266. [CrossRef] [PubMed]
8. Alizadeh, E.; Chakraborty, D.; Ptasinska, S. Low-Energy Electron Generation for Biomolecular Damage Inquiry: Instrumentation and Methods. *Biophysica* **2022**, *2*, 475–497. [CrossRef]
9. Ptasinska, S. A missing puzzle in dissociative electron attachment to biomolecules: The detection of radicals. *Atoms* **2021**, *9*, 77. [CrossRef]
10. Cipriani, M.; Ingólfsson, O. HF Formation through Dissociative Electron Attachment—A Combined Experimental and Theoretical Study on Pentafluorothiophenol and 2-Fluorothiophenol. *Int. J. Mol. Sci.* **2022**, *23*, 2430. [CrossRef] [PubMed]
11. Kopyra, J.; Wierzbicka, P.; Tulwin, A.; Thiam, G.; Bald, I.; Rabilloud, F.; Abdoul-Carime, H. Experimental and theoretical studies of dissociative electron attachment to metabolites oxaloacetic and citric acids. *Int. J. Mol. Sci.* **2021**, *22*, 7676. [CrossRef] [PubMed]
12. Arthur-Baidoo, E.; Ameixa, J.; Ončák, M.; Denifl, S. Ring-selective fragmentation in the tirapazamine molecule upon low-energy electron attachment. *Int. J. Mol. Sci.* **2021**, *22*, 3159. [CrossRef] [PubMed]
13. Cipriani, M.; Svavarsson, S.; Ferreira da Silva, F.; Lu, H.; McElwee-White, L.; Ingólfsson, O. The role of low-energy electron interactions in cis-pt(Co)2 br2 fragmentation. *Int. J. Mol. Sci.* **2021**, *22*, 8984. [CrossRef] [PubMed]
14. Kopyra, J.; Koenig-Lehmann, C.; Bald, I.; Illenberger, E. A single slow electron triggers the loss of both chlorine atoms from the anticancer drug cisplatin: Implications for chemoradiation therapy. *Angew. Chem.-Int. Ed.* **2009**, *48*, 7904–7907. [CrossRef] [PubMed]
15. Zdrowowicz, M.; Spisz, P.; Hać, A.; Herman-Antosiewicz, A.; Rak, J. Influence of Hypoxia on Radiosensitization of Cancer Cells by 5-Bromo-2′-deoxyuridine. *Int. J. Mol. Sci.* **2022**, *23*, 1429. [CrossRef] [PubMed]
16. Mendes, M.; Kossoski, F.; Lozano, A.I.; Pereira-da-silva, J.; Rodrigues, R.; Ameixa, J.; Jones, N.C.; Hoffmann, S.V.; Ferreira da Silva, F. Excited states of bromopyrimidines probed by vuv photoabsorption spectroscopy and theoretical calculations. *Int. J. Mol. Sci.* **2021**, *22*, 6460. [CrossRef] [PubMed]
17. Mariano, A.; Bigioni, I.; D'abusco, A.S.; Conrado, A.B.; Maina, S.; Francioso, A.; Mosca, L.; Fontana, M. Pheomelanin effect on UVB radiation-induced oxidation/nitration of L-tyrosine. *Int. J. Mol. Sci.* **2022**, *23*, 267. [CrossRef] [PubMed]
18. Yablonskaya, O.I.; Voeikov, V.L.; Novikov, K.N.; Buravleva, E.V.; Menshov, V.A.; Trofimov, A.V. Effect of Humid Air Exposed to IR Radiation on Enzyme Activity. *Int. J. Mol. Sci.* **2022**, *23*, 601. [CrossRef] [PubMed]
19. Halat, M.; Klimek-Chodacka, M.; Orleanska, J.; Baranska, M.; Baranski, R. Electronic circular dichroism of the cas9 protein and grna:Cas9 ribonucleoprotein complex. *Int. J. Mol. Sci.* **2021**, *22*, 2937. [CrossRef] [PubMed]
20. Kang, D.H.; Oh, S.H.; Ko, J.H.; Lee, K.S.; Kojima, S. Elastic properties of taurine single crystals studied by brillouin spectroscopy. *Int. J. Mol. Sci.* **2021**, *22*, 7116. [CrossRef] [PubMed]

International Journal of *Molecular Sciences*

MDPI

Article

The Role of Low-Energy Electron Interactions in cis-$Pt(CO)_2Br_2$ Fragmentation

Maicol Cipriani [1], Styrmir Svavarsson [1], Filipe Ferreira da Silva [2], Hang Lu [3], Lisa McElwee-White [3] and Oddur Ingólfsson [1,*]

1 Department of Chemistry and Science Institute, University of Iceland, Dunhagi 3, 107 Reykjavik, Iceland; mac31@hi.is (M.C.); styrmir94@gmail.com (S.S.)
2 CEFITEC, Departamento de Física, Faculdade de Ciências e Tecnologia, Universidade NOVA de Lisboa, 2829-516 Caparica, Portugal; fr.silva@fct.unl.pt
3 Department of Chemistry, University of Florida, Gainesville, FL 32611-7200, USA; hlu90@ufl.edu (H.L.); lmwhite@chem.ufl.edu (L.M.-W.)
* Correspondence: odduring@hi.is

Abstract: Platinum coordination complexes have found wide applications as chemotherapeutic anticancer drugs in synchronous combination with radiation (chemoradiation) as well as precursors in focused electron beam induced deposition (FEBID) for nano-scale fabrication. In both applications, low-energy electrons (LEE) play an important role with regard to the fragmentation pathways. In the former case, the high-energy radiation applied creates an abundance of reactive photo- and secondary electrons that determine the reaction paths of the respective radiation sensitizers. In the latter case, low-energy secondary electrons determine the deposition chemistry. In this contribution, we present a combined experimental and theoretical study on the role of LEE interactions in the fragmentation of the Pt(II) coordination compound cis-$PtBr_2(CO)_2$. We discuss our results in conjunction with the widely used cancer therapeutic Pt(II) coordination compound cis-$Pt(NH_3)_2Cl_2$ (cisplatin) and the carbonyl analog $Pt(CO)_2Cl_2$, and we show that efficient CO loss through dissociative electron attachment dominates the reactivity of these carbonyl complexes with low-energy electrons, while halogen loss through DEA dominates the reactivity of cis-$Pt(NH_3)_2Cl_2$.

Keywords: anticancer drugs; cisplatin; platinum (II) halo-carbonyl complexes; low-energy electrons; dissociative electron attachment; dissociative ionization; focused electron beam induced deposition

Citation: Cipriani, M.; Svavarsson, S.; Ferreira da Silva, F.; Lu, H.; McElwee-White, L.; Ingólfsson, O. The Role of Low-Energy Electron Interactions in cis-$Pt(CO)_2Br_2$ Fragmentation. *Int. J. Mol. Sci.* **2021**, 22, 8984. https://doi.org/10.3390/ijms22168984

Academic Editor: Mauro Coluccia

Received: 26 July 2021
Accepted: 16 August 2021
Published: 20 August 2021

1. Introduction

Platinum coordination complexes, such as cisplatin [$Pt(NH_3)_2Cl_2$], have been widely used as chemotherapeutic anticancer drugs since the 1970s [1,2]. Cisplatin-based chemotherapy has proven to be highly effective against testicular cancer [3,4] and other various types of malignancies, such as metastatic melanoma, carcinoma of the head and neck and metastatic breast cancer [5]. The cytotoxicity of this platinum complex lies in its ability to form CDDP–DNA adducts inhibiting replication [1,2]. In fact, after entering the cell, the cisplatin undergoes hydrolysis, and as a result, the two chlorine atoms are lost. The remaining fragment forms the CDDP–DNA adducts by binding to guanine or purine nucleobases that inhibit transcriptions of the cancer cell, blocking its replication. The synchronous combination of platinum complexes and radiation (chemoradiation) has increased the survival probability of cancer patients due to the enhancement of the cell-killing effect of radiation [6–8]. It has been hypothesized that cisplatin, in addition to acting directly as a chemotherapeutic agent, also acts as a radiosensitizer. Apart from the medical/biological applications, platinum-based complexes are also used as precursors for focused electron beam-induced deposition (FEBID) [9–13], an electron-induced materials deposition technique for the fabrication of functional nanostructures [14]. In both applications, low-energy electrons (LEEs) play an important role in dictating the fragmentation pathways through electron capture as well as by electron ionization. In chemoradiation, these low-energy electrons are produced in the interaction of the ionizing radiation with the tissue material and in FEBID by interaction of the high energy electron beam with the substrate surface and the depositing material [15,16]. In both cases, the energy distribution of these secondary electrons peaks at or below 10 eV, and has a contribution close to 0 eV and a long tail extending to higher energies [17]. These LEEs can inflict considerable damage on the DNA [18], causing single and double strand breaks (SSB and DSB) [19]. In a 2008 study by Zheng et al. [7], the authors showed that when cisplatin is covalently bonded to DNA, SSB and DSB induced by LEEs are substantially enhanced. This enhancement has been attributed to bond cleavage triggered by the formation of transient negative ions (TNI) through electron capture, i.e., dissociative electron attachment (DEA). Dissociative electron attachment studies on cisplatin have been performed by Kopyra et al. [20], determining the fragmentation pathways under interaction with low-energy electrons. In this study, it was shown that electrons close to 0 eV can easily fragment this molecule by cleavage of the Pt−Cl bonds, leading to the loss of one or even both Cl atoms with considerable intensity. Hence, one single low-energy electron efficiently triggers the cleavage of both the Pt−Cl bonds. In the FEBID process, gas phase precursors, usually organometallics, are introduced into a high vacuum (HV) chamber in close proximity to a substrate surface where they are subjected to a tightly focused high-energy electron-beam. The high energy electron beam generates a considerable number of low-energy SEs [21,22] that interact with the precursor molecules, initiating chemical reactions through DEA, dissociative ionization (DI), neutral dissociation (ND) and dipolar dissociation (DD). These processes determine the decomposition of the precursor molecule at the substrate surface leading to deposition of the nonvolatile fragments formed, while volatile fragments are pumped away. A description of these processes is given, e.g., in References [23,24]. Ideally, for the creation of metallic deposits, the organometallic precursors should fully dissociate, leaving the metallic atom on the surface while the ligands are pumped away. In recent years, several studies on LEE interactions with organometallic platinum compounds have been carried out in the context of FEBID [25–28], studies that are also relevant to the action of radiation sensitizers where low-energy electrons are expected to play a significant role.

In low-energy electron interactions with $Pt(CO)_2Cl_2$ in the gas phase under single collision conditions, Ferreira da Silva et al. [28] found that, in contrast to cisplatin [20], DEA close to 0 eV electron incident energies leads exclusively to CO loss. This channel is very efficient for both single and double CO loss, while Cl loss is inefficient and only observed at higher energies. Similarly, CO loss is also the main channel in dissociative ionization of this compound, though significant Cl loss is also observed and the bare Pt^+ ion is formed with appreciable intensity. In a UHV surface study where adsorbed monolayers of $Pt(CO)_2Cl_2$ (and $Pt(CO)_2Br_2$) were exposed to 500 eV electrons from a flood gun (and correspondingly the generated low-energy SEs), J. A. Spencer et al. [12] found that these compounds decomposed by rapid CO loss leading to a $PtCl_2$ deposit with the 1:2 stoichiometric ratio of the initial compound. Prolonged electron irradiation then led to nearly quantitative removal of the chlorine. The authors attributed this to an initial DEA step leading to the CO loss. In this context, a comparative electron beam deposition study of $Pt(CO)_2Cl_2$ and $Pt(CO)_2Br_2$ was recently conducted by A. Mahgoub et al. [29]. Interestingly, it was found that while both these compounds behave similarly, the UHV deposits contained a significant portion of the halogen species but little or no carbon, while the deposits created under HV contained only small amounts of halogen species but high carbon content. It is possible that the presence of trace water in the HV experiments leads to the formation of volatile HCl in the irradiation process, decreasing the chlorine content in the deposits. This has been observed by M. Rohdenburg et al. [13,30] for cisplatin in electron-induced intermolecular reactions of the chlorine with hydrogen from the amine ligand and in electron-induced reactions of $(\eta^3\text{-}C_3H_5)\,Ru(CO)_3Cl$ in the presence of NH_3 as processing gas. It is thus clear that the environment plays a critical role in electron-induced decomposition of these compounds, and this is especially true for biological media where water is plentiful.

In the current study, we have performed DEA and DI experiments on $Pt(CO)_2Br_2$ in the gas phase under single collision conditions as well as thermochemical calculations at the DFT and coupled-cluster level of theory for the respective processes. Relaxed potential energy surface scans were computed, and vertical electron attachment energies and the respective MOs were calculated along with the respective electronic excitation energies. For comparison, the vertical transition energies from the anionic ground state to the first excited anionic state are also presented for all the Pt(II) halogen carbonyls; $Pt(CO)_2X$ (X = F, Cl, Br and I). We compare our findings with previous work on electron-induced decomposition of $Pt(CO)_2Cl_2$ and of $Pt(CO)_2Br_2$ as well as cisplatin; $Pt(NH_3)_2Cl_2$.

2. Results and Discussion

Figure 1 shows the DEA ion yield curves for $Pt(CO)Br_2^-$ and $PtBr_2^-$, from $Pt(CO)_2Br_2$, i.e., the energy dependence of the loss of one and two CO ligands, respectively. To better allow comparison, the intensities are normalized with respect to the pressure and the intensity of SF_6^- formation from SF_6 at 0 eV recorded before each measurement. These are the two most efficient DEA processes and both fragments are formed with appreciable intensity close to 0 eV. However, while $[Pt(CO)Br_2]^-$ peaks at 0 eV, the maximum of the low-energy $PtBr_2^-$ contribution is at about 0.07 eV, and both contributions are broad and asymmetric towards higher energies. The loss of both CO ligands, and the formation of $PtBr_2^-$, is also observed through a higher-lying resonance (or resonances), contributing to the ion yield close to 3 eV. This contribution is not observed in the single CO loss ion yield curve. This is similar to the previous observations for $Pt(CO)_2Cl_2$ [28], where DEA contributions through higher lying resonances were observed in the $[PtCl_2]^-$ but not in the $[Pt(CO)Cl_2]^-$ ion yield curves. The single CO loss from $Pt(CO)_2Cl_2$ was found to be exothermic, and it was suggested that the excess energy in the $[Pt(CO)Cl_2]^-$ fragment (i.e., after the first CO loss) makes its survival probability low at the onset of the high energy resonance (or resonances), and further decomposition to $[PtCl_2]^-$ is the predominant process. The losses of one and two CO ligands from $Pt(CO)_2Br_2$ through DEA are also found to be exothermic, and at the DLPNO-CCSD(T) level of theory, we find the threshold for the loss of one and two CO ligands to be −1.57 and −0.48 eV, respectively.

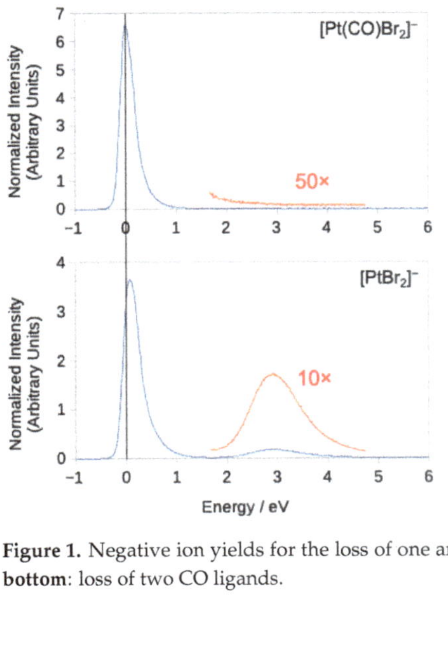

Figure 1. Negative ion yields for the loss of one and two CO ligands. **Top**: loss of one CO ligand; **bottom**: loss of two CO ligands.

This is also clear from the cut through the relative potential energy surfaces (PESs) shown in Figure 2. These are calculated through relaxed energy scans at the wB97X-D3/ma-def2-TZVP level of theory along the OC−Pt(CO)Br_2 and OC−PtBr_2 dissociation coordinates, respectively. For $[Pt(CO)Br_2]^-$, the energy contribution of the CO ligand ($\varepsilon_{(CO)}$) is included in the calculations. The single-point energies obtained from the relaxed energy scans were fitted with Morse potential energy function. From the fitting, we obtained the Pt−CO dissociation energies from the minimum of the potential curve (D_e) and the Pt−CO bond lengths (R_e) for the neutral molecule ($Pt(CO)_2Br_2$), the molecular anion ($[Pt(CO)_2Br_2]^-$), and the $[Pt(CO)Br_2]^-$ fragment. It is noted that these PESs do not include the zero-point vibrational energy (ZPVE). For the neutral molecule, we derived a dissociation energy of 1.5 eV and an equilibrium Pt–CO bond length of 1.9 Å. For the molecular anion, these values were found to be 0.2 eV and 1.9 Å, respectively. The Pt−CO bond length in $[Pt(CO)Br_2]^-$ was found to be 1.8 Å and the dissociation energy was found to be 1.4 eV. The corresponding Pt−CO bond lengths from our geometry optimization at the wB97X-D3/ma-def2-TZVP level of theory are 1.9 Å for $Pt(CO)_2Br_2$, $[Pt(CO)_2Br_2]^-$ and $[Pt(CO)Br_2]^-$, which agrees with these derived from the PES fits.

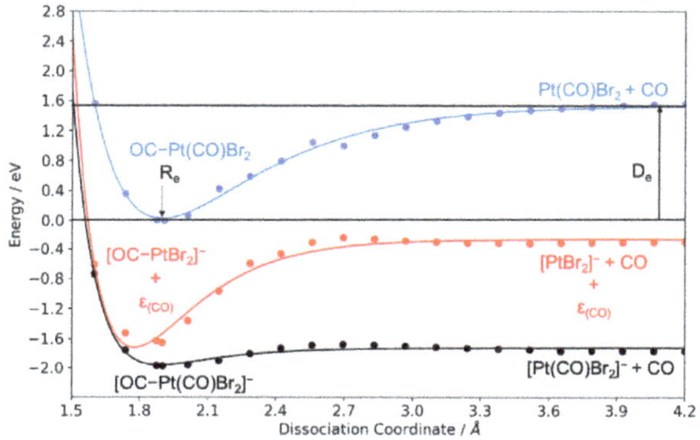

Figure 2. Relaxed potential energy surface scans for OC−Pt(CO)Br_2 dissociation from the neutral parent and OC−Pt(CO)Br_2 and OC−PtBr_2 dissociations from the respective anion. The calculations were performed at wB97X-D3/ma-def2-TZVP level of theory.

As is clear from Figure 2, the PESs for the formation of the anionic fragment $[Pt(CO)Br_2]^-$ and $[PtBr_2]^-$ lie entirely below the ground state of the neutral molecule in the range of the dissociative coordinate 1.4 to 4.2 Å. Thus, the survival probability of $[Pt(CO)Br_2]^-$, with respect to further CO loss, drops rapidly above the threshold. This, in turn, is reflected in the shift and broadening of the low-energy contribution for $[Pt(CO)Br_2]^-$ as compared to $PtBr_2^-$ and the lack of any $[Pt(CO)Br_2]^-$ contribution through the higher lying resonance at around 3 eV.

Similar to $Pt(CO)_2Cl_2$ [28], we attribute the low-energy contributions in the $Pt(CO)_2Br_2$ ion yields to the initial formation of the ground state negative ion and the first excited negative ion state, i.e., electron occupation of the LUMO and the slightly higher lying LUMO+1. Figure 3 shows the electrostatic potential isosurfaces for the corresponding SOMO and SOMO+1. The former of these is a mixture of contributions from the π^* CO orbitals and the Pt 5 d_{xz} p and has antibonding character. The latter is predominantly composed of the Pt $d_{x^2-y^2}$ orbital with σ^* P−L antibonding character (L = CO or Br) and direct CO loss from this excited anion state is given by its repulsive σ^* nature. The same process from the ground anion state, however, is in principle symmetry forbidden and requires effective coupling of the CO π^* orbital with the respective σ^* Pt−L. For $Pt(CO)_2Cl_2$ [28], it has been hypothesized that such effective coupling is provided by the out-of-plane bending of the CO group.

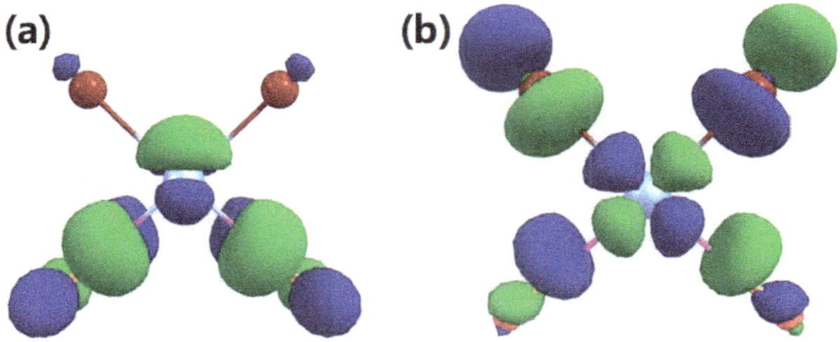

Figure 3. Isosurfaces of the frontier orbitals of the $Pt(CO)_2Br_2$ anion obtained with a contour value of 0.05. (**a**): SOMO. (**b**): SOMO+1.

From our calculation, we derive a dipole moment for $Pt(CO)_2Br_2$ of 5.0 D, which should be well above the limit for a capture into a dipole bound state [31]. For the lower lying (0 eV) resonance, we anticipate that this provides a gateway for the DEA process as has been discussed by, e.g., Sommerfeld for nitromethane, uracil and cyanoacetylene [32]. In this mechanism, the initially formed diffuse dipole-bound state couples with the nuclear motion, channeling the excess energy into the vibrational degrees of freedom. The so formed vibrational Feshbach resonance couples with the respective valence state, in this case the LUMO, leading to a transient negative ion characterized by the excess electron defining the SOMO. At higher energies, where the angular momentum of the electron has *l* components higher than zero, the initial capture may rather be through the formation of the respective shape resonance. A detailed discussion on the actual capture mechanism at these low incident energies exceeds the scope of this paper, but we refer the interested reader to Reference [24] and references therein.

The calculated adiabatic electron affinity of $Pt(CO)_2Br_2$ was found to be 2.31 eV at the ωB97X-D3 level of theory and 2.11 eV at the DLPNO-CCSD(T)/aug- cc-pVQZ level of theory. The vertical attachment energy (VAE) for the anion ground state was found to be −1.14 eV at the ωB97X-D3 level of theory and −0.99 eV for the first excited anion state using a delta-SCF approach at the same level of theory. Similar to $PtCl_2(CO)_2$, the VAEs associated with formation of the ground state and first excited state anions are both negative at the equilibrium geometry of the neutral molecule. The resulting vertical excitation energy for $PtBr_2(CO)_2$ from the anion ground state to the first anionic excited state was found to be 0.15 eV, while for $PtCl_2(CO)_2$, the value was found to be 0.38 eV with delta-SCF at the same level of theory [28]. This reflects the increased destabilization of the σ* SOMO+1 with increasing electronegativity of the halogen ligand and suggests that substitution of the Br ligands with I atoms would further decrease the excitation energy while substitution with F atoms would largely increase the excitation energy. This is confirmed by the vertical transition energies between these anion states, obtained at TDDFT wB97X-D3/ma-def2-TZVP level of theory, shown in Table 1. As can be seen in Table 1, the vertical excitation energy from the anionic ground state to the first anionic excited state decreases following the trend $[Pt(CO)_2F_2]^-$ > $[Pt(CO)_2Cl_2]^-$ > $[Pt(CO)_2Br_2]^-$ > $[Pt(CO)_2I_2]^-$.

Table 1. Calculated vertical excitation energy from the anionic ground state to the first anionic excited at the TDDFT wB97X-D3/ma-def2-TZVP level of theory.

Anions	Vertical Excitation SOMO+1 ← SOMO (eV)
$[Pt(CO)_2F_2]^-$	0.58
$[Pt(CO)_2Cl_2]^-$	0.44
$[Pt(CO)_2Br_2]^-$	0.27
$[Pt(CO)_2I_2]^-$	0.13

This is also reflected in the more structured low-energy contributions from $Pt(CO)_2Cl_2$ as compared to the high energy tail observed in $Pt(CO)_2Br_2$. To visualize this, Figure 4 shows a fit of two Gaussian functions to the low-energy contributions in the ion yield curves for $[Pt(CO)Br_2]^-$ and $PtBr_2^-$. An excellent fit with an R^2 value of 0.99 is achieved with a fairly narrow lower energy contribution peaking at about 0 eV electron energy and a broader higher energy component peaking at 0.18 and 0.33 eV, respectively. These values for the higher energy contributions are in both cases slightly below the corresponding VAE as is to be expected due to the intrinsic competition between autodetachment and dissociation. The lower value for $[Pt(CO)Br_2]^-$ as compared to $[PtBr_2]^-$ is also in line with the expected energy dependence of the survival probability of that fragment with respect to further CO loss to form $[PtBr_2]^-$.

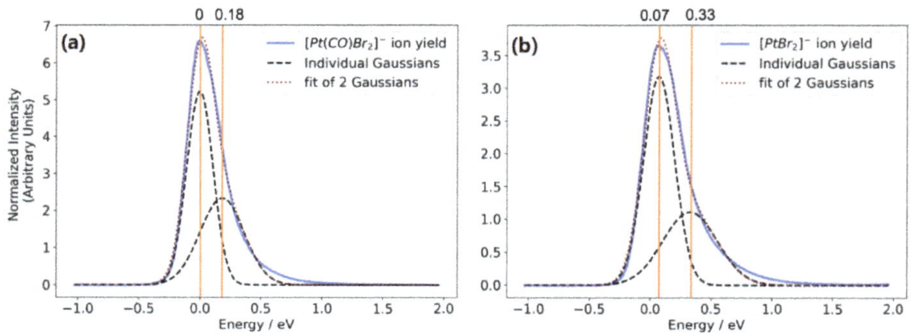

Figure 4. Fit of two Gaussian functions to the $[Pt(CO)Br_2]^-$ ion yield (**a**) and the $[PtBr_2]^-$ ion yield (**b**). The first represents contributions from dissociation from the anionic ground state, while the latter represents these from the first excited anionic state.

In addition to the CO loss fragments, DEA to $Pt(CO)_2Br_2$ also leads to the formation of $[Pt(CO)Br]^-$, $[PtBr]^-$ and Br^-, though with considerably lower intensity. The ion yield traces for these are shown in Figure 5, and Table 2 lists the threshold values for all fragments observed from $Pt(CO)_2Br_2$, calculated at the wB97X-D3/ma-def2-TZVP and DLPNO-CCSD(T)/aug-cc-pVQZ level of theory. For comparison, the onsets of individual contributions estimated from the ion yield curves, i.e., the appearance energies (AEs), are also shown in Table 2. With the exception of the formation of $[PtBr]^-$, all DEA channels observed from $Pt(CO)_2Br_2$ are found to be exothermic. Similar to $[PtBr_2]^-$, the ion yield curve for $[Pt(CO)Br]^-$ shows two contributions, one that peaks close to 0.5 eV and one with considerably higher intensity peaking close to 3 eV. We attribute the former of these to dissociation from the first excited anionic state, though contributions from the high energy tail from the ground state transient negative ion (TNI) cannot be excluded. The 3 eV contribution is shifted to slightly higher energies as compared to the double CO loss, which is likely rooted in the competition between these channels falling in favor of the more exothermic double CO loss at lower energies. The situation is very similar for the Br^- formation, which also appears through two contributions peaking at around 0.15 and 3.4 eV, respectively, and here we also attribute the low-energy contribution to the σ^*, first anionic excited state. Interestingly we do not observe any $[Pt(CO)_2Br]^-$ contributions in DEA to $Pt(CO)_2Br_2$, suggesting that $[Pt(CO)Br]^-$ is formed through initial CO loss, i.e., through Br loss from $[Pt(CO)Br_2]^-$. This may be rooted in the synergistic back-bonding Br−Pt−CO interaction through the contribution of Br lone pair electron density through the metal d orbitals into the π^*(CO). If the CO is lost first, then the partial extra bond from Br−Pt is gone, and the Br that was trans to the now missing CO may be more disposed to dissociation.

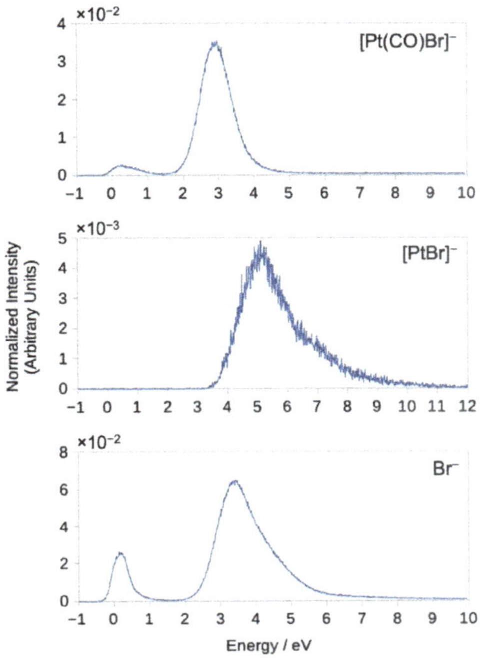

Figure 5. Negative ion yields for $[Pt(CO)Br]^-$, $[PtBr]^-$ and Br^-.

Table 2. Appearance energies and calculated thermochemical thresholds for all observed DEA fragments calculated at the wB97X-D3/ma-def2-TZVP and DLPNO-CCSD(T)/aug-cc-pVQZ level of theory. Threshold energies include the thermal energy of the neutral at 50 °C.

Fragments	AEs (eV)	Threshold Energy wB97X-D3/ma-def2-TZVP (eV)	Threshold Energy DLPNO-CCSD(T)/aug-cc-pVQZ (eV)
$[Pt(CO)Br_2]^-$	0.0	−2.06	−1.57
$[Pt(CO)Br]^-$	0.0–2.0	−0.20	−0.017
$[PtBr_2]^-$	0.0	−0.70	−0.48
$[PtBr]^-$	3.8	3.56	3.80
$[Br]^-$	0.0–2.3	−0.93	−0.59

Finally, $[PtBr]^-$ is formed with an onset close to its threshold at 3.8 eV and a maximum at about 5 eV. This fragment, which is formed through the loss of three ligands, appears with low intensity, and we anticipate that it is formed through the high energy tail of the resonance (or resonances), contributing to the $[PtBr_2]^-$, $[Pt(CO)Br]^-$ and Br^- formation at around 3 eV.

At the wB97X-D3/ma-def2-TZVP level of theory, taking into account the ZPVE and the thermal energy corrections at room temperature, we find the halogen−Pt bond energies for $Pt(CO)_2Cl_2$ and $Pt(CO)_2Br_2$ to be 3.3 and 2.8 eV, respectively. The calculated threshold for the formation of $[Pt(CO)Br]^-$ was found to be −0.017 eV at the DLPNO-CCSD(T) level of theory, while the threshold for $[Pt(CO)Cl]^-$ was found to be 0.19 eV at the same level of theory. Similarly, the calculated threshold for the formation of Br^- was found to be −0.59 eV at the DLPNO-CCSD(T) level of theory, while the calculated threshold for Cl^- formation from $Pt(CO)_2Cl_2$ was found to be −0.51 eV at the same level of theory. It is noted that the $[Pt(CO)Br]^-$ calculations include the thermal energy of the neutral at 50 °C, while those for $Pt(CO)_2Cl_2$ include the thermal energy of the neutral at 85 °C.

As compared to $Pt(CO)_2Cl_2$ and $Pt(CO)_2Br_2$, DEA to cisplatin $[Pt(NH_3)_2Cl_2]$ shows a very different behavior. Here, the dominant DEA channels are the formation of $[Pt(NH_3)_2Cl]^-$, Cl^- and $[Pt(NH_3)_2]^-$ [20], i.e., the cleavage of the Pt–halogen bonds. This is interesting, as similar to the Pt–CO bond, the Pt–NH_3 bond is much weaker than the Pt–Cl bond, and in their study, Kopyra et al. [20] calculated the respective bond energies to be 1.5 eV for the Pt-NH_3 bond and 3.3 eV for the Pt–Cl bond using DFT. The LUMO, involved in LEE attachment to cisplatin, however, has a repulsive σ^* character in all the ligands [20], which is more similar to the LUMO+1 of the current compound, through which effective Pt–Br bond cleavage is observed. Thus, at very low energies, where the electron attachment cross sections are the highest, direct relaxation of $[Pt(NH_3)_2Cl_2]^-$ by lengthening of the Pt–Cl bond beyond its bonding distance is allowed while this process is in principle symmetry forbidden from the π^*_{CO} character LUMO of $[Pt(CO)_2Cl]^-$ and $[Pt(CO)_2Br]^-$.

Dissociative ionization of $Pt(CO)_2Br_2$ leads to much more extensive fragmentation than DEA. Figure 6 shows the positive ion DI spectrum of $Pt(CO)_2Br_2$ recorded at 70 eV electron energy, and Table 3 lists the relative contributions of individual fragments normalized to the contribution of the parent ion as well as the efficiency of CO and Br removal per incident electron. The DI spectrum shows all the fragments associated with the breakdown of the $Pt(CO)_2Br_2$. As for $Pt(CO)_2Cl_2$, the dominant contribution is from the parent ion $[Pt(CO)_2Br_2]^+$, and the bare Pt^+ ion is observed with significant intensity. The loss of one and two CO, i.e., the formation of $[Pt(CO)Br_2]^+$ and $[PtBr_2]^+$, is also appreciable, while the loss of one Br and two Br, i.e., $[Pt(CO)_2Br]^+$ and $[Pt(CO)_2]^+$, is much less significant. The formation of $[PtBr]^+$ is appreciable, while other fragments appear with marginal intensity.

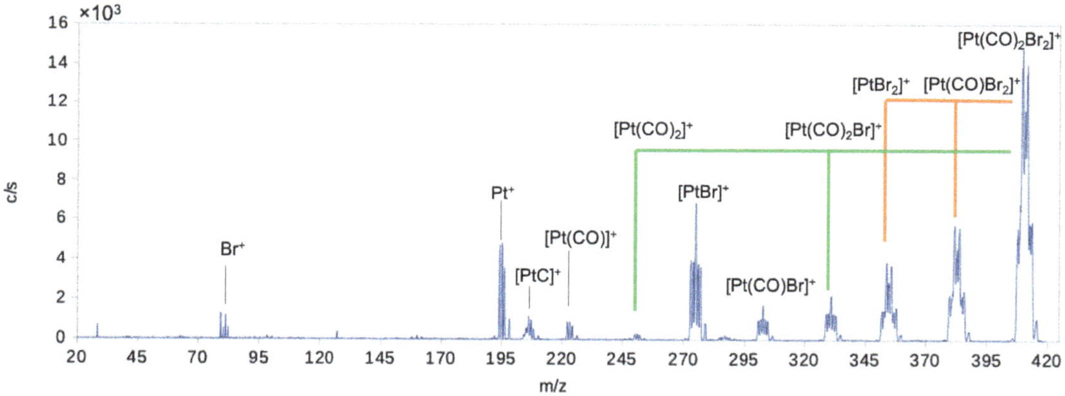

Figure 6. Positive ion mass spectrum of $Pt(CO)_2Br_2$ recorded at 70 eV incident electron energy. Green lines show the sequential loss of the two Br ligands, while orange lines show the sequential loss of the two CO ligands.

Finally, for comparison with the electron-induced decomposition of $Pt(CO)_2Br_2$ in FEBID and at surfaces, we have calculated the average bromine and carbonyl loss per incident in DEA and DI (see Table 3). The average CO loss per incident in DEA was calculated by taking the integral intensities of all ion yield curves of CO-loss fragments from about 0 to 10 eV (see Figure 1) and weighing these by the number of carbonyls lost. In a similar way, the average Br loss through DEA was obtained by weighing the integral intensities of the $[Pt(CO)Br]^-$, $[PtBr]^-$ and Br^- ion yield curves by the number by bromines lost. For DI, the average CO and Br losses per incident were obtained in a similar way by integrating over the isotope distribution of the respective fragment peaks in the positive ion mass spectrum and weighing these by the number of CO and Br lost in the respective processes. In DEA we found an average CO loss of 1.4 per incident through DEA, and an average Br loss per incident of only 0.04. For DI, we found an average CO loss of 0.7 per incident and an average Br loss of 0.3. This is comparable to the observations made for $Pt(CO)_2Cl_2$ [28], where the average CO loss in DEA is also found to be 1.4 and the Br loss negligible, while DI leads to an average of 0.6 CO and 0.5 Br lost per incident, respectively. In electron-induced decomposition of $Pt(CO)_2Cl_2$ at surfaces [12], CO loss was found to be the dominating process at low electron doses, leading to an average CO loss of around 1–2. Similarly, in UHV electron-induced deposition experiments, the deposits made with $Pt(CO)_2Br_2$ and $Pt(CO)_2Cl_2$ were found to retain the nearly 1:2 platinum:halogen ratio of the precursor compounds [29]. These surface and deposition studies suggest that DEA rather than DI is dominating in the initial decomposition step. However, as mentioned in the introduction, depositions made with $Pt(CO)_2Cl_2$ and $Pt(CO)_2Br_2$ under HV conditions were found to contain very little of the respective halogen species with carbon being the main contaminant [29].

Table 3. Relative yields of the fragments formed by DI and DEA to $Pt(CO)_2Br_2$, average weighted CO loss per DEA and DI incident and average weighted Br loss per DI incident.

Fragments	Relative DI Contributions	Relative DEA Contributions
$[Pt(CO)_2Br_2]^+$	1	-
$[Pt(CO)Br_2]^{+/-}$	0.38	1
$[PtBr_2]^{+/-}$	0.23	0.76
$[Pt(CO)_2Br]^+$	0.10	-
$[Pt(CO)Br]^{+/-}$	0.07	0.02
$[PtBr]^{+/-}$	0.25	0.004
$[Pt(CO)_2]^+$	0.01	-
$[Pt(CO)]^+$	0.02	-
$[PtC]^+$	0.04	-
Pt^+	0.11	-
$Br^{+/-}$	0.01	0.05
Average CO loss	0.7	1.4
Average Br loss	0.3	0.04

For electron irradiation of cisplatin at surfaces [13], it has been shown that intermolecular reactions of the chlorine with hydrogen from the amino ligands readily produce HCl that desorbs and effectively reduces the Cl content in the adsorbate. Similarly, the use of NH_3 as a processing gas in electron-induced decomposition of $(\eta^3\text{-}C_3H_5)Ru(CO)_3Cl$ at surfaces has proven effective in Cl removal through HCl formation [30]. We speculate that residual water may have the same effect in FEBID of $Pt(CO)_2Cl_2$ and $Pt(CO)_2Cl_2$ under HV conditions and thus explain the very different observations under HV and UHV conditions. This is important when considering electron-induced decomposition in a biological environment, where water is omnipresent.

3. Materials and Methods

3.1. Experimental Setup

Low-energy electron interactions with $Pt(CO)_2Br_2$ were studied in a crossed electron-molecule beam apparatus. The experimental setup has been covered previously [33], and only a short description will be given here. The instrument consists of a quadrupole mass spectrometer (HIDEN EPIC1000), a trochoidal electron monochromator (TEM) and an effusive gas inlet system. The TEM and the ion extraction elements are maintained at 120 °C during measurements to avoid target gas condensation on the electrical lens components. A quasi mono-energetic electron beam, generated with the TEM, crosses an effused molecular beam of the target gas. The ions formed in the crossed beam region are then extracted from the reaction region by a weak electric field (<1 V/cm) and analyzed by the mass spectrometer. Both positive and negative ions can be studied, and ion yield curves are recorded by scanning through the electron energy at a fixed mass (m/z), while mass spectra are recorded by scanning through the relevant m/z range at fixed energies.

The background pressure inside the chamber is on the order of 10^{-6} Pa and the pressure during measurements typically about 10^{-5} Pa to ensure single collision conditions. The energy scale was calibrated by the well documented SF_6^- formation from SF_6 at 0 eV [34] recorded before and after each measurement. The energy resolution was estimated from the FWHM of that signal and was found to be 100–150 meV for the current measurements. $PtBr_2(CO)_2$ is solid at room temperature and was sublimed at ~50 °C in the gas inlet system.

3.2. $PtBr_2(CO)_2$ Synthesis

$PtBr_2(CO)_2$ was synthesized at the University of Florida as previously reported [29] and characterized by comparison to literature data [35].

3.3. Computational Method

Ab initio calculations were performed with the quantum chemistry package ORCA 4.2.1 [36]. All geometries were first optimized using a range-separated hybrid functional ωB97X-D3 [37,38] with minimal augmented triple zeta basis set, ma-def2-TZVP [39,40] and the def2 effective core potential (ECP) [41] for platinum core electrons. Harmonic vibrational frequencies of the molecule and fragments were calculated at the same level of theory. They were confirmed to be positive, i.e., all structures were stationary points on the potential energy surface, and were used to yield zero-point energy contributions for the molecule and all the fragments as well as thermal energy correction for the neutral parent at 50 °C. Final threshold energies were calculated using the coupled cluster approach at the DLPNO-CCSD(T) [42–45] level of theory on the ωB97X-D3 optimized geometries. A large and diffuse aug-cc-pVQZ basis set [46–48] (aug-cc-pVQZ-PP basis set and associated pseudopotential for Pt) [49] was used. All thresholds include zero-point vibrational energies for all fragments and the thermal energy of the neutral molecule at 50 °C (ωB97X-D3/ma-def2-TZVP level of theory). Vertical excitation energies for the first excited anion states were calculated using time-dependent DFT (TDDFT) at the ωB97X-D3/ma-def2-TZVP level of theory.

4. Conclusions

Dissociative electron attachment to $Pt(CO)_2Br_2$ in the energy range from about 0 to 12 eV and dissociative ionization at 70 eV were studied in a crossed electron-molecule beam experiment. The thermochemical thresholds for all DEA channels were calculated, and relaxed potential energy surface scans were computed for the main channels. The vertical electron attachment energies and the respective MOs were calculated for the lowest lying anionic states as well as the vertical transition energy from the anionic ground state to the first excited anionic state for the Pt(II) halogen carbonyls $Pt(CO)_2X$ (X = F, Cl, Br and I). In DI at 70 eV, the main contributions are from the parent ion $[Pt(CO)_2Br_2]^+$, but the loss of one and two CO, $[Pt(CO)Br_2]^+$ and $[PtBr_2]^+$, and the formation of $PtBr^+$ and the bare Pt^+ ion are also significant. The loss of one and two Br, $[Pt(CO)_2Br]^+$ and $[Pt(CO)_2]^+$, the formation of $[Pt(CO)Br]^+$, and the platinum carbide PtC^+ and Br^+ are also apparent, though with lesser intensities. All DEA channels, except the formation of $PtBr^-$ were found to be exothermic, and the dominating DEA channels are the loss of one and two CO leading to the formation of $[Pt(CO)Br_2]^-$ and $[PtBr_2]^-$, while Br loss is insignificant. The CO loss appears predominantly through fairly broad contributions in the respective ion yields, peaking close to 0 eV and markedly asymmetric towards higher energies. We attribute these to overlapping contributions from the anionic ground state of a mixed π^* CO, Pt 5 d_{xz} p and the first electronically excited anionic state, which is of a σ^*, Pt $d_{x^2-y^2}$ character. The overall behavior of $[Pt(CO)_2Br_2]$ with respect to electron-induced dissociation is similar to what has been observed for $[Pt(CO)_2Cl_2]$ [28], and the differences observed can at large be attributed to the weaker Pt−Br bond as compared to Pt−Cl and the resulting lower dissociation thresholds. Compared to the widely used cancer therapeutic cisplatin, the behavior of the Pt(II) carbonyl halides with respect to low-energy electron-induced dissociation is very different. While the dominating DEA channels in the Pt(II) carbonyl halides are the loss of one and two CO and halogen loss is negligible, the main fragments observed in DEA to $[Pt(NH_3)_2Cl_2]$ are $[Pt(NH_3)_2Cl]^-$, Cl^- and $[Pt(NH_3)_2]^-$ [20]. Hence, cleavage of the Pt−halogen bonds dominates. From the thermochemical point of view, this is not expected as the Pt−NH_3 bond is significantly weaker than the Pt−Cl bond (1.5 eV vs. 3.3 eV in cisplatin) [20]. However, the anionic ground state of cisplatin has a repulsive σ^* character unlike the π^*_{CO} character of the anionic ground state of the Pt(II) carbonyl halides. Relaxation of $[Pt(NH_3)_2Cl_2]^-$ from its anionic ground state, through Pt−Cl bond rupture along the σ^* coordinates is thus a direct process. This is not the case for the π^*, CO character anionic ground state of $[Pt(CO)_2Cl_2]^-$ and $[Pt(CO)_2Br_2]^-$ where this process is symmetry forbidden and effective coupling of the π^*_{CO} component with the respective σ^* states is necessary to effectuate dissociation.

Author Contributions: Conceptualization, F.F.d.S., O.I. and L.M.-W.; compound synthesis, L.M.-W. and H.L.; experimental data acquisition, S.S., F.F.d.S. and O.I.; theoretical calculations, M.C.; data curation, M.C.; writing—original draft preparation, M.C.; writing—review and editing, O.I., M.C., F.F.d.S., L.M.-W. and H.L.; visualization, M.C. All authors have read and agreed to the published version of the manuscript.

Funding: This project was conducted within the framework of ELENA, a Horizon 2020 research and innovation program under Marie Sklodowska-Curie Innovative Training Network, under the grant agreement No. 722149. M.C. and O.I. acknowledge support from the Icelandic Center of Research (RANNIS), grant no. 13049305(1−3). M.C. acknowledges a doctoral grant from the University of Iceland Research Fund. H.L. and L.M.-W. thank the National Science Foundation for support under grants CHE-1607547 and CHE-1904802. F.F.d.S. acknowledges the Portuguese National Funding Agency for Science Research and Technology (FCT-MCTES), through the research grants PTDC/FIS-AQM/31215/2017 and UIDB/00068/2020 (CEFITEC).

Institutional Review Board Statement: Not applicable.

Informed Consent Statement: Not applicable.

Data Availability Statement: The data underlying this article will be shared on reasonable request from the corresponding author.

Conflicts of Interest: The authors declare no conflict of interest. The funders had no role in the design of the study; in the collection, analyses, or interpretation of data; in the writing of the manuscript, or in the decision to publish the results.

References

1. Wilson, G.D.; Bentzen, S.M.; Harari, P.M. Biologic Basis for Combining Drugs With Radiation. *Semin. Radiat. Oncol.* **2006**, *16*, 2–9. [CrossRef] [PubMed]
2. Kelland, L. The resurgence of platinum-based cancer chemotherapy. *Nat. Rev. Cancer* **2007**, *7*, 573–584. [CrossRef] [PubMed]
3. Hanna, N.; Einhorn, L.H. Testicular Cancer: A Reflection on 50 Years of Discovery. *J. Clin. Oncol.* **2014**, 32, 3085–3092. [CrossRef] [PubMed]
4. Einhorn, E.H. Testicular cancer: An oncological success story. *Clin. Cancer Res.* **1997**, *3*, 2630–2632. [PubMed]
5. Glover, D.; Grabelsky, S.; Fox, K.; Weiler, C.; Cannon, L.; Glick, J. Clinical trials of WR-2721 and cis-platinum. *Int. J. Radiat. Oncol.* **1989**, *16*, 1201–1204. [CrossRef]
6. Brabec, V. Chemistry and Structural Biology of 1,2-Interstrand Adducts of Cisplatin. In *Platinum-Based Drugs in Cancer Therapy*; Kelland, L.R., Farrell, N.P., Eds.; Humana Press: Totowa, NJ, USA, 2000; Volume 1, pp. 37–63. [CrossRef]
7. Zheng, Y.; Hunting, D.J.; Ayotte, P.; Sanche, L. Role of Secondary Low-Energy Electrons in the Concomitant Chemoradiation Therapy of Cancer. *Phys. Rev. Lett.* **2008**, *100*, 198101. [CrossRef]
8. Dąbkowska, I.; Rak, J.; Gutowski, M. DNA strand breaks induced by concerted interaction of H radicals and low-energy electrons. *Eur. Phys. J. D* **2005**, *35*, 429–435. [CrossRef]
9. Landheer, K.; Rosenberg, S.G.; Bernau, L.; Swiderek, P.; Utke, I.; Hagen, C.W.; Fairbrother, D.H. Low-Energy Electron-Induced Decomposition and Reactions of Adsorbed Tetrakis(trifluorophosphine)platinum $[Pt(PF_3)_4]$. *J. Phys. Chem. C* **2011**, *115*, 17452–17463. [CrossRef]

10. Hedhili, M.N.; Bredehöft, J.H.; Swiderek, P. Electron-Induced Reactions of $MeCpPtMe_3$ Investigated by HREELS. *J. Phys. Chem. C* **2009**, *113*, 13282–13286. [CrossRef]
11. Wnuk, J.D.; Gorham, J.M.; Rosenberg, S.G.; van Dorp, W.F.; Madey, T.E.; Hagen, C.W.; Fairbrother, D.H. Electron Induced Surface Reactions of the Organometallic Precursor Trimethyl(methylcyclopentadienyl)platinum(IV). *J. Phys. Chem. C* **2009**, *113*, 2487–2496. [CrossRef]
12. Spencer, J.A.; Wu, Y.-C.; McElwee-White, L.; Fairbrother, D.H. Electron Induced Surface Reactions of cis -$Pt(CO)_2Cl_2$: A Route to Focused Electron Beam Induced Deposition of Pure Pt Nanostructures. *J. Am. Chem. Soc.* **2016**, *138*, 9172–9182. [CrossRef] [PubMed]
13. Rohdenburg, M.; Martinović, P.; Ahlenhoff, K.; Koch, S.; Emmrich, D.; Gölzhäuser, A.; Swiderek, P. Cisplatin as a Potential Platinum Focused Electron Beam Induced Deposition Precursor: NH_3 Ligands Enhance the Electron-Induced Removal of Chlorine. *J. Phys. Chem. C* **2019**, *123*, 21774–21787. [CrossRef]
14. Utke, I.; Hoffmann, P.; Melngailis, J. Gas-assisted focused electron beam and ion beam processing and fabrication. *J. Vac. Sci. Technol. B Microelectron. Nanom. Struct.* **2008**, *26*, 1197. [CrossRef]
15. LaVerne, J.A.; Pimblott, S.M. Electron Energy-Loss Distributions in Solid, Dry DNA. *Radiat. Res.* **1995**, *141*, 208. [CrossRef] [PubMed]
16. Cobut, V. Monte Carlo simulation of fast electron and proton tracks in liquid water—I. physical and physicochemical aspects. *Radiat. Phys. Chem.* **1998**, *51*, 229–243. [CrossRef]
17. Pimblott, S.M.; LaVerne, J.A. Production of low-energy electrons by ionizing radiation. *Radiat. Phys. Chem.* **2007**, *76*, 1244–1247. [CrossRef]
18. Sanche, L. Low energy electron-driven damage in biomolecules. *Eur. Phys. J. D* **2005**, *35*, 367–390. [CrossRef]
19. Boudaïffa, B.; Cloutier, P.; Hunting, D.; Huels, M.A.; Sanche, L. Resonant formation of DNA strand breaks by low-energy (3 to 20 eV) electrons. *Science* **2000**, *287*, 1658–1660. [CrossRef]
20. Kopyra, J.; Koenig-Lehmann, C.; Bald, I.; Illenberger, E. A Single Slow Electron Triggers the Loss of Both Chlorine Atoms from the Anticancer Drug Cisplatin: Implications for Chemoradiation Therapy. *Angew. Chem. Int. Ed.* **2009**, *48*, 7904–7907. [CrossRef]
21. Schaefer, J.; Hoelzl, J. A contribution to the dependence of secondary electron emission from the work function and fermi energy. *Thin Solid Films* **1972**, *13*, 81–86. [CrossRef]
22. Ohya, K.; Harada, A.; Kawata, J.; Nishimura, K. Monte Carlo Simulation of Yield and Energy Distribution of Secondary Electrons Emitted from Metal Surfaces. *Jpn. J. Appl. Phys.* **1996**, *35*, 6226–6232. [CrossRef]
23. Utke, I.; Moshkalev, S.; Russell, P. *Nanofabrication Using Focused Ion and Electron Beams: Principles and Applications (Oxford Series in Nanomanufacturing Book 1)*; Oxford University Press Inc.: New York, NY, USA, 2012.
24. Ingólfsson, O. *Low-Energy Electrons: Fundamentals and Applications*; Pan Stanford Publishing Pte. Ltd.: Singapore, 2019.
25. Engmann, S.; Stano, M.; Matejčík, Š.; Ingólfsson, O. Gas phase low energy electron induced decomposition of the focused electron beam induced deposition (FEBID) precursor trimethyl (methylcyclopentadienyl) platinum(iv) ($MeCpPtMe_3$). *Phys. Chem. Chem. Phys.* **2012**, *14*, 14611. [CrossRef] [PubMed]
26. May, O.; Kubala, D.; Allan, M. Dissociative electron attachment to $Pt(PF_3)_4$ —A precursor for Focused Electron Beam Induced Processing (FEBIP). *Phys. Chem. Chem. Phys.* **2012**, *14*, 2979–2982. [CrossRef] [PubMed]
27. Zlatar, M.; Allan, M.; Fedor, J. Excited States of $Pt(PF_3)_4$ and Their Role in Focused Electron Beam Nanofabrication. *J. Phys. Chem. C* **2016**, *120*, 10667–10674. [CrossRef]
28. Ferreira da Silva, F.; Thorman, R.M.; Bjornsson, R.; Lu, H.; McElwee-White, L.; Ingólfsson, O. Dissociation of the FEBID precursor *cis*-$Pt(CO)_2Cl_2$ driven by low-energy electrons. *Phys. Chem. Chem. Phys.* **2020**, *22*, 6100–6108. [CrossRef]
29. Mahgoub, A.; Lu, H.; Thorman, R.M.; Preradovic, K.; Jurca, T.; McElwee-White, L.; Fairbrother, H.; Hagen, C.W. Electron beam-induced deposition of platinum from $Pt(CO)_2Cl_2$ and $Pt(CO)_2Br_2$. *Beilstein J. Nanotechnol.* **2020**, *11*, 1789–1800. [CrossRef]
30. Rohdenburg, M.; Boeckers, H.; Brewer, C.R.; McElwee-White, L.; Swiderek, P. Efficient NH_3-based process to remove chlorine from electron beam deposited ruthenium produced from $(\eta^3\text{-}C_3H_5)Ru(CO)_3Cl$. *Sci. Rep.* **2020**, *10*, 10901. [CrossRef]
31. Desfrançois, C.; Abdoul-Carime, H.; Schermann, J.-P. Ground-state dipole-bound anions. *Int. J. Mod. Phys. B* **1996**, *10*, 1339–1395. [CrossRef]
32. Sommerfeld, T. Dipole-bound states as doorways in (dissociative) electron attachment. *J. Phys. Conf. Ser.* **2005**, *4*, 245–250. [CrossRef]
33. Bjarnason, E.H.; Ómarsson, B.; Engmann, S.; Ómarsson, F.H.; Ingólfsson, O. Dissociative electron attachment to titatinum tetrachloride and titanium tetraisopropoxide. *Eur. Phys. J. D* **2014**, *68*, 121. [CrossRef]
34. Braun, M.; Marienfeld, S.; Ruf, M.-W.; Hotop, H. High-resolution electron attachment to the molecules CCl 4 and SF_6 over extended energy ranges with the (EX)LPA method. *J. Phys. B At. Mol. Opt. Phys.* **2009**, *42*, 125202. [CrossRef]
35. Bagnoli, F.; Belli Dell'Amico, D.; Calderazzo, F.; Englert, U.; Marchetti, F.; Merigo, A.; Ramello, S. Halo-carbonyl complexes of platinum(II) and palladium(II). *J. Organomet. Chem.* **2001**, *622*, 180–189. [CrossRef]
36. Neese, F. Software update: The ORCA program system, version 4.0. *WIREs Comput. Mol. Sci.* **2018**, *8*. [CrossRef]
37. Chai, J.-D.; Head-Gordon, M. Long-range corrected hybrid density functionals with damped atom–atom dispersion corrections. *Phys. Chem. Chem. Phys.* **2008**, *10*, 6615. [CrossRef] [PubMed]
38. Grimme, S.; Antony, J.; Ehrlich, S.; Krieg, H. A consistent and accurate ab initio parametrization of density functional dispersion correction (DFT-D) for the 94 elements H-Pu. *J. Chem. Phys.* **2010**, *132*, 154104. [CrossRef] [PubMed]

39. Weigend, F.; Ahlrichs, R. Balanced basis sets of split valence, triple zeta valence and quadruple zeta valence quality for H to Rn: Design and assessment of accuracy. *Phys. Chem. Chem. Phys.* **2005**, *7*, 3297. [CrossRef] [PubMed]
40. Zheng, J.; Xu, X.; Truhlar, D.G. Minimally augmented Karlsruhe basis sets. *Theor. Chem. Acc.* **2011**, *128*, 295–305. [CrossRef]
41. Andrae, D.; Häußermann, U.; Dolg, M.; Stoll, H.; Preuß, H. Energy-adjustedab initio pseudopotentials for the second and third row transition elements. *Theor. Chim. Acta* **1990**, *77*, 123–141. [CrossRef]
42. Riplinger, C.; Neese, F. An efficient and near linear scaling pair natural orbital based local coupled cluster method. *J. Chem. Phys.* **2013**, *138*, 034106. [CrossRef]
43. Riplinger, C.; Pinski, P.; Becker, U.; Valeev, E.F.; Neese, F. Sparse maps—A systematic infrastructure for reduced-scaling electronic structure methods. II. Linear scaling domain based pair natural orbital coupled cluster theory. *J. Chem. Phys.* **2016**, *144*, 024109. [CrossRef]
44. Riplinger, C.; Sandhoefer, B.; Hansen, A.; Neese, F. Natural triple excitations in local coupled cluster calculations with pair natural orbitals. *J. Chem. Phys.* **2013**, *139*, 134101. [CrossRef]
45. Saitow, M.; Becker, U.; Riplinger, C.; Valeev, E.F.; Neese, F. A new near-linear scaling, efficient and accurate, open-shell domain-based local pair natural orbital coupled cluster singles and doubles theory. *J. Chem. Phys.* **2017**, *146*, 164105. [CrossRef] [PubMed]
46. Dunning, T.H. Gaussian basis sets for use in correlated molecular calculations. I. The atoms boron through neon and hydrogen. *J. Chem. Phys.* **1989**, *90*, 1007–1023. [CrossRef]
47. Woon, D.E.; Dunning, T.H. Gaussian basis sets for use in correlated molecular calculations. III. The atoms aluminum through argon. *J. Chem. Phys.* **1993**, *98*, 1358–1371. [CrossRef]
48. Kendall, R.A.; Dunning, T.H.; Harrison, R.J. Electron affinities of the first-row atoms revisited. Systematic basis sets and wave functions. *J. Chem. Phys.* **1992**, *96*, 6796–6806. [CrossRef]
49. Figgen, D.; Peterson, K.A.; Dolg, M.; Stoll, H. Energy-consistent pseudopotentials and correlation consistent basis sets for the 5d elements Hf–Pt. *J. Chem. Phys.* **2009**, *130*, 164108. [CrossRef]

International Journal of
Molecular Sciences

MDPI

Article

HF Formation through Dissociative Electron Attachment—A Combined Experimental and Theoretical Study on Pentafluorothiophenol and 2-Fluorothiophenol

Maicol Cipriani and Oddur Ingólfsson *

Department of Chemistry and Science Institute, University of Iceland, Dunhagi 3, 107 Reykjavik, Iceland; mac31@hi.is
* Correspondence: odduring@hi.is

Abstract: In chemoradiation therapy, dissociative electron attachment (DEA) may play an important role with respect to the efficiency of the radiosensitizers used. The rational tailoring of such radiosensitizers to be more susceptive to DEA may thus offer a path to increase their efficiency. Potentially, this may be achieved by tailoring rearrangement reactions into the DEA process such that these may proceed at low incident electron energies, where DEA is most effective. Favorably altering the orbital structure of the respective molecules through substitution is another path that may be taken to promote dissociation up on electron capture. Here we present a combined experimental and theoretical study on DEA in relation to pentafluorothiophenol (PFTP) and 2-fluorothiophenol (2-FTP). We investigate the thermochemistry and dynamics of neutral HF formation through DEA as means to lower the threshold for dissociation up on electron capture to these compounds, and we explore the influence of perfluorination on their orbital structure. Fragment ion yield curves are presented, and the thermochemical thresholds for the respective DEA processes are computed as well as the minimum energy paths for HF formation up on electron capture and the underlying orbital structure of the respective molecular anions. We show that perfluorination of the aromatic ring in these compounds plays an important role in enabling HF formation by further lowering the threshold for this process and through favorable influence on the orbital structure, such that DEA is promoted. We argue that this approach may offer a path for tailoring new and efficient radiosensitizers.

Keywords: chemoradiation; radiosensitizers; low-energy electron interaction; perfluorination; HF formation; pentafluorothiophenol; 2-fluorothiophenol; dissociative electron attachment

Citation: Cipriani, M.; Ingólfsson, O. HF Formation through Dissociative Electron Attachment—A Combined Experimental and Theoretical Study on Pentafluorothiophenol and 2-Fluorothiophenol. *Int. J. Mol. Sci.* **2022**, 23, 2430. https://doi.org/10.3390/ijms23052430

Academic Editor: Henry Chermette

Received: 20 January 2022
Accepted: 22 February 2022
Published: 23 February 2022

1. Introduction

In recent years, appreciable attention has been paid to the interaction of low-energy electrons (LEEs) with DNA and radiosensitizers applied in cancer therapy [1–6]. Although much progress has been made in the fight against cancer, tumor hypoxia still represents an obstacle to traditional cancer therapy. Hypoxia is generally present in solid tumors due to their limited vascularization. The decrease in O_2 availability in tumor masses may make chemotherapy and radiotherapy ineffective [7–10]. A more efficient therapy is the concomitant application of radiation with oxygen-mimetic radiosensitizers, most commonly nitro-imidazoles [11,12]. In these electron-affinic radiosensitizers, the nitro group binds with the DNA free radicals generated by ionizing radiation and consequently induces DNA strand breaks [12]. However, at the microscopic level, low-energy electron (LEE) interaction plays an important role in sensitizing cancer cells to radiation [3,13]. The interaction of the ionizing radiation with a biological tissue generates LEEs (<20 eV) with energy distribution that peaks at or below 10 eV, with appreciable contribution close to 0 eV and a tail extending to higher energies [14]. At electron energies below 10 eV, electron-affinic radiosensitizers can be subjected to LEE induced reactions, which yield radical species that can damage DNA. In an aqueous medium, LEEs solvate on a picosecond scale [15].

However, before solvation, an LEE can occupy a vacant virtual orbital of a radiosensitizer, giving rise to a transient negative ion (TNI). If reaction channels are available at these electron energies, a TNI can undergo fragmentation via dissociative electron attachment (DEA), generating a negatively charged ion and neutral counter-fragment(s) [16,17]. This process is in competition with the relaxation of the TNI through autodetachment, i.e., the re-emission of the electron without fragmentation. The electron attachment process, which is the formation of the TNI, is most efficient at low energies, and the autodetachment lifetime decreases rapidly with increasing electron energies. Thus, DEA is most efficient at very low incident electron energies, given that the respective dissociation channel is thermochemically accessible. Hence, exothermic DEA processes, which may proceed close to 0 eV electron energy, are most efficient. The important role of LEEs and DEA in DNA radiolysis was shown by Boudaïffa et al. [1] already in the year 2000, in a study where the authors demonstrated that DEA processes can trigger single and double-strand breaks. This work triggered manifold studies on LEE interaction with DNA components and the fragmentation of negative ions of DNA components, with the bulk of this work being reviewed, for example, in references [18–22]. Furthermore, LEE interaction with radiosensitizers has also been investigated in a number of studies [3–6,23–25]. In these studies, it has, e.g., been shown for 5-halouracils that halogenation increases the DEA cross-sections and such halogenated uracils present sharp peaks with high cross sections in the 0–4 eV electron energy range [23]. In this context, it has also been shown that substitution of thymine with the higher electron affinity 5-halouraci significantly sensitizes DNA to radiation [26]. Similarly, Rackwitz et al. [24] have shown enhanced strand-brake efficiency through DEA to DNA oligonucleotedes when replacing adenine with 2-fluoro adenine, the active component in the chemotherapeutic fludarabine that has also been considered for use in chemoradiation therapy [27–29]. Rackwitz et al. [24] associate the observed strand brakes to resonances they observe in gas phase DEA to 2-fluoro adenine at around 5.5 eV and note that these are shifted towards lower energies when compared to DEA to the native adenine. In addition, DEA to the oxygen-mimetic radiosensitizers 2-nitroimidazole and 4(5)-nitroimidazole has been shown to effetely fragment these molecules [5,6].

Dissociative electron attachment is not limited to single bond ruptures but may also involve the rupture of multiple bonds and the formation of new bonds. The generation of new chemical bonds provides additional energy to the system and can thus promote DEA and open new reaction channels otherwise inaccessible at low electron energies. In recent years, dissociative electron attachment (DEA) reactions leading to neutral HF formation from perfluorinated benzene derivatives have been the object of several studies [30–33]. The formation of HF feeds 5.9 eV into the system, i.e., the bond energy of HF, and can promote reaction channels that involve the rupture of multiple bonds. Ómarsson et al. [30,31] conducted detailed experimental and theoretical investigation of HF formation through DEA to pentafluorotoluene (PFT), pentafluoroaniline (PFA), and pentafluorophenol (PFP). There it was shown that the polarization of the X–H bond plays a determining role in neutral HF formation through the promotion of the formation of an intermediate intramolecular hydrogen bond, X–H···F. In these studies, the authors correlated the different magnitude of the polarization of the X–H bond for X = C, N, and O, with the stabilization of the intermediate X–H···F leading to the HF loss in the respective DEA processes. With respect to the thermochemistry, the HF formation upon DEA to PFP was found to be exothermic, partly attributed the subsequent rearrangement of the charge retaining fragment, while in the case of PFA and PFT, the HF formation was found to be endothermic. In fact, it was also pointed out by Rackwitz et al. [24] that the neutral HF formation may provide the thermochemical prerequisite for the fragmentation effectuated in DEA to 2-fluoro adenine and that such neutral halogen acid formation is frequently observed in DEA to halo-nucleobases.

Motivated by the possibility to promote reaction channels in radiosensitizers through HF formation upon DEA, we extend the previous investigations and compare the two compounds, pentafluorothiophenol (PFTP) and 2-fluorothiophenol (2-FTP). A comparison between PFTP and 2-FTP is informative as both these compounds may form the intermediate X–H···F to a fluorine in the ortho position of the aromatic ring and thus dissociate by neutral HF loss up on electron attachment. However, the perfluorination of PFTP changes the order of the involved molecular orbitals and also favorably influences the thermochemistry of the process. Furthermore, the comparison of PFTP to PFP is interesting because S and O atoms have similar electron configurations, being neighbors in the same group within the periodic table. Because S is less electronegative than N and O, the process involving the HF formation will not be as well supported by the polarization of the S–H bond. However, the orbital structure of S is more extended than that of O (principal quantum number (n) = 3 as compared to 2 for oxygen), so the S–H bond is longer and weaker than the O–H bond. Thus, R–SH is a stronger acid than R–OH. For the HF formation through DEA to take place close to 0 eV incident electron energy, where the attachment cross section is highest, the electron affinity of the biradical [M-HF] must compensate the energy difference between the cleavage of the two bonds (M–F and M–H) and the formation of the new H–F bond. In addition, 2-FTP has also been the object of a near ultraviolet photodissociation study in regard to the S–H bond cleavage [34], which in turn is a prerequisite for HF formation.

Here we present a combined theoretical and experimental study where we use PFTP and 2-FTP as model compounds to explore the potential of substitution to enhance the susceptibility of such compounds towards low energy electrons. We present ion yield curves for all DEA fragments observed from these compounds and we explore the influence of fluorination on the relative energies of the respective low-lying anionic states in conjunction with the thermochemistry and reaction paths leading to fragmentation up on electron capture. Specifically, we focus on HF formation as a potential means to supply additional energy into the DEA channels in order to move the fragmentation threshold close to 0 eV, where the attachment cross sections are highest. In this study, we show that the perfluorination of the molecule is not only important with respect to the attachment cross-section but also plays an important role with respect to the orbital structure and the thermochemistry behind the HF formation. We discuss the nature of the SOMOs involved in the electron attachment processes, calculate the thermochemical thresholds of these processes, and compute the minimum energy paths for HF loss for both compounds. We argue that such molecular functionalization may serve as a basis for the design of more efficient radiosensitizers.

2. Results and Discussion

Figure 1 shows negative ion yield curves for observed fragments formed through DEA to PFTP (left) and 2–FBT (right), respectively. The ion yield curves are shown for the incident electron energy range from approximately 0 to 10 eV and are normalized to the respective target gas pressure and the relative cross-section of SF_6^- formation from SF_6 at 0 eV incident electron energy.

The most pronounced DEA channel for PFTP leads to neutral HF loss from the transient negative ion (TNI) formed in the initial attachment process, that is the formation of $[M\text{-}HF]^-$. This channel is most significant at threshold, i.e., at 0 eV, but has a higher-lying contribution centered at around 4.7 eV, which is approximately three orders of magnitude lower in intensity. The contribution peaking at approximately 0 eV is distinctly asymmetric towards higher energies, and we anticipate that this is due to overlapping contributions from two distinct resonances. Hydrogen loss is also observed from PFTP at low energies (at approximately 0.3 eV) and through a higher-lying resonance appearing in the ion yield curves at approximately 4.5 eV. The relative, maximum cross-section for the hydrogen loss from this molecule, that is the $[M\text{-}H]^-$ formation at 0.3 eV, is two orders of magnitude lower than that for the HF formation. However, the relative cross-section for the $[M\text{-}H]^-$ contribution from PFTP at approximately 4.5 eV is three orders of magnitude higher than that for the $[M\text{-}HF]^-$ formation at approximately 4.7 eV. This is understandable, as the attachment cross-section is significantly higher close to 0 eV as compared to 0.3 eV; however, at energies significantly above threshold, at approximately 4.5 eV in this case, direct dissociation such as the hydrogen loss is expected to be more efficient as compared to rearrangement processes such as the HF formation. Finally, DEA to PFTP also leads to the loss of SH, i.e., the observation of the anionic fragment $[M-SH]^-$. Similar to the HF formation and the H loss ion yield curves, the $[M-SH]^-$ ion yield curve has a low energy contribution with an onset at approximately 0 eV and peak intensity at approximately 0.8 eV and a less intense second contribution at higher energy that is centered at approximately 4 eV. The maximum relative cross section for the $[M-SH]^-$ is three orders of magnitude lower than that for the $[M-HF]^-$ formation, i.e., an order of magnitude lower than that for $[M-H]^-$. This channel leads to the formation of the stable pentafluorbenzenide anion.

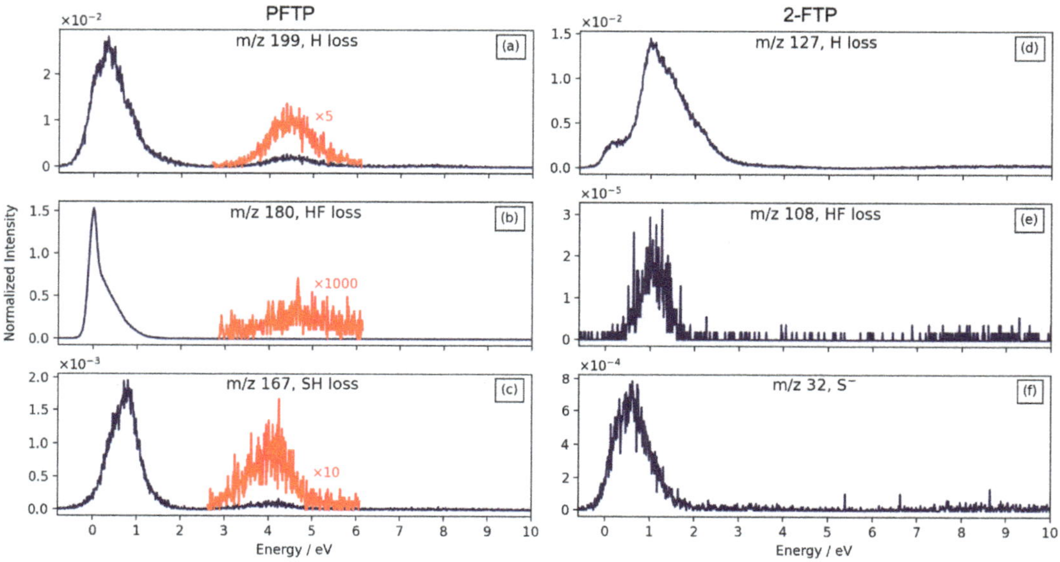

Figure 1. Left: DEA ion yield curves from PFTP for the channels: (**a**) H loss, (**b**) HF loss, and (**c**) SH loss. Right: DEA ion yield curves from 2-FTP for the channels: (**d**) H loss, (**e**) HF loss, (**f**) S^- formation. The intensities are normalized with respect to the target gas pressure and the formation of SF_6^- from SF_6 at 0 eV incident electron energy.

Dissociative electron attachment to 2–FBT also leads to the formation of $[M-H]^-$ and $[M-HF]^-$ through resonances at low incident electron energies. These contributions both have their maxima at approximately 1.0 eV. However, unlike PFTP, the relative cross-section for the HF loss from 2-FTP, i.e., the formation of $[M-HF]^-$, is three orders of magnitude lower than that for the direct hydrogen loss, $[M-H]^-$. The low energy contribution to the $[M-H]^-$ formation from 2-FTP is composed of a contribution at approximately 0 eV, appearing as a low energy shoulder on the main contribution that peaks at approximately 1 eV and is asymmetric towards high energies. We attribute this 0 eV shoulder to 'hot -band transitions' or I^- ($m/z = 127$) from some iodine containing compound residual in the 2-FTP sample or possibly in our inlet system. The assignments of the resonance reflected in the low energy $[M-H]^-$ contribution is discussed in more details below. In addition to the $[M-H]^-$ and $[M-HF]^-$ channels, the formation of S^- is also observed in DEA to 2-FTP. This channel is, similarly to the others, most efficient at low energies, with an onset at approximately 0 eV and a maximum cross-section at approximately 0.6 eV. The maximum relative cross-section for the S^- formation from 2-FTP is approximately two orders of magnitude lower than that for the hydrogen loss from this compound.

Hence, while the most pronounced DEA channel for 2-FTP is direct hydrogen loss, HF loss is the dominating DEA channel from PFTP. In fact, the relative cross section for neutral

HF formation from 2-FTP is five orders of magnitude lower than that for HF formation from 2-FTP. Furthermore, the onset of the HF loss from 2-FTP is at approximately 0.5 eV, indicating that this channel is endothermic, while the cross section for HF loss from PFTP peaks at approximately 0 eV, as would be expected for an exothermic process.

In order to elucidate the thermochemistry and the dynamics of the DEA processes for PFTP and 2-FTP, we have calculated the 0K reaction enthalpies (ΔH_{0K}) at the B3LYP D3BJ/aug-cc-pVTZ level of theory for all the observed fragments. These are given in Table 1 along with the thermally corrected values (ΔE_{th}) derived by adding the thermal energy correction at room temperature to the parent molecule. This approach is taken as we expect thermal equilibrium for the parent molecules within the inlet system, but not for the DEA fragments formed under single collision conditions. For the HF formation, we additionally considered a rearrangement of the aromatic ring whereby the 6-membered benzene ring is rearranged to a 5-memebered ring with an exocyclic–CS moiety: C_5F_4–CS^- and C_5H_4–CS^-, respectively, see Figure 2.

Table 1. Calculated 0K reaction enthalpies (ΔH_{0K}) and thermally corrected thresholds (ΔE_{th}) for the fragments observed in DEA to PFTP and 2-FTP. For the HF loss, the values are shown for both the direct process, $[M\text{-}HF]^-/C_6F_4S^-$ or $C_6H_4S^-$ and for the rearrangement process shown in Figure 2, $[M\text{-}HF]^-/C_5F_4CS^-$ or $C_6F_4S^-$. At the bottom of the table, the same values are calculated for the HF formation upon DEA to PFP. The calculations are performed at the B3LYP D3BJ/aug-cc-pVTZ level of theory.

m/*z*	Fragment	ΔH_{0K}	ΔE_{th}
PFTP			
199	$[M\text{-}H]^-/C_6F_5S^-$	0.42	0.13
180	$[M\text{-}HF]^-/C_6F_4S^-$	−0.0084	−0.29
180	* $[M\text{-}HF]^-/C_5F_4CS^-$	−0.066	−0.35
167	$[M\text{-}SH]^-/C_6F_5^-$	0.61	0.32
2-FTP			
127	$[M\text{-}H]^-/C_6H_4FS^-$	1.03	0.84
108	$[M\text{-}HF]^-/C_6H_4S^-$	0.62	0.42
108	* $[M\text{-}HF]^-/C_5H_4CS^-$	0.47	0.28
32	S^-	0.33	0.14
PFP			
164	$[M\text{-}HF]^-/C_6F_4O^-$	−0.066	−0.33
164	* $[M\text{-}HF]^-/C_5F_4CO^-$	−0.27	−0.54

* Calculated threshold considering the rearrangement of the ring after the HF formation.

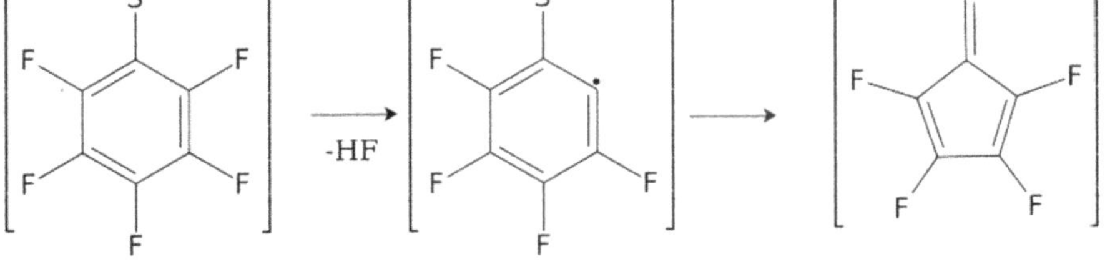

Figure 2. Considered rearrangement of the phenyl ring subsequent to the HF loss upon DEA to PFTP and 2-FTP. In the figure, this is shown for PFTP as an example. In this process, after the HF loss, the 6-membered benzene ring rearranges into a 5-membered ring with exocyclic–CS.

This rearrangement was proposed by Ómarsson et al. [30,31] in their studies on DEA to PFP, PFA, and PFT. Similar to PFTP, effective HF loss from PFP was observed at 0 eV in those studies, while the direct HF loss from PFP, calculated at the B2PLYP/aug-pc-2 level of theory, was found to be endothermic by 0.59 eV. However, in better agreement with the experimental results, a rearrangement leading to a 5-membered ring structure of the anion resulted in a threshold at −0.19 eV. For comparison, we have also calculated the thresholds (ΔE_{th}) and 0K reaction enthalpies (ΔH_{0K}) for HF formation upon DEA to PFP at the B3LYP D3BJ/aug-cc-pVTZ level of theory and, in fact, at this level of theory, we find the direct HF loss to be exothermic. In addition to the B3LYP D3BJ/aug-cc-pVTZ calculations for PFTP and 2-FTP, shown in Table 1, we have also performed calculations at the ωB97X-D3/aug-cc-pVTZ, ωB97X-D3/aug-cc-pVQZ, and DLPNO-CCSD(T)/aug-cc-pVQZ levels of theory for HF formation upon DEA to PFTP and PFP; these give qualitatively the same results and are presented as Supporting Material in Table S1.

According to our calculations (shown in Table 1), the direct HF formation from PFTP is exothermic by 0.29 eV, and rearrangement of the charge-retaining ring only lowers the threshold to −0.35 eV. For the H and SH losses from PFTP, the calculated thermochemical thresholds were found to be 0.13 and 0.32 eV, respectively. This is in good agreement with our experimental results, where the peak intensities for PFTP are found to be at 0.0 eV for the exothermic HF loss, while the endothermic H and the HS losses are shifted to slightly higher energies. Furthermore, the high relative cross-section for the $[M\text{-}HF]^-$ formation is consistent with the higher attachment cross-section expected at threshold (~0.0 eV) [35].

For 2-FTP, at the B3LYP D3BJ/aug-cc-pVTZ level of theory, we found the thermochemical threshold for the direct HF loss, Table 1, to be 0.42 eV and considering the rearrangement of the ring we found the threshold to be 0.28 eV. The hydrogen loss is found to be energetically less favorable, i.e., endothermic by 0.84 eV. In fact, this is a 0.61 eV higher threshold than the respective threshold for hydrogen loss from PFTP. This is mainly a result of the perfluorination increasing the electron affinity of the charge-retaining fragment C_6F_4–S as compared to that for C_6H_4F–S. At the B3LYP D3BJ/aug-cc-pVTZ level of theory, we found the 0K adiabatic electron affinities of C_6F_5–S and C_6H_4F–S to be 2.18 eV and 1.69 eV, respectively.

Despite the fact that the HF loss from 2-FTP is energetically more favorable than the H loss, the H loss dominates the ion yields observed upon DEA to this compound. The significantly higher cross-section for H loss as compared to HF loss must thus be rooted in the dynamics of these processes. To further explore the dynamics of this process, we have performed NEB-TS calculations at the B3LYP D3BJ/aug-cc-pVTZ level of theory to compute the reaction paths for the HF loss from both PFTP and 2–FBT, including the potential rearrangement of the charge retaining phenyl ring as discussed here above. Figures 3 and 4 show the calculated minimum energy paths, on the B3LYP potential energy surface (PES) for the HF formation from PFTP and 2-FTP (step 3) and the subsequent rearrangement of the aromatic ring to form C_5F_4–CS^- and C_5H_4–CS^-(step 9), respectively, from the anionic ground states of PFTP and 2-FTP (step 1). The total energy of the neutral parent molecule, calculated at the B3LYP D3BJ/aug-cc-pVTZ level of theory, is set at 0 eV, and the black line extending from the *y*-axis marks the relative energy of the neutral ground states. The open circles correspond to the calculated single point energies of the system along the reaction paths, but the blue line is only meant to guide the eye.

Similar to what appears in the minimum energy path for the HF formation from the ground state of the PFP anion, calculated by Ómarsson et al. [31], the formation of the HF, hydrogen-bonded intermediate in step 3 is favored over the molecular anion. Both in the case of PFTP and 2-FTP, see step 2 in Figures 3 and 4, this process (from step 1 to 3) proceeds with an energy barrier. For PFTP it is approximately 0.4 eV and for 2-FTP it is approximately 0.55 eV, relative to the single point energies of the respective relaxed anionic ground states. However, the relaxed PFTP anionic ground state is already 0.85 eV below the respective relaxed neutral ground state of PFTP. The barrier in step 2 and the relaxed $C_6F_4S^-$ anion, shown in step 4, are thus 0.45 eV and 0.29 eV below the relaxed neutral ground state of PFTP, respectively. Hence, with respect to the neutral, this is a barrierless exothermic reaction and may thus proceed at the 0.0 eV incident electron energy. This is what is observed in the experiments. For 2-FTP, on the other hand, the anionic ground state is 0.09 eV above the neutral ground state, and the activation barrier, shown in step 2, and the relaxed $C_6H_4S^-$ anion, shown in step 4, lies above the neutral ground state. The saddle point for this process, in step 2, is 0.64 eV above the neutral ground state. Hence, with respect to the neutral, this is a barrierless exothermic reaction and may thus proceed at the 0.0 eV incident electron energy. This is what is observed in the experiments. Furthermore, the energy barrier for the HF loss, step 2 in Figure 4, is comparable to the threshold energy for the direct hydrogen loss. The HF loss can thus only proceed at higher energy, i.e., above the threshold, and this in turn favors the faster, direct hydrogen loss, as is observed in the respective ion yields.

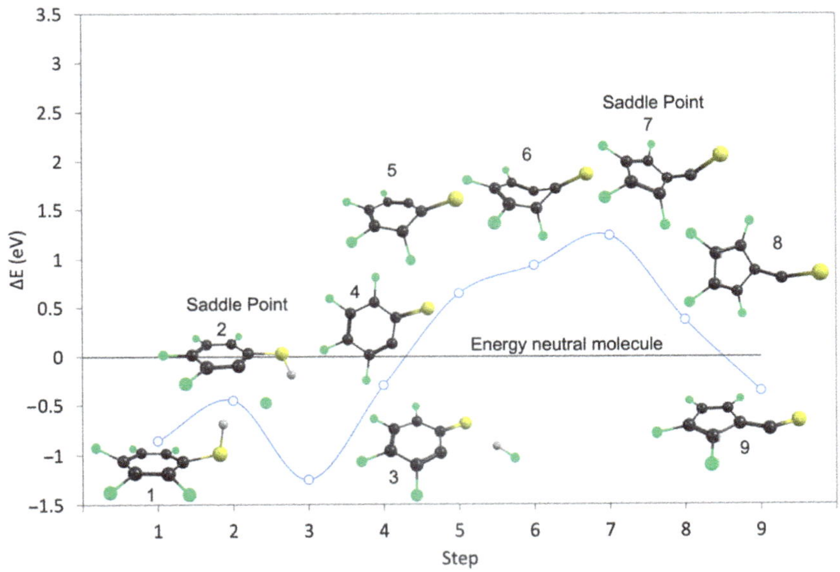

Figure 3. Minimum energy path for the direct HF loss (steps 1–4) and the formation of the 5-membered ring C_5F_4–CS^- (steps 4–9) from the anionic ground state of PFTP (step 1) calculated using the NEB-TS method at the B3LYP D3BJ/aug-cc-pVTZ level of theory. Step 4 corresponds to the direct HF loss, i.e., without the rearrangement of the ring. The single point energies (open circles) along the reaction paths are relative to the energy of the neutral parent molecule, which is set to 0 eV (horizontal black line). ZPVEs and thermal energy correction for the neutral molecule were taken into account.

From step 4, the minimum energy path was further calculated considering rearrangement of the ring to form C_5F_4–CS^- and C_5H_4–CS^- from PFTP and 2-FTP, respectively. The same procedure was applied in Ómarsson et al. [31]. In both PFTP and 2-FTP, the ring rearrangement to the pentagonal structure (from steps 4 to 9) proceeds through a deformation of the ring with a high energy barrier. For PFTP, this reaction is slightly more exothermic than the direct HF loss; however, as can be seen in Figure 3, there is a 1.24 eV reaction barrier on this path for PFTP (from step 4 to 9). This shows that the HF formation from PFTP at 0 eV threshold energy is direct and proceeds without rearrangement of the aromatic ring. Similarly, we find a reaction barrier of approximately 2.3 eV on this reaction path for 2-FTP, showing that the low energy contribution in the [M-HF]$^-$ ion yield from 2-FTP must also be attributed to direct HF loss.

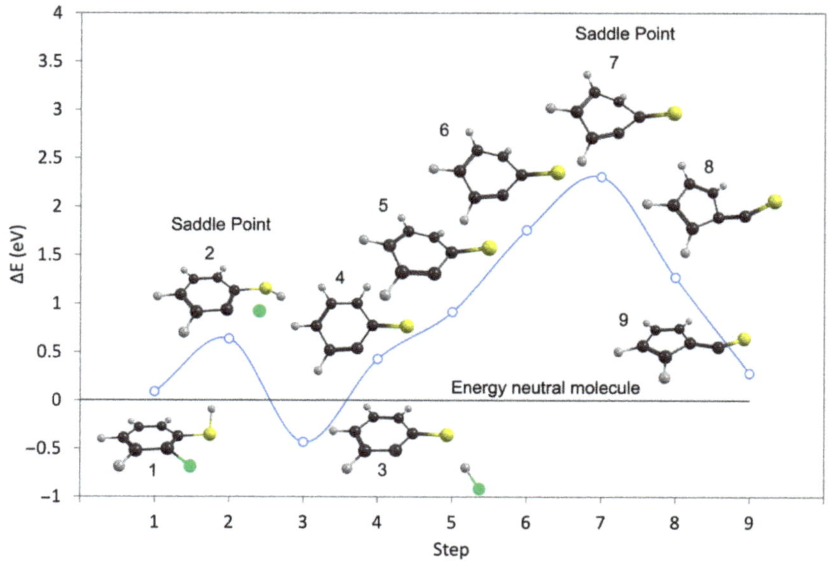

Figure 4. Minimum energy path for the direct HF loss (steps 1–4) and the formation of the 5-membered ring C_5F_4–CS^- (steps 4–9) from the anionic ground state of 2-FTP (step 1), calculated using the NEB-TS method at the B3LYP D3BJ/aug-cc-pVTZ level of theory. Step 4 corresponds to the direct HF loss, i.e., without the rearrangement of the ring. The single point energies (open circles) along the reaction paths are relative to the energy of the neutral parent molecule, which is set to 0 eV (horizontal black line). ZPVEs and thermal energy correction for the neutral molecule were taken into account.

Similarly, our threshold calculations for the $[M\text{-}HF]^-$ from PFP, at the B3LYP D3BJ/aug-cc-pVTZ level of theory show that the direct HF loss is also exothermic here (−0.33 eV) and on the minimum energy path for the ring rearrangement in this molecule, Ómarsson et al. [31] found the rearrangement barrier to be close to 2 eV. Therefore, it is reasonable to assume that the $[M\text{-}HF]^-$ formation from PFP observed at the 0 eV threshold also occurs as a direct process without involving the rearrangement of the ring.

It is clear from the experiments and the calculations presented above that the perfluorination in PFTP makes the HF loss in DEA energetically more favorable as compared to 2-FTP. Hence, fluorination may potentially be used to sensitize such molecules with respect to DEA by lowering the thermochemical thresholds for these reactions and thus enabling them to proceed at very low energies where the attachment cross section is highest.

With respect to the orbital structure associated with the resonant attachment processes reflected in the ion yields of these compounds, it is worth looking at that of benzene and substituted benzenes. In electron attachment to benzene, the X^2E_{2u} anionic ground state is formed in the gas phase at 1.15 eV through single electron occupation of the doubly degenerate LUMO $e_{2u}(\pi^*)$, as has been assigned through electron transmission spectroscopy [36]. This radical anion distorts due to the Jahn–Teller effect (JT), and the symmetry of the molecule is lowered from D_{6h} to D_{2h}, splitting the degenerated $e_{2u}(\pi^*)$ LUMO into two components: ^2A_u and ^2B_u [37]. Similarly, the D_{6h} symmetry of the neutral benzene is also broken by substitution at the ring. A single substitution removes the degeneracy of the e_{2u} orbital and lowers the D_{6h} symmetry to C_{2v}, whereby the doubly degenerate $e_{2u}(\pi^*)$ molecular orbital (MO) splits into the components, a_2 (π^*) and b_1 (π^*) [38–40]. At the carbon carrying the substituent, the B_1-type orbital displays maximum electron density, whilst the A_2-type has a node at this point. The magnitude of the splitting of these orbitals is influenced by the different combination of the mesomeric and the inductive effect of the respective substituent [41]. While the inductive effect stabilizes both the A_2 and the B_1 anion states, the mesomeric effect destabilizes the B_1 state but generally does not affect the A_2 significantly. Fluorination of aromatic rings moderately lowers the energy of the π^* MOs, but strongly lowers the σ^* MOs due to the strong inductive effect of fluorine as compared to its mesomeric effect [42,43]. This is commonly referred to as the perfluoro effect [38,39]. The geometrical structures and the nature of the ground and excited states of fluoro-substituted benzene anions have been studied, both experimentally and theoretically, for example, with electron-spin resonance techniques [44,45] and electron transmission, inner-shell electron energy loss, and magnetic circular dichroism spectroscopy [46] as well as INDO and Hartree Fock calculations [47,48]. Generally, the findings have been that the energy level of the low-lying σ^* MO decreases with increasing fluorination and, in the case of C_6F_6, the lowest virtual MO is found to be the σ^* MO. This is visualized informatively in an energy diagram shown in reference [46]. Similar trends have also been observed in heavily fluorinated pyridine anions [49]. Furthermore, in the theoretical studies [47,48], the authors argued that the structure of the polyfluorinated benzene anions undergoes a distortion due to the pseudo-Jahn–Teller effect (pJT), resulting in a planar carbon structure with C–F bonds out of plane. The extra electron occupies a pseudo-π orbital formed by the mixing of the π^* and σ^* orbitals. The $Q_{(b1)}$ pJT distortion [47] is given by the vibronic interaction between the totally symmetric σ^* state and $^2B_1$–π state. As may be seen in comparison to the schematic representation in [47], we note that the $Q_{(b1)}$ pJT distortion correlates well with the relaxed structure of the PFTP anion optimized at the B3LYP D3BJ/aug-cc-pVTZ level, shown in Figure 5.

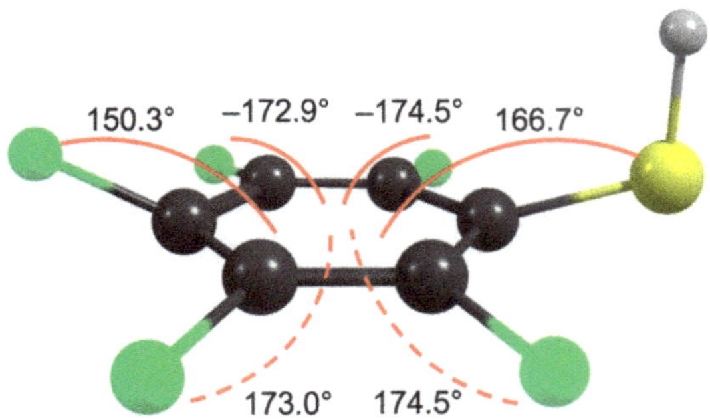

Figure 5. Relaxed geometry of the PFTP anion optimized at the B3LYP D3BJ/aug-cc-pVTZ level of theory. The angles shown in the figure are the angles between the out-of-plane C–F and C–S bonds and the plane of the ring.

Both the influence of fluorination on the order of the lowest lying π^* and σ^* orbitals and the JT distortion up on electron capture is important in DEA to these compounds as a direct dissociation along the substituent's σ^* bond to the aromatic ring is symmetry forbidden from the π^* MOs in the C_{2v} point group [50]. Hence, effective coupling between the respective π^* and σ^* states is required for such dissociation to take place. Occupation of the σ^* orbital, on the other hand, can lead to direct dissociation. This may influence the dissociation cross-section significantly, especially where there is strong competition with autodetachment, and the survival probability of the initially formed TNI is determined.

Figure 6 shows the LUMO, LUMO + 1, and LUMO + 2 of PFTP, along with the respective vertical electron attachment energies calculated using the EOM-EA-CCSD method with the B3LYP orbitals and aug-cc-pVTZ basis set. We note that their values are strongly dependent on the basis set while the order is reliable. Adhering to Jordan et al. [40], we labelled the π^* orbitals according to the C_{2V} point group. The LUMO of PFTP was found to have an σ^* character and is anti-bonding along the C–F and C–S coordinates, and there is a polarization along the S–H bond. From a hydrogen bonded S···H···F intermediate, this provides preferential conditions for HF loss from PFTP and the formation of $[M\text{-}HF]^-$.

This is consistent with the high efficiency of the $[M\text{-}HF]^-$ formation at approximately 0 eV, assuming that it will proceed from the σ* electronic ground state of the anion. The vertical attachment energy to this state was found to be 0.15 eV. The LUMO + 1 of PFTP has a π* character and correlates with the b_1 (π*) MO. The vertical attachment energy to this orbital is approximately 0.73 eV greater than the respective vertical attachment energy to the σ* electronic ground state of the anion. This may be explained by the strong inductive effect through the perfluorination stabilizing the σ* significantly stronger than the π*. Additionally, the mesomeric effect of the S atom is not strong enough to destabilize the b_1 (π*) MO and push it above the a_2 (π*) MO. This is due to the poor overlap of the $3p_x$ orbital of the S atom with the b_1 (π) and b_1 (π*) orbitals of the benzene ring. We attribute the asymmetry of the low energy peak in the $[M\text{-}HF]^-$ ion yield from PFTP to dissociation through single electron occupation of π* LUMO + 1. The significantly higher intensity through the σ* ground state may in part be due to the direct dissociation from the σ* state, as compared to the required coupling of the π* with the σ* coordinate, even though such coupling should be promoted by the pJT, causing π*–σ* mixing through the out of plain bending of the fluorine and –SH substituents, as shown in Figure 4. However, the energy dependency of the autodetachment lifetime will also play a significant role. In fact, these effects are intertwined as the DEA cross-section is defined as the product of the electron-attachment cross-section and the survival probability of the TNI [51,52]. With less coupling and increased energy, the autodetachment process becomes more significant, reducing the survival probability with respect to dissociation, which in turn is reflected in lower DEA cross-sections at higher energies. This affects the shape of the peak in the ion yield curve, which appears asymmetric with a long tail on the right side.

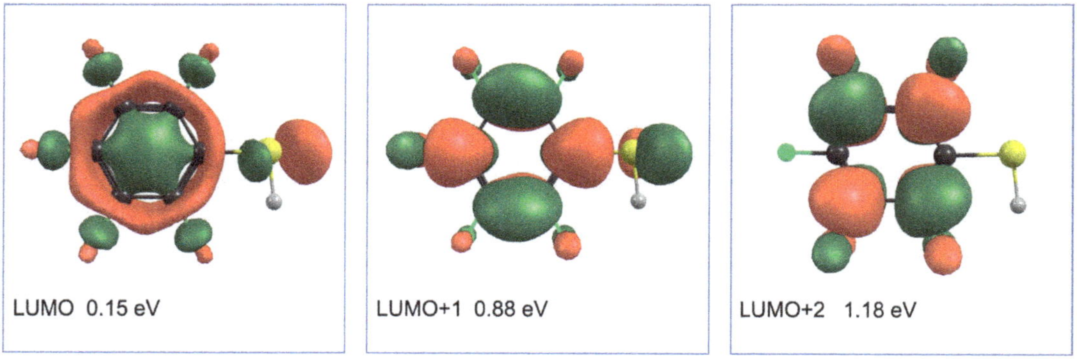

Figure 6. Contour plots of the LUMO, LUMO + 1, and LUMO + 2 of PFTP (B3LYP orbitals). The respective vertical electron attachment energy calculated using the EOM-EA-CCSD method with the B3LYP orbitals and aug-cc-pVTZ basis set are shown for each orbital. The LUMO has a σ* character, while LUMO + 1 and LUMO + 2 have a π* character and correlate with b_1 (π*) and a_2 (π*) MOs, respectively.

This interpretation is demonstrated in Figure 7, where we present a fitting of the low energy contribution in the negative ion yield curve for neutral HF loss from PFTP upon DEA using a combined fit of normal and skewed gaussian curves. The fitting has been carried out with a python script using the LMFIT library [53]. For the lower energy contribution, the energy dependence of the autodetachment lifetime is neglected (hence, the normal Gaussian) and the natural width of the underlying resonance is considered to be well below the instrumental energy resolution. The FWHM of this contribution in the ion yield should thus reflects the energy resolution of the instrument, but in praxis it is approximately 250 meV. The skewed gaussian curve is chosen for the higher energy component to take into account the asymmetry of the peaks due to the energy dependence of the attachment process and the autodetachment lifetime [16]. With this approach, where we consider contributions from both the singly occupied LUMO and LUMO + 1, an excellent fit to the low-energy contribution in the [M-HF]$^-$ ion yield from PFTP is obtained.

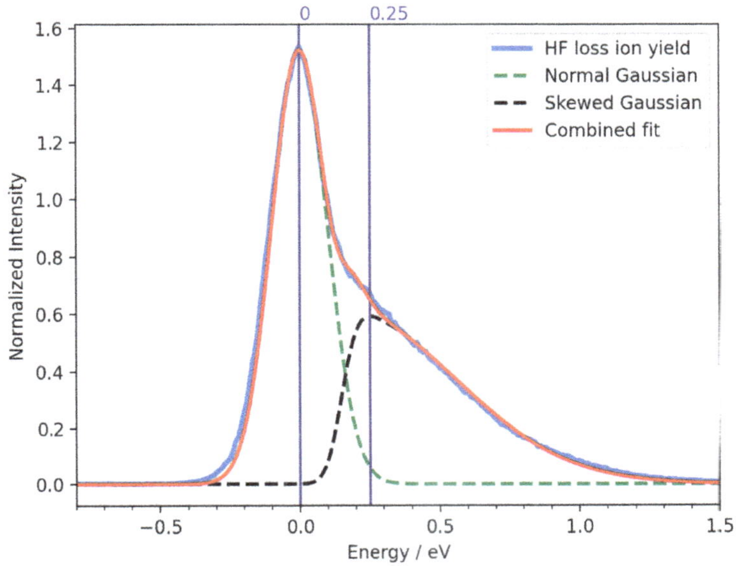

Figure 7. Combined fit of a Normal Gaussian (green dashed line) and a Skewed Gaussian (black dashed line) of the negative ion yield curve (blue line) for neutral HF loss and formation of [M-HF]$^-$ upon DEA from PFTP. The resulting fit is represented by the red line.

This is consistent with the picture in the first low-energy resonance that appears with a peak intensity at 0 eV in the ion yield curves; the unpaired electron is temporarily accommodated in the σ* MO. In the second resonance, with a maximum contribution at 0.25 eV in the ion yield curve, the extra electron is temporarily placed in the $b_1(\pi^*)$ MO. Here, the autodetachment is significant, and the long tail on the right side of this contribution reflects the lower survival probability at higher attachment energies, due to the shorter lifetime of the respective temporary anion state.

With respect to the LUMO + 2, shown in Figure 5, this correlates with the a_2 (π^*) MO. This A_2 state has no electron density on the SH substituent, and a S···H···F hydrogen bond formation from this state is not to be expected. Correspondingly we do not expect a contribution from the A_2 TNI to the [M-HF]$^-$ formation. Furthermore, we expect both the [M-H]$^-$ and [M-SH]$^-$ formations to be direct channels that compete with the [M-HF]$^-$ formation. These channels are slightly endothermic, as discussed above, and thus comparatively more efficient at higher energies. The ion yields for these fragments are correspondingly expected to derive their intensity from the high energy side of the σ* resonance and the $b_1(\pi^*)$ resonance, either directly or through vibrational energy redistribution. Finally, the high-energy contribution at approximately 4.5–4.7 eV in the [M-HF]$^-$, [M-H]$^-$, and [M-SH]$^-$ ion yield curves are most likely routed from the same resonance(s).

Figure 8 shows the LUMO, LUMO + 1, and LUMO + 2 of the 2-FTP, along with the respective vertical attachment energies calculated using the EOM-EA-CCSD method with the B3LYP orbitals and the aug-cc-pVTZ basis set. Different from PFTP, both the LUMO and LUMO + 1 in 2-FTP have a π* character and, in analogy to the nomenclature used for PFTP, they correlate with the $a_2(\pi^*)$ and $b_1(\pi^*)$ MOs, respectively. From these, the LUMO + 1 is anti-bonding along the C–F coordinate, providing a favorable condition for HF loss and the formation of [M-HF]$^-$. However, different from the direct HF formation from the σ* SOMO in PFTP, this process is symmetry forbidden from the π* LUMO + 1 of 2-FTP and requires effective π*–σ*coupling. Calculated at the B3LYP D3BJ/aug-cc-pVTZ level of theory, we find the threshold for this process to be 0.42 eV, and we anticipate that the low relative cross section for the HF formation from 2-FTP is due to inefficient coupling of the LUMO + 1 with the respective C–F σ* state, in combination with the high threshold for this process. Hence, at these energies, autodetachment, and conceivably S^- formation, prevail over the HF formation.

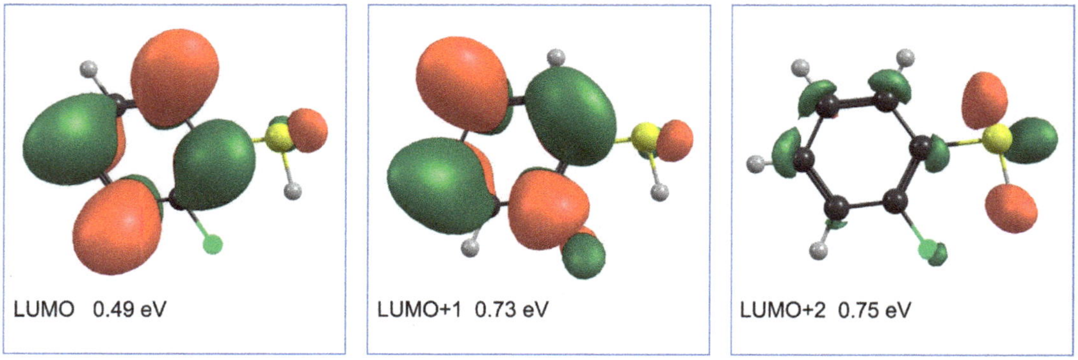

Figure 8. Contour plots of the LUMO, LUMO + 1, and LUMO + 2 of PFTP (B3LYP orbitals). The respective vertical electron attachment energy calculated using the EOM-EA-CCSD method with the B3LYP orbitals and aug-cc-pVTZ basis set are shown for each orbital. The LUMO and LUMO + 1 have a π* character and correlate with a_2 (π*) and b_1 (π*) MOs, respectively, while LUMO + 2 is a partial diffuse orbital with some electron density on the S and H atoms.

The hydrogen loss from 2-FTP is by far the most efficient DEA channel for this molecule and is characterized in the ion yields by a broad asymmetric contribution peaking at 0.88 eV and tailing off towards higher energies. A shoulder at approximately 0 eV in the ion yield curve is also observed, which we attribute to 'hot -band transitions' or I^- ($m/z = 127$) formation from iodine containing contaminations. The thermochemical threshold for the hydrogen loss, calculated at the B3LYP D3BJ/aug-cc-pVTZ level of theory, is found to be 0.82 eV, and we anticipate that this process proceeds predominantly from the partial diffuse LUMO + 2 orbital, which has some electron density on the S and H atoms. This assignment is also supported by the single contribution fit to the ion yield curve shown in Figure 9, where an excellent agreement is obtained by a fit of a single skewed Gaussian to the hydrogen loss ion yield. For completeness, a Gaussian contribution peaking at 0.2 eV is also included to reproduce the 0 eV impurity contribution. In principle, all conditions for HF formation could also proceed from the LUMO + 2; however, at these energies the hydrogen-bonded intermediate is not stable and the direct hydrogen loss prevails as the most efficient channel.

In a sense, DEA can be compared to photo dissociation as both are effectuated by a single electron occupation of previously unoccupied antibonding orbitals. In this context, we note a recent study by Marchetti et al. [34] on near ultraviolet spectroscopy and the photodissociation dynamics of 2- and 3-substituted thiophenols. There it was shown for 2-FTP that the repulsive S–H $^1n\sigma^*$ state crosses the $^1\pi\pi^*$ state close to its vibrational ground state. Population transfer from the π* to the repulsive S–H σ* may thus proceed through non-adiabatic coupling above the respective vibrational ground state, but tunnelling would be required from the ground state. In the current terminology, this may offer an alternative path for HF formation from the π* LUMO of the TNI formed in the initial attachment process.

It is clear from the current experiments and calculations that the perfluorination in PFTP, as compared to 2-FTP, does not only lower the thermochemical threshold for the HF loss in DEA, but also lowers the lowest σ* MO below the respective π* MOs, providing a very favorable condition for the HF loss. Potentially, this may be taken advantage of to promote the interaction of radiosensitizers with low-energy electrons, thus increasing their efficiency.

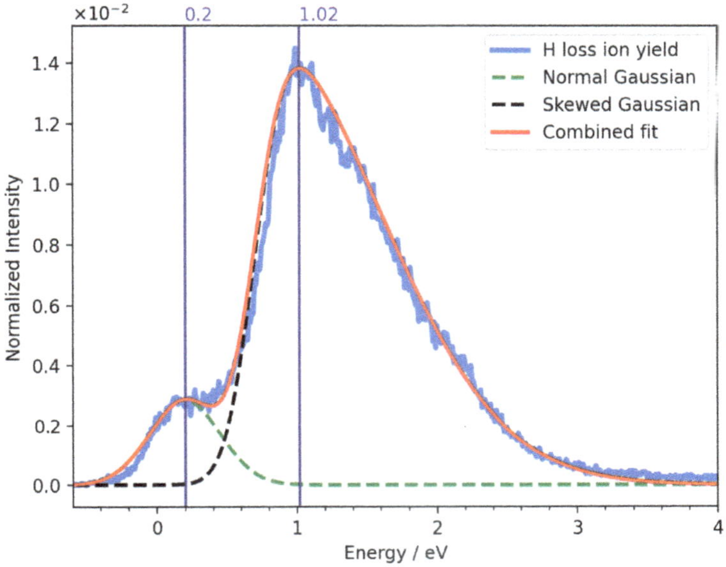

Figure 9. Single fit of a Skewed Gaussian (black dashed line) of the negative ion yield curve (blue line) for the H loss upon DEA from 2-FTP. Additionally, a Normal Gaussian peaking at 0.2 eV (green dashed line) is included to reproduce the impurity contribution at 0 eV. The red line represents the resulting fit.

3. Materials and Methods

3.1. Experimental Setup

Negative ion yield curves were recorded by means of a crossed electron-molecular beam apparatus. The experimental setup has been described in detail previously [54] and only a short description will be given here. The instrument is composed of a trochoidal electron monochromator (TEM), an effusive gas inlet system, and a quadrupole mass spectrometer (Hiden EPIC1000). The monochromator was heated to 120 °C with two internal halogen lamps to avoid condensation of the target gas on the electrical lens components. A quasi mono-energetic electron beam, generated with the TEM, crosses a molecular beam of the target gas obtained through the effusive gas inlet system. The ions resulting from the electron-molecule interactions in the collision region are then extracted through a small electric field (<1 V/cm) and analyzed by mass spectrometer. The DEA ion yield curves are recorded by scanning through the relevant electron energy at a fixed mass (m/z). The compounds were purchased from ABCR GMBH & Co. (Karlsruher, Germany), with a stated purity of 97% for PFTP and 98% for 2-FTP.

The electron energy was calibrated to the well-known 0 eV resonance for SF_6^- formation from SF_6, and the energy resolution of the electron beam at 0 eV was determined from the FWHM and was in the range of 120–140 meV. The background pressure inside the chamber was approximately 3×10^8 mbar, while the sample gas pressure was in the range of (2–10) $\times 10^7$ mbar for PFTP and (1–4) $\times 10^6$ mbar for 2-FTP.

3.2. Theoretical Procedures

All quantum chemical calculations were carried out using ORCA version 4.2.1 [55].

Geometry optimization of all the charged fragments and neutral molecule were performed at the B3LYP [56–58] level of theory, using the aug-cc-pVTZ basis set [59,60] and D3(BJ) dispersion correction [61,62]. For the closed shell systems, the Restricted Kohn–Sham (RKS) formalism was used, while the Unrestricted Kohn–Sham (RKS) formalism was used for open-shell systems. Harmonic vibrational frequencies were calculated at the same level of theory to derive zero-point vibrational energy (ZPVE) and thermal energy corrections. All threshold calculations refer to the single point energy of the relaxed structures, and they have been performed at the same level of theory as the geometry optimization. The reaction enthalpies at 0K (ΔH_{0K}) were calculated by subtracting the total energy of all fragments from the total energy of the parent molecule, including the respective ZPVEs. The thermally corrected thresholds (E_{th}) were obtained by subtracting the thermal energy of the parent molecule at room temperature from the reaction enthalpies at 0K.

In addition, for HF formation from PFP and PFTP, geometry optimizations and harmonic vibrational frequencies calculation were also performed at the ωB97X-D3 [61,63]/aug-cc-pVTZ level of theory, and single-point energy calculations were performed at the wB97X-D3/aug-cc-pVQZ [59,60] and DLPNO-CCSD(T) [64–67]/aug-cc-pVQZ level on the respective ωB97X-D3 optimized geometries. In the DLPNO-CCSD(T) calculations, for open shell systems, quasi-restricted orbitals (QROs) [68] were used as a reference determinant from the UHF orbitals. ZPVEs and thermal energy correction for the neutral were obtained from the ωB97X-D3 vibrational frequencies calculation. These calculations are given as Supplementary Information in Table S1.

Vertical electron attachment energies to the virtual orbitals of PFTP and 2-FTP were calculated using the EOM-EA-CCSD method with B3LYP orbitals and aug-cc-pVTZ basis set.

The minimum energy path for the HF formation and rearrangement mechanism was calculated using the Nudged Elastic Band method with transition state (TS) optimization (NEB-TS) [69] at the B3LYP D3BJ level, using the aug-cc-pVTZ basis set. For the reactants and products, we used the optimized geometry at the B3LYP D3BJ/aug-cc-pVTZ level of theory. The transition states were checked through the calculation of harmonic vibrational frequencies, and confirmed as effective saddle points, and it was found that they had only one negative frequency (imaginary mode).

Generally, basis set superposition errors (BSSE) are small in DFT as compared to methods based on wave-function theory (WFT), due to the lower basis set sensitivity. With respect to the current study, such errors are well within the experimental energy resolution and are not taken into account.

Finally, neither spin orbit coupling nor relativistic effects were taken into account. These effects are small for the current systems and well within the limit of our experimental accuracy.

4. Conclusions

Here we presented a combined theoretical and experimental study on DEA in relation to PFTP and 2-FTP, exploring the influence of perfluorination on the susceptibility of these compounds to DEA. We reported the energy dependence of the relative DEA cross-sections for the observed fragments and the thermochemical thresholds calculated for the respective DEA processes, as well as the reaction paths computed for the formation of neutral HF up on DEA to these compounds. We also showed the nature of the LUMOs involved in the initial electron attachment process and the respective transitions energies.

We found that the perfluorination in PFTP, as compared to 2-FTP, influences significantly the DEA processes. While the dominant DEA channel in PFTP is the HF loss, direct H loss is the dominating DEA process in 2-FTP, and HF loss is insignificant. We attribute this to the exothermic nature of the HF formation from PFTP, provided not only by the energy gain through the HF formation, but also by the perfluorination. Hence, the perfluorination increases the electron affinity of the charge-retaining fragment, providing additional energy in the process. This is reflected in the respective ion yield curve for the HF loss from PFTP, which is characterized by high relative cross sections already at the 0 eV threshold. In fact, it is 5 orders of magnitude higher than the relative cross section for HF formation from

2-FTP. The key point here is that, due to the very high electron attachment cross sections at very low energies, DEA may be made very efficient by tailoring exothermic reaction channels into the respective molecules. The perfluorination also influences the molecular orbital structure, and specifically the energy ordering of the lowest lying MOs. This is due to the dominating inductive effect of fluorine providing significantly stronger stabilization for the lowest lying σ^* orbitals than the respective π^* orbitals. This is very important with respect to the DEA efficiency as effective coupling of the lowest lying orbitals with the respective dissociation coordinates is essential for DEA to be effective at low electron attachment energies. For PFTP, this is provided because the lowest lying virtual orbital is of a σ^* character, it is anti-bonding along the C–F and C–S coordinates and is polarized along the S–H bond. Hence, single electron occupation of this orbital provides all prerequisites for "direct" HF formation at very low attachment energies. The two lowest lying virtual MOs of 2-FTP, on the other hand, are antibonding π^* orbitals. Different from the direct HF formation from single electron occupation of the σ^* LUMO in PFTP, this process is symmetrically forbidden from these π^* orbitals and requires effective π^*–σ^*coupling to proceed. Aided by the slight endothermicity, this puts relaxation through dissociation at a disadvantage compared to relaxation through re-emission of the electron, rendering the low energy DEA processes for 2-FTP inefficient as compared to PFTP.

It is clear from the current study, as well as from many previous DEA studies in the literature, that perfluorination enhances the susceptibility of many compounds towards low energy electrons, and rearrangement reactions such as HF formation may be used to open up exothermic DEA channels. In the context of the role of DEA in the functionality of radio sensitizers, these may be seen as important tools to promote efficient DEA reactions at low electron energies, and we argue that such tools may be valuable for a bottom-up approach in the design of efficient radiosensitizers.

Supplementary Materials: The following supporting information can be downloaded at: https://www.mdpi.com/article/10.3390/ijms23052430/s1.

Author Contributions: Conceptualization, M.C. and O.I.; methodology, M.C. and O.I.; validation, M.C. and O.I.; investigation, M.C. and O.I.; resources, O.I.; experimental data acquisition, M.C.; data curation, M.C.; theoretical calculations, M.C.; writing—original draft preparation, M.C.; writing—review and editing, M.C. and O.I.; visualization, M.C.; supervision, O.I.; funding acquisition, O.I. All authors have read and agreed to the published version of the manuscript.

Funding: M.C. and O.I. acknowledge support from the Icelandic Center of Research (RANNIS), grant no. 13049305(1–3). M.C. acknowledges a doctoral grant from the University of Iceland Research Fund.

Institutional Review Board Statement: Not applicable.

Informed Consent Statement: Not applicable.

Data Availability Statement: The data underlying this article will be shared on reasonable request from the corresponding author.

Conflicts of Interest: The authors declare no conflict of interest.

References

1. Boudaıffa, B.; Cloutier, P.; Hunting, D.; Huels, M.A.; Sanche, L. Resonant Formation of DNA Strand Breaks by Low-Energy (3 to 20 eV) Electrons. *Science* **2000**, *287*, 1658–1660. [CrossRef] [PubMed]
2. Kumar, A.; Becker, D.; Adhikary, A.; Sevilla, M.D. Reaction of Electrons with DNA: Radiation Damage to Radiosensitization. *Int. J. Mol. Sci.* **2019**, *20*, 3998. [CrossRef] [PubMed]
3. Ameixa, J.; Arthur-Baidoo, E.; Meißner, R.; Makurat, S.; Kozak, W.; Butowska, K.; Ferreira da Silva, F.; Rak, J.; Denifl, S. Low-energy electron-induced decomposition of 5-trifluoromethanesulfonyl-uracil: A potential radiosensitizer. *J. Chem. Phys.* **2018**, *149*, 164307. [CrossRef]
4. Meißner, R.; Feketeová, L.; Bayer, A.; Limão-Vieira, P.; Denifl, S. Formation of negative and positive ions in the radiosensitizer nimorazole upon low-energy electron collisions. *J. Chem. Phys.* **2021**, *154*, 074306. [CrossRef] [PubMed]
5. Tanzer, K.; Feketeová, L.; Puschnigg, B.; Scheier, P.; Illenberger, E.; Denifl, S. Reactions in Nitroimidazole Triggered by Low-Energy (0–2 eV) Electrons: Methylation at N1-H Completely Blocks Reactivity. *Angew. Chem. Int. Ed.* **2014**, *53*, 12240–12243. [CrossRef]

6. Ribar, A.; Fink, K.; Probst, M.; Huber, S.E.; Feketeová, L.; Denifl, S. Isomer Selectivity in Low-Energy Electron Attachment to Nitroimidazoles. *Chem. A Eur. J.* **2017**, *23*, 12892–12899. [CrossRef] [PubMed]
7. Harris, A.L. Hypoxia—A key regulatory factor in tumour growth. *Nat. Rev. Cancer* **2002**, *2*, 38–47. [CrossRef]
8. Hockel, M.; Vaupel, P. Tumor Hypoxia: Definitions and Current Clinical, Biologic, and Molecular Aspects. *JNCI J. Natl. Cancer Inst.* **2001**, *93*, 266–276. [CrossRef]
9. Wojtkowiak, J.W.; Verduzco, D.; Schramm, K.J.; Gillies, R.J. Drug Resistance and Cellular Adaptation to Tumor Acidic pH Microenvironment. *Mol. Pharm.* **2011**, *8*, 2032–2038. [CrossRef]
10. Graham, K.; Unger, E. Overcoming tumor hypoxia as a barrier to radiotherapy, chemotherapy and immunotherapy in cancer treatment. *Int. J. Nanomed.* **2018**, *13*, 6049–6058. [CrossRef]
11. Bonnet, M.; Hong, C.R.; Wong, W.W.; Liew, L.P.; Shome, A.; Wang, J.; Gu, Y.; Stevenson, R.J.; Qi, W.; Anderson, R.F.; et al. Next-Generation Hypoxic Cell Radiosensitizers: Nitroimidazole Alkylsulfonamides. *J. Med. Chem.* **2018**, *61*, 1241–1254. [CrossRef] [PubMed]
12. Wardman, P. Chemical Radiosensitizers for Use in Radiotherapy. *Clin. Oncol.* **2007**, *19*, 397–417. [CrossRef] [PubMed]
13. Park, Y.; Polska, K.; Rak, J.; Wagner, J.R.; Sanche, L. Fundamental Mechanisms of DNA Radiosensitization: Damage Induced by Low-Energy Electrons in Brominated Oligonucleotide Trimers. *J. Phys. Chem. B* **2012**, *116*, 9676–9682. [CrossRef] [PubMed]
14. Pimblott, S.M.; LaVerne, J.A. Production of low-energy electrons by ionizing radiation. *Radiat. Phys. Chem.* **2007**, *76*, 1244–1247. [CrossRef]
15. Farhataziz; Rodgers, M.A.J. *Radiation Chemistry-Principles and Applications*; VCH: New York, NY, USA, 1987.
16. Ingólfsson, O. Low Energy Electron–Induced Dissociation. In *Low-Energy Electrons*; Jenny Stanford Publishing: Delhi, India, 2019; pp. 47–120.
17. Christophorou, L.G.; Olthoff, J.K. Electron interactions with plasma processing gases: Present status and future needs. *Appl. Surf. Sci.* **2002**, *192*, 309–326. [CrossRef]
18. Kohanoff, J.; McAllister, M.; Tribello, G.A.; Gu, B. Interactions between low energy electrons and DNA: A perspective from first-principles simulations. *J. Phys. Condens. Matter* **2017**, *29*, 383001. [CrossRef]
19. Alizadeh, E.; Orlando, T.M.; Sanche, L. Biomolecular Damage Induced by Ionizing Radiation: The Direct and Indirect Effects of Low-Energy Electrons on DNA. *Annu. Rev. Phys. Chem.* **2015**, *66*, 379–398. [CrossRef]
20. Baccarelli, I.; Bald, I.; Gianturco, F.A.; Illenberger, E.; Kopyra, J. Electron-induced damage of DNA and its components: Experiments and theoretical models. *Phys. Rep.* **2011**, *508*, 1–44. [CrossRef]
21. Bald, I.; Denifl, S. The Role of Low-Energy Electrons in DNA Radiation Damage. In *Low-Energy Electrons*; Jenny Stanford Publishing: Delhi, India, 2019; pp. 285–340, ISBN 9780429058820.
22. Flosadóttir, H.D.; Ómarsson, B.; Bald, I.; Ingólfsson, O. Metastable decay of DNA components and their compositions—A perspective on the role of reactive electron scattering in radiation damage. *Eur. Phys. J. D* **2012**, *66*, 13. [CrossRef]
23. Scheer, A.M.; Aflatooni, K.; Gallup, G.A.; Burrow, P.D. Bond Breaking and Temporary Anion States in Uracil and Halouracils: Implications for the DNA Bases. *Phys. Rev. Lett.* **2004**, *92*, 068102. [CrossRef]
24. Rackwitz, J.; Kopyra, J.; Dąbkowska, I.; Ebel, K.; Ranković, M.L.; Milosavljević, A.R.; Bald, I. Sensitizing DNA towards Low-Energy Electrons with 2-Fluoroadenine. *Angew. Chem. Int. Ed.* **2016**, *55*, 10248–10252. [CrossRef] [PubMed]
25. Kopyra, J.; Keller, A.; Bald, I. On the role of fluoro-substituted nucleosides in DNA radiosensitization for tumor radiation therapy. *RSC Adv.* **2014**, *4*, 6825. [CrossRef]
26. Lawrence, T.S.; Davis, M.A.; Maybaum, J.; Stetson, P.L.; Ensminger, W.D. The effect of single versus double-strand substitution on halogenated pyrimidine-induced radiosensitization and DNA strand breakage in human tumor cells. *Radiat. Res.* **1990**, *123*, 192–198. [CrossRef] [PubMed]
27. McGinn, C.J.; Shewach, D.S.; Lawrence, T.S. Radiosensitizing Nucleosides. *JNCI J. Natl. Cancer Inst.* **1996**, *88*, 1193–1203. [CrossRef]
28. Nitsche, M.; Christiansen, H.; Lederer, K.; Griesinger, F.; Schmidberger, H.; Pradier, O. Fludarabine combined with radiotherapy in patients with locally advanced NSCLC lung carcinoma: A phase I study. *J. Cancer Res. Clin. Oncol.* **2012**, *138*, 1113–1120. [CrossRef]
29. Grégoire, V.; Ang, K.K.; Rosier, J.-F.; Beauduin, M.; Garden, A.S.; Hamoir, M.; Hittelman, W.N.; Humblet, Y.; Khuri, F.R.; Milas, L.; et al. A phase I study of fludarabine combined with radiotherapy in patients with intermediate to locally advanced head and neck squamous cell carcinoma. *Radiother. Oncol.* **2002**, *63*, 187–193. [CrossRef]
30. Ómarsson, B.; Bjarnason, E.H.; Ingólfsson, O.; Haughey, S.; Field, T.A. Chemical control through dissociative electron attachment—A study on pentafluorotoluene, pentafluoroaniline and pentafluorophenol. *Chem. Phys. Lett.* **2012**, *539–540*, 7–10. [CrossRef]
31. Ómarsson, B.; Bjarnason, E.H.; Haughey, S.A.; Field, T.A.; Abramov, A.; Klüpfel, P.; Jónsson, H.; Ingólfsson, O. Molecular rearrangement reactions in the gas phase triggered by electron attachment. *Phys. Chem. Chem. Phys.* **2013**, *15*, 4754. [CrossRef]
32. Ómarsson, B.; Bjornsson, R.; Ingólfsson, O. Proton Shuttling and Reaction Paths in Dissociative Electron Attachment to *o*- and *p*-Tetrafluorohydroquinone, an Experimental and Theoretical Study. *J. Phys. Chem. A* **2017**, *121*, 5580–5585. [CrossRef]
33. Dąbkowska, I.; Flosadóttir, H.D.; Orzol, M.; Ptasinska, S.; Bald, I.; Ingólfsson, O.; Illenberger, E. Reactions in gas phase and condensed phase C6F5X (X = NCO, CH2CN) triggered by low energy electrons. *Phys. Chem. Chem. Phys.* **2009**, *11*, 5323. [CrossRef]

34. Marchetti, B.; Karsili, T.N.V.; Cipriani, M.; Hansen, C.S.; Ashfold, M.N.R. The near ultraviolet photodissociation dynamics of 2- and 3-substituted thiophenols: Geometric vs. electronic structure effects. *J. Chem. Phys.* **2017**, *147*, 013923. [CrossRef] [PubMed]
35. Klar, D.; Ruf, M.-W.; Hotop, H. Dissociative electron attachment to CCl_4 molecules at low electron energies with meV resolution. *Int. J. Mass Spectrom.* **2001**, *205*, 93–110. [CrossRef]
36. Sanche, L.; Schulz, G.J. Electron transmission spectroscopy: Resonances in triatomic molecules and hydrocarbons. *J. Chem. Phys.* **1973**, *58*, 479–493. [CrossRef]
37. Chowdhury, S.; Grimsrud, E.P.; Heinis, T.; Kebarle, P. Electron affinities of perfluorobenzene and perfluorophenyl compounds. *J. Am. Chem. Soc.* **1986**, *108*, 3630–3635. [CrossRef]
38. Hinde, A.L.; Poppinger, D.; Radom, L. Ab initio study of the benzene radical anion. *J. Am. Chem. Soc.* **1978**, *100*, 4681–4685. [CrossRef]
39. Birch, A.J.; Hinde, A.L.; Radom, L. A theoretical approach to the Birch reduction. Structures and stabilities of the radical anions of substituted benzenes. *J. Am. Chem. Soc.* **1980**, *102*, 3370–3376. [CrossRef]
40. Jordan, K.D.; Michejda, J.A.; Burrow, P.D. Electron transmission studies of the negative ion states of substituted benzenes in the gas phase. *J. Am. Chem. Soc.* **1976**, *98*, 7189–7191. [CrossRef]
41. Christophorou, L.G.; McCorkle, D.L.; Carter, J.G. Compound-negative-ion-resonant states and threshold-electron-excitation spectra of monosubstituted benzene derivatives. *J. Chem. Phys.* **1974**, *60*, 3779–3786. [CrossRef]
42. Brundle, C.R.; Robin, M.B.; Kuebler, N.A.; Basch, H. Perfluoro effect in photoelectron spectroscopy. I. Nonaromatic molecules. *J. Am. Chem. Soc.* **1972**, *94*, 1451–1465. [CrossRef]
43. Brundle, C.R.; Robin, M.B.; Kuebler, N.A. Perfluoro effect in photoelectron spectroscopy. II. Aromatic molecules. *J. Am. Chem. Soc.* **1972**, *94*, 1466–1475. [CrossRef]
44. Yim, M.B.; Wood, D.E. Free radicals in an adamantane matrix. XII. EPR and INDO study of σ^*-π^* crossover in fluorinated benzene anions. *J. Am. Chem. Soc.* **1976**, *98*, 2053–2059. [CrossRef]
45. Wang, J.T.; Williams, F. 13C Hyperfine interaction in the hexafluorobenzene radical anion. *Chem. Phys. Lett.* **1980**, *71*, 471–475. [CrossRef]
46. Hitchcock, A.P.; Fischer, P.; Gedanken, A.; Robin, M.B. Antibonding σ^* valence MOs in the inner-shell and outer-shell spectra of the fluorobenzenes. *J. Phys. Chem.* **1987**, *91*, 531–540. [CrossRef]
47. Shchegoleva, L.N.; Bilkis, I.I.; Schastnev, P.V. Geometrical and electronic structure of fluoro-substituted benzene radical anions based on quantum chemical analysis of hyperfine interactions. *Chem. Phys.* **1983**, *82*, 343–353. [CrossRef]
48. Shchegoleva, L.N.; Beregovaya, I.V.; Schastnev, P.V. Potential energy surface of $C_6F_6^-$ radical anion. *Chem. Phys. Lett.* **1999**, *312*, 325–332. [CrossRef]
49. Yim, M.B.; DiGregorio, S.; Wood, D.E. Free radicals in an adamantane matrix. 13. Electron paramagnetic resonance study of .sigma.*-.pi.* orbital crossover in fluorinated pyridine anions. *J. Am. Chem. Soc.* **1977**, *99*, 4260–4263. [CrossRef]
50. Skalický, T.; Chollet, C.; Pasquier, N.; Allan, M. Properties of the π^* and σ^* states of the chlorobenzene anion determined by electron impact spectroscopy. *Phys. Chem. Chem. Phys.* **2002**, *4*, 3583–3590. [CrossRef]
51. Schulz, G.J.; Asundi, R.K. Isotope Effect in the Dissociative Attachment in H_2 at Low Energy. *Phys. Rev.* **1967**, *158*, 25–29. [CrossRef]
52. Krishnakumar, E.; Denifl, S.; Čadež, I.; Markelj, S.; Mason, N.J. Dissociative Electron Attachment Cross Sections for H_2 and D_2. *Phys. Rev. Lett.* **2011**, *106*, 243201. [CrossRef]
53. Newville, M.; Stensitzki, T.; Allen, D.B.; Ingargiola, A. LMFIT: Non-Linear Least-Square Minimization and Curve-Fitting for Python (0.8.0). *Zenodo* **2014**. [CrossRef]
54. Bjarnason, E.H.; Ómarsson, B.; Engmann, S.; Ómarsson, F.H.; Ingólfsson, O. Dissociative electron attachment to titatinum tetrachloride and titanium tetraisopropoxide. *Eur. Phys. J. D* **2014**, *68*, 121. [CrossRef]
55. Neese, F. Software update: The ORCA program system, version 4.0. *WIREs Comput. Mol. Sci.* **2018**, *8*, e1327. [CrossRef]
56. Becke, A.D. Density-functional thermochemistry. III. The role of exact exchange. *J. Chem. Phys.* **1993**, *98*, 5648–5652. [CrossRef]
57. Lee, C.; Yang, W.; Parr, R.G. Development of the Colle-Salvetti correlation-energy formula into a functional of the electron density. *Phys. Rev. B* **1988**, *37*, 785–789. [CrossRef]
58. Stephens, P.J.; Devlin, F.J.; Chabalowski, C.F.; Frisch, M.J. Ab Initio Calculation of Vibrational Absorption and Circular Dichroism Spectra Using Density Functional Force Fields. *J. Phys. Chem.* **1994**, *98*, 11623–11627. [CrossRef]
59. Kendall, R.A.; Dunning, T.H.; Harrison, R.J. Electron affinities of the first-row atoms revisited. Systematic basis sets and wave functions. *J. Chem. Phys.* **1992**, *96*, 6796–6806. [CrossRef]
60. Woon, D.E.; Dunning, T.H. Gaussian basis sets for use in correlated molecular calculations. III. The atoms aluminum through argon. *J. Chem. Phys.* **1993**, *98*, 1358–1371. [CrossRef]
61. Grimme, S.; Antony, J.; Ehrlich, S.; Krieg, H. A consistent and accurate ab initio parametrization of density functional dispersion correction (DFT-D) for the 94 elements H-Pu. *J. Chem. Phys.* **2010**, *132*, 154104. [CrossRef] [PubMed]
62. Grimme, S.; Ehrlich, S.; Goerigk, L. Effect of the damping function in dispersion corrected density functional theory. *J. Comput. Chem.* **2011**, *32*, 1456–1465. [CrossRef] [PubMed]
63. Chai, J.-D.; Head-Gordon, M. Long-range corrected hybrid density functionals with damped atom–atom dispersion corrections. *Phys. Chem. Chem. Phys.* **2008**, *10*, 6615. [CrossRef]

64. Riplinger, C.; Neese, F. An efficient and near linear scaling pair natural orbital based local coupled cluster method. *J. Chem. Phys.* **2013**, *138*, 034106. [CrossRef]
65. Riplinger, C.; Pinski, P.; Becker, U.; Valeev, E.F.; Neese, F. Sparse maps—A systematic infrastructure for reduced-scaling electronic structure methods. II. Linear scaling domain based pair natural orbital coupled cluster theory. *J. Chem. Phys.* **2016**, *144*, 024109. [CrossRef]
66. Riplinger, C.; Sandhoefer, B.; Hansen, A.; Neese, F. Natural triple excitations in local coupled cluster calculations with pair natural orbitals. *J. Chem. Phys.* **2013**, *139*, 134101. [CrossRef]
67. Saitow, M.; Becker, U.; Riplinger, C.; Valeev, E.F.; Neese, F. A new near-linear scaling, efficient and accurate, open-shell domain-based local pair natural orbital coupled cluster singles and doubles theory. *J. Chem. Phys.* **2017**, *146*, 164105. [CrossRef]
68. Neese, F. Importance of Direct Spin−Spin Coupling and Spin-Flip Excitations for the Zero-Field Splittings of Transition Metal Complexes: A Case Study. *J. Am. Chem. Soc.* **2006**, *128*, 10213–10222. [CrossRef]
69. Ásgeirsson, V.; Birgisson, B.O.; Bjornsson, R.; Becker, U.; Neese, F.; Riplinger, C.; Jónsson, H. Nudged Elastic Band Method for Molecular Reactions Using Energy-Weighted Springs Combined with Eigenvector Following. *J. Chem. Theory Comput.* **2021**, *17*, 4929–4945. [CrossRef]

International Journal of *Molecular Sciences*

MDPI

Article

Experimental and Theoretical Studies of Dissociative Electron Attachment to Metabolites Oxaloacetic and Citric Acids

Janina Kopyra [1,*], Paulina Wierzbicka [1], Adrian Tulwin [1], Guillaume Thiam [2], Ilko Bald [3,*], Franck Rabilloud [2] and Hassan Abdoul-Carime [4]

1 Faculty of Exact and Natural Sciences, Siedlce University of Natural Sciences and Humanities, 3 Maja 54, 08-110 Siedlce, Poland; paulina.wierzbicka@uph.edu.pl (P.W.); adrian.tul@interia.pl (A.T.)
2 Université de Lyon, Université Claude Bernard Lyon 1, CNRS, Institut Lumière Matière, UMR5306, F-69622 Villeurbanne, France; guillaume.thiam@univ-lyon1.fr (G.T.); franck.rabilloud@univ-lyon1.fr (F.R.)
3 Institute of Chemistry, University of Potsdam, Karl-Liebknecht-Str. 24-25, 14476 Potsdam, Germany
4 Institut de Physique des 2 Infinis, Université Lyon 1, Université de Lyon, CNRS/IN2P3, UMR5822, F-69003 Lyon, France; hcarime@ipnl.in2p3.fr
* Correspondence: janina.kopyra@uph.edu.pl (J.K.); bald@uni-potsdam.de (I.B.)

Abstract: In this contribution the dissociative electron attachment to metabolites found in aerobic organisms, namely oxaloacetic and citric acids, was studied both experimentally by means of a crossed-beam setup and theoretically through density functional theory calculations. Prominent negative ion resonances from both compounds are observed peaking below 0.5 eV resulting in intense formation of fragment anions associated with a decomposition of the carboxyl groups. In addition, resonances at higher energies (3–9 eV) are observed exclusively from the decomposition of the oxaloacetic acid. These fragments are generated with considerably smaller intensities. The striking findings of our calculations indicate the different mechanism by which the near 0 eV electron is trapped by the precursor molecule to form the transitory negative ion prior to dissociation. For the oxaloacetic acid, the transitory anion arises from the capture of the electron directly into some valence states, while, for the citric acid, dipole- or multipole-bound states mediate the transition into the valence states. What is also of high importance is that both compounds while undergoing DEA reactions generate highly reactive neutral species that can lead to severe cell damage in a biological environment.

Keywords: dissociative electron attachment; negative ions; oxaloacetic acid; citric acid; mass spectrometry

Citation: Kopyra, J.; Wierzbicka, P.; Tulwin, A.; Thiam, G.; Bald, I.; Rabilloud, F.; Abdoul-Carime, H. Experimental and Theoretical Studies of Dissociative Electron Attachment to Metabolites Oxaloacetic and Citric Acids. *Int. J. Mol. Sci.* **2021**, *22*, 7676. https://doi.org/10.3390/ijms22147676

Academic Editor: Bice Conti

Received: 7 June 2021
Accepted: 9 July 2021
Published: 18 July 2021

1. Introduction

Oxaloacetic acid (OAA, $C_4H_4O_5$, m/z 132) and citric acid (CA, $C_6H_8O_7$, m/z 192) are important metabolites found in aerobic organisms. Within the citrate (or Krebs) cycle, citrate is decomposed into oxaloacetate to release chemically stored energy. OAA possesses two carboxyl groups and a carbonyl group, while CA possesses three carboxyl groups and a hydroxyl group (Figure 1). The presence of these oxygen-rich functional groups is key to their electron-accepting properties, which are pivotal for the electron transfer and redox reactions taking place within the metabolism in the Krebs cycle. Although the electron transfer chains established in the organism are well-balanced, the delicate equilibria might be disturbed by the presence of "free" electrons generated, for example, by ionizing radiation through water radiolysis [1–3].

The last two decades have seen the implication of low energy electrons in various application fields as diverse as radiation chemotherapy, chemistry or nano-lithography [4–8]. In the irradiated material, the primary energetic particles or radiation generate a very large number of ballistic secondary electrons with an energy distribution below 10 eV [9]. These slow particles are now known to be able to induce efficient fragmentation to molecules as large as DNA [10]. Previous studies have shown that DNA is not the only critical target in biological radiation damage. In addition to DNA a decomposition of pivotal metabolites such as CA and OAA can also lead to dysfunction of the metabolism and ultimately to cell deaths. Identification of such critical damage pathways could trigger the development of more targeted therapies or radiation protection strategies, e.g., development of new radiosensitizers [11,12].

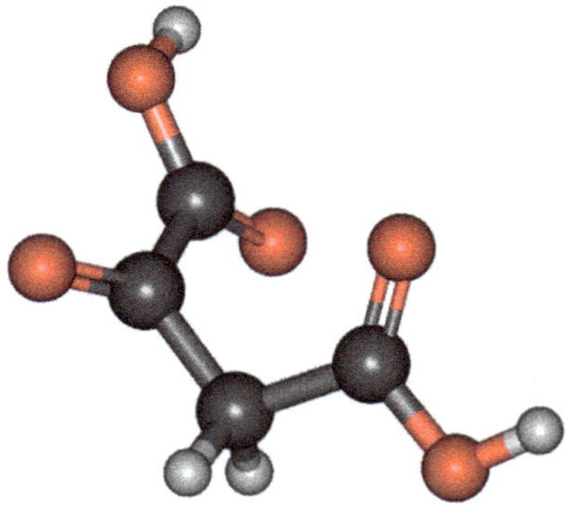

(**a**)

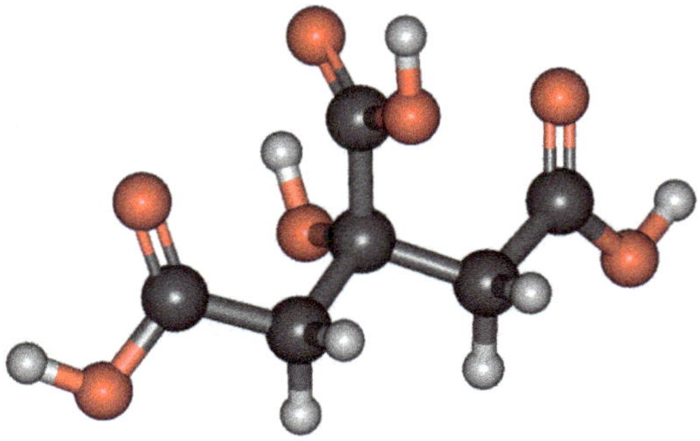

(**b**)

Figure 1. Optimized structure of the (**a**) neutral cis-enol oxaloacetic acid, (**b**) citric acid. The molecules were optimized at the ωB97x/cc-pvtz level of theory at 0 K.

In addition to its biological role, CA is used as a reducing agent for nanoparticle synthesis and citrate serves as a stabilizing agent for example for gold nanoparticles. Gold nanoparticles are interesting due to their optical properties associated with the excitation of their surface plasmon resonance. Recent studies have shown that this excitation can lead to the formation of hot electrons within the nanoparticles, which can be transferred to adsorbed molecules [13,14] such as the stabilizing citrate molecules. Indeed, it was demonstrated that irradiation of citrate stabilized gold nanoparticles with visible light can lead to decomposition of the citrate [15]. In this context it is very important to characterize the interaction of citrate or the protonated form of citrate, i.e., CA, with low-energy electrons. Nanoparticles are also suggested as radiosensitizers in radiation therapy because they increase the local dose of LEEs upon irradiation [16]. In order to assess their mode of action, it is also important to characterize the interaction of LEEs with the stabilizing coating of nanoparticles.

In this work, we demonstrate that both oxaloacetic and citric acid are highly reactive towards electrons with very low energy, i.e., close to 0 eV. However, the involved processes at these energies differ according to the nature of the acids. Additionally, we observe the formation of highly reactive neutral species that can lead to severe cell damage in a biological environment.

2. Methodology

2.1. Experimental Procedure

The interaction of electrons with OAA and CA was studied by means of electron-molecular crossed-beam setup consisting of a cathode being a source of electrons, an oven, and a quadrupole mass analyzer (QMS) [17]. All elements of the experimental setup are located in an ultra-high vacuum chamber with a base pressure of 10^{-9} mbar. The working pressure was three (6.5×10^{-6} mbar) or two ($1.6\text{–}2 \times 10^{-7}$ mbar) orders of magnitude higher with respect to the base pressure for OAA and CA, respectively. A quasi-monoenergetic electron beam is generated from a trochoidal electron monochromator (FWHM of about 200 meV, electron current of 15-20 nA) and perpendicularly crosses with an effusive molecular beam of OAA or CA. The molecular beam emanates from the container containing the sample of the investigated compounds (Sigma Aldrich) heated by halogen bulbs located around the spectrometer. CA decomposes at temperatures above 434 K, as it has been shown by thermogravimetry and differential scanning calorimetry measurements [18]. OAA can adopt two conformations, the cis-enol and the trans-enol forms. The melting point of the cis-enol form is 425 K, and that of the trans-enol forms is higher, i.e., 457 K [19]. Since the experiments were undertaken at 400 K (for CA) and 379 K (for OAA), the electrons are likely interacting with unaltered intact molecules. OAA is likely to be produced mainly in the cis-enol structure. The negative ions formed from the integration of the electrons with the investigated compounds are extracted from the reaction zone by small draw-out-field (<1.0 Vcm^{-1}), analyzed by the QMA and detected by single pulse counting technique. In the present experiment we use SF_6 as a reference gas to calibrate the energy scale since it produces a near 0 eV $[SF_6]^-$ resonance. It should be noted that the experiments are done without the presence of the reference gas to avoid unwanted reactions of dissociative electron transfer with the target compounds [20].

2.2. Theoretical Method

Calculations were performed in the framework of the density-functional theory (DFT) using the Gaussian16 suite of programs [21]. We used the long-range-corrected hybrid density functional ωB97x [22] together with the Gaussian atomic basis set cc-pvtz and aug-cc-pvtz [23].

Two structures were optimized for the neutral OAA corresponding to the cis-enol form (Figure 1a) and the trans-enol (Figure S1), the former being more stable than the second one by 0.05 eV. The vertical electron affinity (VEA) defined as the energy difference between the neutral and lowest anion state, each in the optimized structure of the neutral molecule, is calculated at ωB97x/aug-cc-pvtz level. It is found to be positive by 0.20 eV (cis-enol) and 0.50 eV (trans-enol). The dipole moment for each structure is calculated to be 1.87 D and 2.25 D, respectively. The average quadrupole moment and the isotropic polarizability are calculated to be −49.08 D. Å and 9.43 $Å^3$ for the cis-enol configuration and −49.80 D.Å and 9.57 $Å^3$ for trans-enol configuration. The structure of the neutral CA is shown in Figure 1b. In contrast to OAA, the VEA of CA is calculated to be negative (−1.52 eV). The average dipole and quadrupole moments and isotropic polarizability are estimated to be 3.33 D, −71.12 D. Å and 14.08 $Å^3$, respectively.

The prediction of resonant attachment energies can be investigated with approaches ranging from empirical methods to high-level ab initio calculation coupled with the stabilization method [24]. Here, we use an alternative, and less cumbersome, methodology to the stabilization method. The attachment energies of the incoming electron are calculated as the sum of the VEA and the excitation energies of the anion computed at TDDFT level (time-dependent DFT). For describing the valence-type excitations, TDDFT calculations are performed at ωB97x/cc-pvtz level, where the use of the relatively small basis set cc-pvtz (without any diffuse function) serves as a confinement for transitory states and is well suited to describe valence-type excitations [24]. However, for OAA, the long range bound states ("dipole bound" states) are estimated with the large basis set aug-cc-pvtz.

Finally, the enthalpies and the Gibbs free energies of the fragmentation are calculated at ωB97x/aug-cc-pvtz level at 298 K and 400 K.

3. Results and Discussion

Figures 2 and 3 present the recorded anion fragment yields, for OAA and CA, respectively, as a function of the energy of the incident electrons. Table 1 reports the resonance energies observed from these figures. The detected fragments were tentatively assessed by stoichiometry to the species listed in Table 1. The yield functions exhibit structures indicative of dissociative electron attachment (DEA), which is the most efficient mechanism for the molecular fragmentation at the electron energies below 10 eV. In DEA, the colliding electron is resonantly captured by the target molecule to form a transitory negative ion (TNI). If the lifetime of this TNI is longer than the dissociation time, the transitory negative ion dissociates into a negative fragment and one or more neutral counterparts [25]. Otherwise, the metastable parent anion may survive with a sufficiently long time to be detected mass spectrometrically as a parent anion, or the excess electron may auto-detach producing the neutral precursor molecule possibly in some excited state. The observed DEA signals in Figures 2 and 3 are convolutions of the electron capture processes (or cross section) (as predicted by the calculated resonant states shown in Figure 4, Figure S2) and the survival probabilities (i.e., dissociation dynamics).

We will first consider the mechanism of the electron attachment for the formation of the TNI before considering the energetics for the production of the negative species and their neutral counterparts.

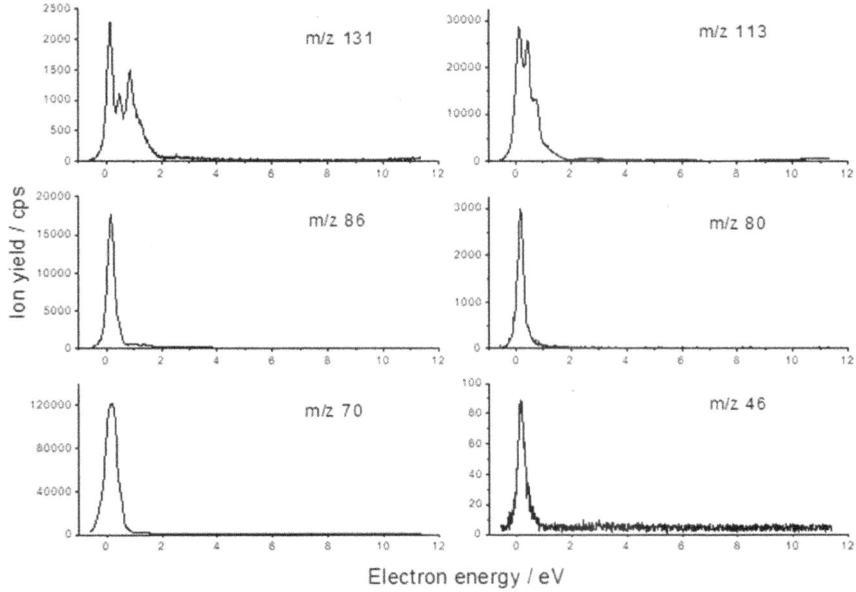

(**a**)

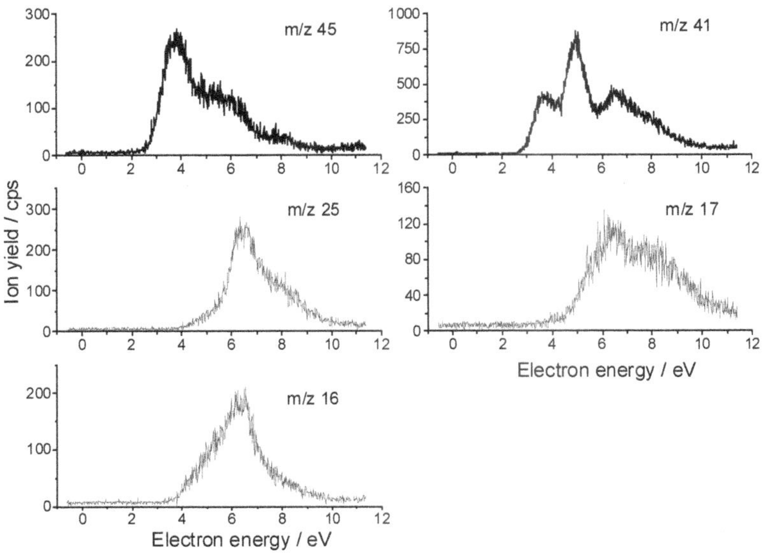

(**b**)

Figure 2. Anion yield functions of the oxaloacetic acid (**a**) fragment anions with m/z 131, 113, 86, 80, 70 and 46, and (**b**) fragment anions with m/z 45, 41, 25, 17, and 16.

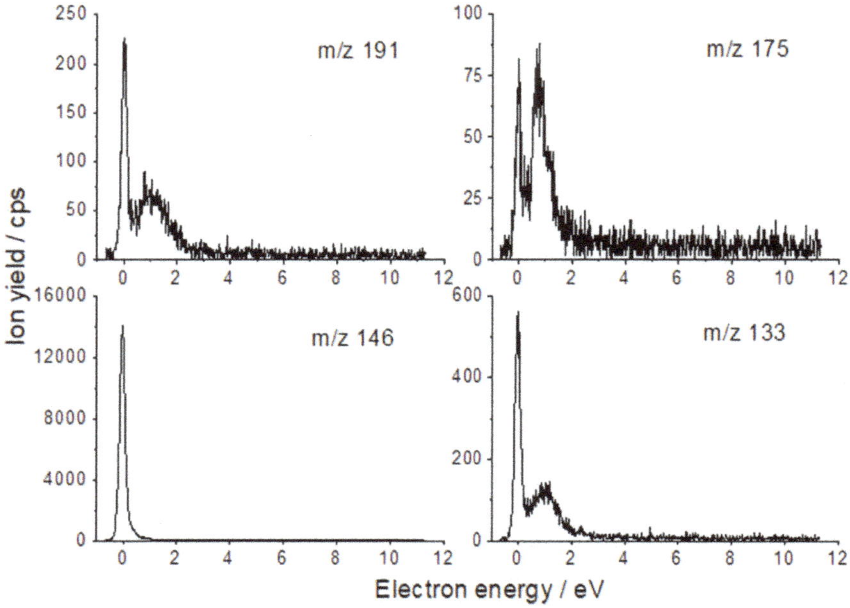

(a)

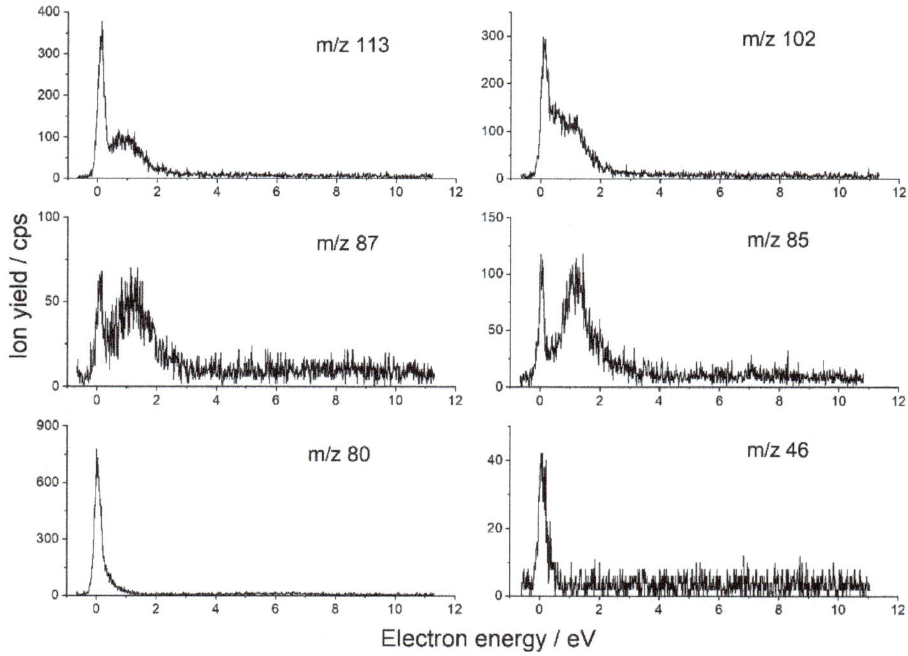

(b)

Figure 3. Anion yield functions of the citric acid: (**a**) fragment anions with m/z 191, 175, 146, and 133; (**b**) fragment anions with m/z 113, 102, 87, 85, 80 and 46.

Table 1. List of the detected m/z negative species created by DEA to oxaloacetic acid (OAA) (left) and to citric acid (CA) (right). The resonance energies are obtained from Figures 2 and 3. The suggested associated radical to the negative fragments and its formation energy are discussed in the text. (* weak peak). In blue the same formula is given in terms of neutral loss channels.

Anion Fragment from OAA m/z	Resonance Energies (eV)	Anion Fragment from CA m/z	Resonance Energies (eV)
131 $[C_4H_3O_5]^-$ $[OAA\text{-}H]^-$	0.14 0.48 0.88 2.78 *	191 $[C_6H_7O_7]^-$ $[CA\text{-}H]^-$	0.01 0.99
113 $[C_4HO_4]^-$ $[OAA\text{-}H_2O\text{-}H]^-$	0.15 0.44 0.71 1.13 2.75 * 4.8 *	175 $[C_6H_7O_6]^-$ $[CA\text{-}OH]^-$	0.0 0.85
86 $[C_3H_2O_3]^-$ $[OAA\text{-}HCOOH]^-$	0.14 1.3 * 3.1 * 6.4 *	146 $[C_5H_6O_5]^-$ $[CA\text{-}HCOOH]^-$	0.0
80 $[C_4O_2]^-$ $[OAA\text{-}H_2O\text{-}H_2O_2]^-$	0.16 0.56 1.4 *	133 $[C_4H_5O_5]^-$ $[CA\text{-}C_2H_3O_2]^-$	0.0 0.94
70 $[C_2H_2O_2]^-$ $[OAA\text{-}HOCOOH]^-$	0.17 0.53 * 1.3 2.83 * 4.4 *	113 $[C_5H_5O_3]^-$ $[CA\text{-}H_2O_2\text{-}HCO_2]^-$	0.12 0.89
46 $[HCOOH]^-$	0.17	102 $[C_4H_6O_3]^-$ $[CA\text{-}2(HCO_2)]^-$	0.11 0.41 0.98
45 $[HCOO]^-$	3.7 5.0 6.3 7.0	87 $[C_3H_3O_3]^-$	0.09 1.22
41 $[C_2OH]^-$	3.7 4.95 6.6 ~8.0	85 $[C_3HO_3]^-$	0.04 1.20
25 $[C_2H]^-$	6.5 8.0	80 $[C_4O_2,]^-$	0.05
17 $[OH]^-$	5.3(4) * 6.3 8.1	46 $[HCOOH]^-$	0.09
16 $[O]^-$	5.2 6.3 7.0		

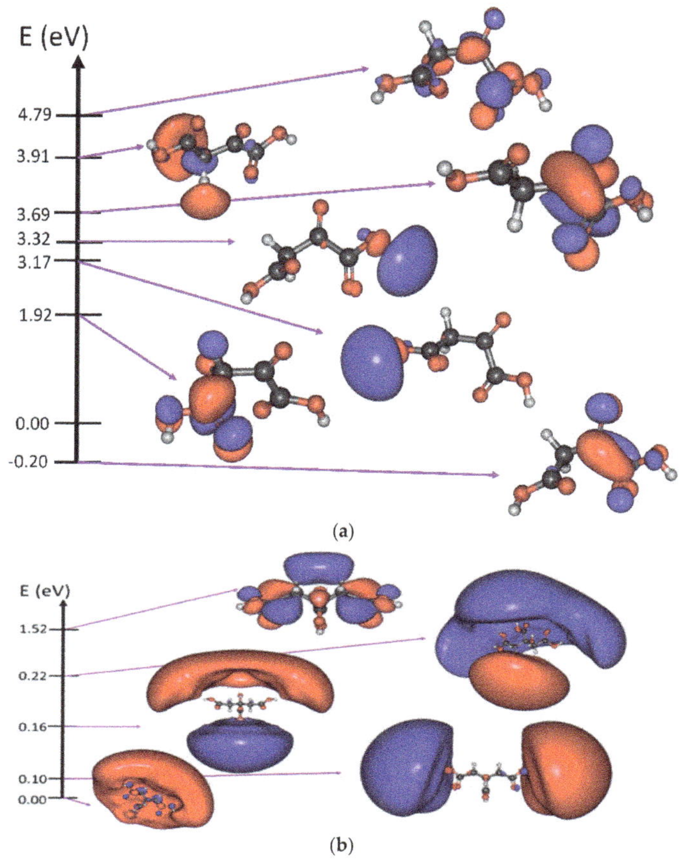

Figure 4. Calculated resonance energies (eV) and molecular orbitals associated to the attachment of the electron for (**a**) cis-enol oxaloacetic and (**b**) citric acids. Results obtained for the trans-enol OA are given in Figure S2. The cis-eno oxaloacetic acid was calculated at the ωB97x/cc-pvtz level at 0 K and the citric acid at the ωB97x/aug-cc-pvtz level of theory at 0 K since the presence of diffuse orbitals.

3.1. Resonant Electron Attachment and Formation of the Transitory Negative Anion

At electron energies below 10 eV, it is now established that the formation of the TNI may arise via three mechanisms, known as (a) the shape resonance, (b) the core-excited resonance [25] and (c) the dipole-bound mediated vibrational Feshbach resonance [26]. For the shape resonance, the incoming electron is trapped by the shape of the electron-molecule potential and occupies a usually unoccupied or a virtual molecular orbital. In the case of a core-excited resonance, the colliding electron excites one of the core-electrons into some electronic excited state, before being trapped by the positive core. Finally, the electron may transfer energy to some vibrational modes of the molecule, before being captured into a "dipole- or multipole-" bound state with a low binding energy, for which the excess electron occupies a very diffuse orbital outside the molecular frame [27]. If such a weakly bound state couples to some dissociative molecular valence orbital, the transitory anion undergoes dissociation [26]. All these mechanisms may arise in the 0–10 eV energy range [25,26,28,29]. Figure 4a,b exhibits the calculated resonance energies and the main molecular orbitals (MOs) associated with the attachment of the electron for OAA and CA, respectively. They represent the potential states into which the excess electron may be trapped to form the TNI.

For the OAA, the calculated MOs indicate that the excess electron occupies pure valence orbitals (Figure 4a) prior to the fragmentation of the TNI via competitive channels as shown in Figure 2. In the stable anion (energy of −0.20 eV in Figure 4a), the extra electron is located in a pi-type orbital at the COCOOH moiety of the cis-enol OAA. In contrast, the electron is located on the other moiety (CH_2COOH group) of the molecule for excitation at 1.92 eV. Localization of the electron on peripheral hydroxyl groups occurs according to our calculations at 3.17 and 3.32 eV. The excitation energies of the trans-enol configuration (Figure S2) are somewhat similar to those of the cis-enol form. The calculated states presented in Figure 4a are associated with the shape resonance, except the resonance at 3.69 eV for which the attachment of the incoming electron with a spin alpha is concomitant with the HOMO-to-LUMO transition (spin beta).

The case of CA is interestingly different. The three lowest molecular orbitals exhibit a strong contribution of a very diffuse MO associated to a "multipole-"bound anion (due to the high values of the dipole and quadrupole moments and polarizability of the CA) with some contribution of valence molecular orbitals. Particularly at ~0 eV, the diffuse orbital encloses all three carboxyl groups (Figure 4b). The next diffuse MO localized at 0.1 eV is found to be close to the side carboxyl groups while at 0.16 eV, the excess electron density is localized near the central hydroxyl group. This diffuse "multipole-"bound state may then couple to the valence MOs, leading to the opening of very rich molecular dissociation decay channels, as observed in Figure 3. At the energy of 1.52 eV, the excess electron occupies pure valence MOs and the DEA is associated with the shape resonance.

3.2. Fragmentation of the Transitory Negative Anion

In DEA, the fragmentation of the TNI produces a negative ion fragment, which is detected by mass spectrometry and one or more neutral counterpart(s). The different fragmentation pathways are depicted in Schemes 1 and 2. In a recent study by Pshenichnyuk et al., DEA to OAA was already reported, and a stable parent anion was observed [30]. The lifetime of the parent anion was estimated to be 25 µs, which is however too short to be observed in the presently used experimental setup. For molecules containing a hydroxyl or a carboxyl group, the dehydrogenation of the parent anion arises from the O-H cleavage near the energy threshold [31–33]. For the formic acid, this threshold has been reported at 1.15 eV [31]. The reaction enthalpy of electron-induced dehydrogenation of HCOOH can be described as a balance between the electron affinity of the HCOO (i.e., 3 17 eV [31]) and the HCOO-H bond dissociation energy (i.e., 4.47 eV [31]). In contrast, for the acetic acid, the pyruvic acid, the benzoic acid and the bromopyruvic acid this threshold is considerably below 1 eV [32–35]. In the present case, for both molecules, we observe the threshold for the molecular dehydrogenation near 0 eV (Figure 2a, m/z 131 and Figure 3a, m/z 191). We performed quantitative calculations on the energetics for the production of the hydrogen atom associated with the [M-H]- anion (M: OAA or CA), and also for the loss of the HCOOH moiety, which has been observed for both OAA and CA. The dissociation channel associated with the loss of a hydrogen atom has been observed for various molecular systems. Table 2 presents the calculated dissociation threshold for the formation of the dehydrogenated parent anion. For both molecules, it has been found that, at 0 K, the dissociation is not accessible near 0 eV, since the endothermicity of the reaction for the cis-enol structure of OAA and for CA (i.e., by 0.95 eV and 0.89 eV, respectively). The Gibbs free energies were also calculated at 400 K for both molecules. The dehydrogenation reaction from cis-enol OAA is found to be 0.19 eV, and the reaction is even more favorable for the trans-enol OAA since the dissociation threshold is calculated to be 0.05 eV. These results, for which the loss of the hydrogen atom arises at the carboxyl site, are in relatively good agreement with the experimental observations (Figure 2a, m/z 131). Furthermore, the near 0 eV resonance observed in the yield function of the [OAA-H]- anion suggests the contribution of the trans-enol form produced in the molecular beam. In the case of CA, the dissociation threshold at 400 K is calculated to be 0.17 eV' again, in relatively good agreement with the experimental observation (Figure 3a, m/z 191). It is noteworthy that the loss of the hydrogen atom arises from the middle site carboxyl groups, while for the side carboxyl and the hydroxyl groups, the reaction energy is higher, i.e., 0.22 eV and 1 eV, respectively. At 298 K, the calculated Gibbs free energies of the $[M\text{-}H]^-$ anions (M: CA or OAA) are found to be slightly higher in energy by about 0.10 eV (Table 2). Thus, we can predict that at room temperature some dissociation channels might be inactivated due to an increase of the thermodynamic threshold.

$$C_4H_4O_5\ (m/z\ 132) + e^- \rightarrow [C_4H_3O_5]^-\ (m/z\ 131) + H$$
$$\rightarrow [C_4HO_4]^-\ (m/z\ 113) + H + H_2O$$
$$\rightarrow [C_3H_2O_3]^-\ (m/z\ 86) + HCOOH$$
$$\rightarrow [C_4O_2]^-\ (m/z\ 80) + H_2O + H_2O_2$$
$$\rightarrow [C_3H_2O_2]^-\ (m/z\ 70) + HOCOOH$$
$$\rightarrow [HCOOH]^-\ (m/z\ 46) + C_3H_2O_3$$
$$\rightarrow [HCOO]^-\ (m/z\ 45) + C_3H_3O_3$$
$$\rightarrow [C_2OH]^-\ (m/z\ 41) + C_2H_3O_4$$
$$\rightarrow [C_2H]^-\ (m/z\ 25) + C_2H_3O_5$$
$$\rightarrow [OH]^-\ (m/z\ 17) + C_4H_3O_4$$
$$\rightarrow [O]^-\ (m/z\ 16) + C_4H_4O_4$$

Scheme 1. Fragmentation reactions of oxaloacetic acid upon interaction with low-energy electrons.

$$C_6H_8O_7\ (m/z\ 192) + e^- \rightarrow [C_6H_7O_7]^-\ (m/z\ 191) + H$$
$$\rightarrow [C_6H_7O_6]^-\ (m/z\ 175) + OH$$
$$\rightarrow [C_5H_6O_5]^-\ (m/z\ 146) + HCOOH$$
$$\rightarrow [C_4H_5O_5]^-\ (m/z\ 133) + C_2H_3O_2$$
$$\rightarrow [C_5H_5O_3]^-\ (m/z\ 113) + HCO_2 + H_2O_2$$
$$\rightarrow [C_4H_6O_3]^-\ (m/z\ 102) + 2HCO_2$$
$$\rightarrow [C_3H_3O_3]^-\ (m/z\ 87) + C_3H_5O_4$$
$$\rightarrow [C_3HO_3]^-\ (m/z\ 85) + C_3H_7O_4$$
$$\rightarrow [C_4O_2]^-\ (m/z\ 80) + C_2H_8O_5$$
$$\rightarrow [HCOOH]^-\ (m/z\ 46) + C_5H_6O_5$$

Scheme 2. Fragmentation reactions of citric acid upon interaction with low-energy electrons.

Table 2. Calculated fragmentation thresholds (DFT level of theory) of cis-enol oxaloacetic (OA) and citric (CA) acids: enthalphy (0 K) and Gibbs free energy at the ωB97x/aug-cc-pvtz level at 400 K. Energies in eV. * Values for trans-enol OAA are given in parentheses. The values in red are Gibbs free energies calculated at 298 K.

	M = OAA *		M = CA	
	Enthalpy	**Gibbs Free Energy**	**Enthalpy**	**Gibbs Free Energy**
[M-H]⁻	0.95 (0.84)	0.19 (0.05) 0.31 (0.17)	0.89	0.17 0.27
[M-HCOO-H]⁻	4.78 (4.73)	3.05 (3.02) 3.38 (3.34)	4.72	2.98 3.31
[M-HCOOH]⁻	0.22 (0.17)	−0.73 (−0.76) −0.53 (−0.57)	0.16	−0.80 −0.60
[M-HCOO-OH]⁻	1.06 (1.01)	0.13 (0.10) 0.68 (0.66)		

We have also examined the channel leading to the loss of either (H + COOH) or (HCOOH) for both molecules. For the citric acid, producing neutral H + COOH requires ~4.7 eV at 0K and ~3 eV at 400 K (Table 2). On the contrary, the cleavage of the central COOH and the H atom from the side carboxyl group concomitantly to the H + COOH recombination is exothermic by 0.80 eV (at 400 K). Since this channel has been observed near 0 eV (Figure 3a, m/z 146), we suggest the recombination to arise. The energetic scenario for the molecular dissociation is found to be similar for the oxaloacetic acid (Table 2, Figure 2a, m/z 86). At 298 K, the reactions mentioned above are found to be still exothermic (Table 2). In the case of OAA, the most intense peak observed at 0.17 eV is measured for the decay channel leading to the loss of a m/z 70 fragment (Figure 2a, m/z 70). The latter is likely to be associated with the loss of neutral (HOCOCH) to form the [OAA-HOCOOH]⁻ negative ion fragment. The reaction energies require 1.06 eV (1.01 eV) at 0 K, but only 0.13 eV (0.10 eV) at 400 K for the cis-enol (trans-enol) configuration (Table 2). At 298 K, the calculated Gibbs free energy is found to be 0.68 eV, i.e., 0.55 eV higher than that calculated at 400 K. Thus, this dissociation channel observed at 0.17 eV at 400 K might not be favorable at 298 K. It has to be noted that the carbonic acid, HOCOOH, is known to be an unstable species which rapidly undergoes dissociation to CO_2 and H_2O [36].

In addition to the loss of the hydrogen atom or the neutral formic acid, further decay channels lead to the dissociation of either a single (e.g., H_2O_2) or multiple (2xOH) neural product(s), as summarized in Table 1 and Schemes 1 and 2. They can arise through a very complex fragmentation scheme (e.g., $C_4O_2^-$ anion fragment observed from the decomposition of the CA TNI), as a single or multiple steps [37].

The fragmentation of the OAA is mainly driven by the electrons with energies below 1eV (99%) as shown in Figure 2. We observe strong resonances already close to zero eV (i.e., 0.15 eV), and the decomposition is most likely driven by the formation of rather stable neutral molecules. Near threshold energies, we can estimate the branching ratio for the formation of the $[C_4H_3O_5]^-$, $[C_4HO_4]^-$, $[C_3H_2O_3]^-$, $[C_4O_2]^-$, [OAA-HOCOOH]⁻, and [HCOOH]⁻ anion fragments and their associated counterpart to be 3.8%, 13.6%, 4.0%, 1.9%, 76,7% and 0.05%, respectively. These branching ratios reflect the dynamics of the dissociation of the TNI at 400 K. Although the loss of the (HOCOOH) neutral species resulting in the anionic fragment at m/z 70 is by far the most dominant dissociation channel, the production of other neutral radicals such as H_2O_2 (or 2 OH) and ballistic hydrogen atoms, represents almost 23% of the decay channels. These radicals may then play a role in the modification of the medium under irradiation when containing OAA molecules. In the case of CA, the molecular fragmentation is induced by electrons with energies below 1 eV, for which 84% arise from the near 0 eV electrons dominated by the loss of the (HCOOH) neutral(s) associated with m/z 146 (Figure 3a), and 16% from ~1 eV electrons (Figure 3b).

4. Conclusions

The present work reports a comparative study of dissociative electron attachment to oxaloacetic and citric acid. The two molecular systems show fragmentation of the transitory negative ion by near 0 eV electron attachment to the neutral parent molecule. However, the process is completely different in both molecules. Indeed, for the oxaloacetic acid, the dissociation operates via the attachment of the excess electron into some valence molecular orbital while for the citric acid, it is very likely to arise initially from the formation of a "dipole-multipole-"bound anion as a gateway for the dissociation of the covalent TNI. The decomposition of both metabolites by low-energy electrons could be associated with a disturbance of a cell's metabolism and therefore the results found here may represent important pathways of radiation induced cell damage.

Nevertheless, in the condensed phase situation the molecules are surrounded by some environment (e.g., H_2O for biological type of environment). In this case it is well known that the electron attachment process may change [38,39]. For instance, it has been demonstrated that the dipole bound character may disappear for the benefit of a valence bound anion, when the molecule is solvated by a couple of water molecules from the environment [40]. Thus, the citric acid might not be sensitive to the electron attacks in contrast to the oxaloacetic acid in condensed media. In contrast, the decomposition of the oxaloacetic acid by low energy ballistic electrons may produce reactive species such as hydrogen peroxide contributing to the alteration of nucleobases [41], the formic acid involved in the glycosidic bond cleavage or the hydrolysis of DNA [42] or the CO_2 as a mediating agent for DNA damage [43].

Supplementary Materials: The following are available online at https://www.mdpi.com/article/10.3390/ijms22147676/s1, Figure S1: Optimized geometry of the trans-enol oxaloacetic acid at 0 K, Figure S2: Calculated resonance energies (eV) and molecular orbitals associated with the attachment of the electron for the trans-enol oxaloacetic acid, calculated at the ωB97x/cc-pvtz level of theory at 0 K.

Author Contributions: J.K. designed the research. J.K., P.W. and A.T. did all experiments. G.T. and F.R. did all calculations. J.K., I.B. and H.A.-C. analyzed the data and wrote the manuscript. All authors have read and agreed to the published version of the manuscript.

Funding: J.K. acknowledges support by the statutory activity subsidy (No 25/20/B) from the Polish Ministry of Science and Higher Education. H.A.-C, G.T. and F.R. thank the ANR-PRC BAMBI (grant number 18-CE30-0009-03) for support of their contribution to this work. In addition, G.T. and F.R. thank the GENCI-IDRIS for supporting the computing time through the grant (A0090807662). I.B. acknowledges support by the European Research Council (ERC; consolidator grant no. 772752). This collaborative work has been performed under the MD-GAS COST Action CA18212. We acknowledge the support of the Deutsche Forschungsgemeinschaft and Open Access Publishing Fund of University of Potsdam.

Data Availability Statement: Data is available from the corresponding authors upon request.

Conflicts of Interest: The authors declare no conflict of interest.

References

1. Pshenichnyuk, A.; Modelli, A.; Komolov, S. Interconnections between dissociative electron attachment and electron-driven biological processes. *Int. Rev. Phys. Chem.* **2018**, *37*, 125–170. [CrossRef]
2. Alizadeh, E.; Orlando, T.M.; Sanche, L. Biomolecular Damage Induced by Ionizing Radiation: The Direct and Indirect Effects of Low-Energy Electrons on DNA. *Annu. Rev. Phys. Chem.* **2015**, *66*, 379–398. [CrossRef]
3. Baccarelli, I.; Bald, I.; Gianturco, F.A.; Illenberger, E.; Kopyra, J. Electron-induced damage of DNA and its components: Experiments and theoretical models. *Phys. Rep.* **2011**, *508*, 1–44. [CrossRef]
4. Sanche, L. Interaction of low energy electrons with DNA: Applications to cancer radiation therapy, Radiat. *Phys. Chem.* **2016**, *128*, 36–43. [CrossRef]
5. Schürmann, R.; Vogel, S.; Ebel, K.; Bald, I. The Physico-Chemical Basis of DNA Radiosensitization: Implications for Cancer Radiation Therapy. *Chem. A Eur. J.* **2018**, *24*, 10271–10279. [CrossRef] [PubMed]

6. Bölher, E.; Wamecke, J.; Swiderek, P. Control of chemical reactions and synthesis by low energy electron. *Chem. Soc. Rev.* **2013**, *42*, 9219–9231.
7. Abdoul-Carime, H.; Bald, I.; Illenberger, E.; Kopyra, J. Selective Synthesis of Ethylene and Acetylene from Dimethyl Sulfide Cold Films Controlled by Slow Electrons. *J. Phys. Chem. C* **2018**, *122*, 24137–24142. [CrossRef]
8. Kopyra, J.; Rabilloud, F.; Wierzbicka, P.; Abdoul-Carime, H. Energy-Selective Decomposition of Organometallic Compounds by Slow Electrons: The Case of Chloro(dimethyl sulfide)gold(I). *J. Phys. Chem. A* **2021**, *125*, 966–972. [CrossRef]
9. Mucke, M.; Braune, M.; Barth, S.; Förstel, M.; Lischke, T.; Ulrich, V.; Arion, T.; Becker, U.; Bradshaw, A.; Hergenhahn, U. A hitherto unrecognized source of low-energy electrons in water. *Nat. Phys.* **2010**, *6*, 143–146. [CrossRef]
10. Boudaiffa, B.; Cloutier, P.; Hunting, D.; Huels, M.A.; Sanche, L. Resonant Formation of DNA Strand Breaks by Low-Energy (3 to 20 eV) Electrons. *Science* **2000**, *287*, 1658–1660.
11. Meißner, R.; Kočišek, J.; Feketeová, L.; Fedor, J.; Fárník, M.; Limão-Vieira, P.; Illenberger, E.; Denifl, S. Low-energy electrons transform the nimorazole molecule into a radiosensitizer. *Nat. Commun.* **2019**, *10*, 2388. [CrossRef]
12. Rak, J.; Chomicz, L.; Wiczk, J.; Westphal, K.; Zdrowowicz, M.; Wityk, P.; Żyndul, M.; Makurat, S.; Golon, Ł. Mechanisms of Damage to DNA Labeled with Electrophilic Nucleobases Induced by Ionizing or UV Radiation. *J. Phys. Chem. B* **2015**, *119*, 8227–8238. [CrossRef]
13. Zhan, C.; Chen, X.-J.; Yi, J.; Li, J.-F.; Wu, D.-Y.; Tian, Z.-Q. From plasmon-enhanced molecular spectroscopy to plasmon-mediated chemical reactions. *Nat. Rev. Chem.* **2018**, *2*, 216–230. [CrossRef]
14. Schürmann, R.; Bald, I. Real-time monitoring of plasmon induced dissociative electron transfer to the potential DNA radiosensitizer 8-bromoadenine. *Nanoscale* **2017**, *9*, 1951. [CrossRef] [PubMed]
15. Schlather, A.E.; Manjavacas, A.; Lauchner, A.; Marangoni, V.S.; DeSantis, C.J.; Nordlander, P.; Halas, N.J. Hot Hole Photoelectrochemistry on Au@SiO2@Au Nanoparticles. *J. Phys. Chem. Lett.* **2017**, *8*, 2060−2067. [CrossRef] [PubMed]
16. Clement, S.; Campbell, J.M.; Deng, W.; Guller, A.; Nisar, S.; Liu, G.; Wilson, B.C.; Goldys, E.M. Mechanisms for Tuning Engineered Nanomaterials to Enhance Radiation Therapy of Cancer. *Adv. Sci.* **2020**, *7*, 2003584. [CrossRef] [PubMed]
17. Kopyra, J. Low energy electron attachment to the nucleotide deoxycytidine monophosphate: Direct evidence for the molecular mechanisms of electron-induced DNA strand breaks. *Phys. Chem. Chem. Phys.* **2012**, *14*, 8287–8289. [CrossRef]
18. Wyrzykowski, D.; Hebanowska, E.; Nowak-Wiczk, G.; Makowski, M.; Chmurzyński, L. Thermal behaviour of citric acid and isomeric aconitic acids. *J. Therm. Anal. Calorim.* **2011**, *104*, 731–735. [CrossRef]
19. Sigma Aldrich Information Product Number 04126.
20. Gohlke, S.; Abdoul-Carime, H.; Illenberger, E. Dehydrogenation of adenine induced by slow (<3 eV) electrons. *Chem. Phys. Lett.* **2003**, *380*, 595–599.
21. Frisch, M.J.; Trucks, G.W.; Schlegel, H.B.; Scuseria, G.E.; Robb, M.A.; Cheeseman, J.R.; Scalmani, G.; Barone, V.; Petersson, G.A.; Nakatsuji, H.; et al. *Gaussian 16, Revision C.01*; Gaussian, Inc.: Wallingford, CT, USA, 2016.
22. Chai, J.-D.; Head-Gordon, M. Systematic optimization of long-range corrected hybrid density functionals. *J. Chem. Phys.* **2008**, *128*, 084106. [CrossRef]
23. Kendall, R.A.; Dunning, T.H.; Harrison, R.J. Electron affinities of the first-row atoms revisited. Systematic basis sets and wave functions. *J. Chem. Phys.* **1992**, *96*, 6796–6806. [CrossRef]
24. Sommerfeld, T.; Weber, R.J. Empirical Correlation Methods for Temporary Anions. *J. Phys. Chem. A* **2011**, *115*, 6675–6682. [CrossRef]
25. Illenberger, E.; Momigny, J. *Gaseous Molecular Ions. An Introduction to Elementary Processes Induced by Ionization*; Baümgartel, H., Franck, E.U., Grünbein, W., Eds.; Springer: Berlin/Heidelberg, Germany, 1992.
26. Sommerfeld, T. Dipole bound states as doorways in (dissociative) electron attachment. *J. Phys. Conf. Ser.* **2005**, *4*, 245–250. [CrossRef]
27. Abdoul-Carime, H.; Desfrançois, C. Electron weakly bound to molecules by dipolar, quadrupolar and polarization forces. *Eur. Phys. J. D* **1998**, *2*, 149–156. [CrossRef]
28. Li, Z.; Ryszka, M.; Dawley, M.M.; Carmichael, I.; Bravaya, K.B.; Ptasińska, S. Dipole-Supported Electronic Resonances Mediate Electron-Induced Amide Bond Cleavage. *Phys. Rev. Lett.* **2019**, *122*, 73002. [CrossRef]
29. Kopyra, J.; Rabilloud, F.; Abdoul-Carime, H. Core-excited resonances initiated by unusually low energy electrons observed in dissociative electron attachment to Ni(II)(bis)acetylacetonate. *J. Chem. Phys.* **2020**, *153*, 124302. [CrossRef]
30. Pshenichnyuk, S.A.; Asfandiarov, N.L. Structural rearrangements as relaxation pathway for molecular negative ions formed via vibrational Feshbach resonance. *Phys. Chem. Chem. Phys.* **2020**, *22*, 16150–16156. [CrossRef]
31. Pelc, A.; Sailer, W.; Scheier, P.; Mason, N.; Märk, T. Low energy electron attachment to formic acid. *Eur. Phys. J. D* **2002**, *20*, 441–444. [CrossRef]
32. Sailer, W.; Pelc, A.; Probst, M.; Limtrakul, J.; Scheier, P.; Illenberger, E.; Märk, T.D. Dissociative electron attachment to acetic acid (CH3COOH). *Chem. Phys. Lett.* **2003**, *378*, 250–256. [CrossRef]
33. Zawadzki, M.; Ranković, M.; Kočišek, J.; Fedor, J. Dissociative electron attachment and anion-induced dimerization in pyruvic acid. *Phys. Chem. Chem. Phys.* **2018**, *20*, 6838–6844. [CrossRef]
34. Zawadzki, M.; Wierzbicka, P.; Kopyra, J. Dissociative electron attachment to benzoic acid (C7H6O2). *J. Chem. Phys.* **2020**, *152*, 174304. [CrossRef]

35. Da Silva, F.F.; Varella, M.T.D.N.; Jones, N.C.; Hoffmann, S.V.; Denifl, S.; Bald, I.; Kopyra, J. Electron-Induced Reactions in 3-Bromopyruvic Acid. *Chem. A Eur. J.* **2019**, *25*, 5498–5506. [CrossRef] [PubMed]
36. Ghoshal, S.; Hazra, M.K. $H_2CO_3 \rightarrow CO_2 + H_2O$ decomposition in the presence of H_2O, HCOOH, CH_3COOH, H_2SO_4 and HO_2 radical: Instability of the gas-phase H_2CO_3 molecule in the troposphere and lower stratosphere. *RSC Adv.* **2015**, *5*, 17623–17635. [CrossRef]
37. Da Silva, F.F.; Matias, C.; Almeida, D.; Garcia, G.; Ingólfsson, O.; Flosadóttir, H.D.; Ómarsson, B.; Ptasinska, S.; Puschnigg, B.; Scheier, P.; et al. NCO^-, a Key Fragment Upon Dissociative Electron Attachment and Electron Transfer to Pyrimidine Bases: Site Selectivity for a Slow Decay Process. *J. Am. Soc. Mass Spectrom.* **2013**, *24*, 1787–1797. [CrossRef] [PubMed]
38. Bald, I.; Langer, J.; Tegeder, P.; Ingólfsson, O. From isolated molecules through clusters and condensates to the building blocks of life. *Int. J. Mass Spectrom.* **2008**, *277*, 4–25. [CrossRef]
39. Schürmann, R.; Tsering, T.; Tanzer, K.; Denifl, S.; Kumar, S.V.K.; Bald, I. Resonant Formation of Strand Breaks in Sensitized Oligonucleotides Induced by Low-Energy Electrons (0.5–9 eV). *Angew. Chem. Int. Ed.* **2017**, *56*, 10952–10955. [CrossRef]
40. Abdoul-Carime, H.; Bouteiller, Y.; Desfrançois, C.; Philippe, L.; Schermann, J.P.; Niinistö, L.; Styring, S.; Tommos, C.; Warncke, K.; Wood, B.R. Excess Electrons in Polar Cluster Anions. *Acta Chem. Scand.* **1997**, *51*, 145–150. [CrossRef]
41. Blakely, W.F.; Fuciarelli, A.F.; Wegher, B.J.; Dizdaroglu, M. Hydrogen peroxide-induced base damage in deoxyribonucleic acid. *Radiat. Res.* **1990**, *121*, 338–343. [CrossRef]
42. Dizaroglu, M.; Gajewski, E. Selected ion mass spectrometry: Assays of oxidative DNA damage. *Methods Enzymol.* **1990**, *186*, 530–544.
43. Yermilov, V.; Yoshie, Y.; Rubio, J.; Ohshima, H. Effects of carbon dioxide/bicarbonate on induction of DNA single-strand breaks and formation of 8-nitroguanine, 8-oxoguanine and base-propenal mediated by peroxynitrite. *FEBS Lett.* **1996**, *399*, 67–70. [CrossRef]

International Journal of *Molecular Sciences*

MDPI

Article

Ring-Selective Fragmentation in the Tirapazamine Molecule upon Low-Energy Electron Attachment

Eugene Arthur-Baidoo [1,2], Joao Ameixa [1,2,3], Milan Ončák [1,*] and Stephan Denifl [1,2,*]

[1] Institute for Ion Physics and Applied Physics, University of Innsbruck, Technikerstrasse 25/3, 6020 Innsbruck, Austria; Eugene.Arthur-Baidoo@uibk.ac.at (E.A.-B.); j.ameixa@campus.fct.unl.pt (J.A.)

[2] Center for Molecular Biosciences Innsbruck, University of Innsbruck, Technikerstrasse 25/3, 6020 Innsbruck, Austria

[3] Atomic and Molecular Collisions Laboratory, Department of Physics, CEFITEC, Universidade NOVA de Lisboa, 2829-516 Caparica, Portugal

* Correspondence: Milan.Oncak@uibk.ac.at (M.O.); Stephan.Denifl@uibk.ac.at (S.D.)

Abstract: We investigate dissociative electron attachment to tirapazamine through a crossed electron–molecule beam experiment and quantum chemical calculations. After the electron is attached and the resulting anion reaches the first excited state, D_1, we suggest a fast transition into the ground electronic state through a conical intersection with a distorted triazine ring that almost coincides with the minimum in the D_1 state. Through analysis of all observed dissociative pathways producing heavier ions (90–161 u), we consider the predissociation of an OH radical with possible roaming mechanism to be the common first step. This destabilizes the triazine ring and leads to dissociation of highly stable nitrogen-containing species. The benzene ring is not altered during the process. Dissociation of small anionic fragments (NO_2^-, CN_2^-, CN^-, NH_2^-, O^-) cannot be conclusively linked to the OH predissociation mechanism; however, they again do not require dissociation of the benzene ring.

Keywords: radiosensitizer; low-energy electron; dissociative electron attachment; roaming; tirapazamine; predissociation

Citation: Arthur-Baidoo, E.; Ameixa, J.; Ončák, M.; Denifl, S. Ring-Selective Fragmentation in the Tirapazamine Molecule upon Low-Energy Electron Attachment. *Int. J. Mol. Sci.* **2021**, *22*, 3159. https://doi.org/10.3390/ijms22063159

Academic Editor: Michael Filatov

Received: 25 February 2021
Accepted: 16 March 2021
Published: 19 March 2021

1. Introduction

Attempts to find innovative methods for applications in cancer radiotherapy and concomitant chemotherapy have been a focus of research for decades. Hypoxic cells in solid tumors, however, pose resistance to radiotherapy, resulting in a reduced tumor response [1–3]. The chemical agents, referred to as radiosensitizers, administered during cancer radiotherapy have the capability of differentially sensitizing hypoxic cells and thus enhancing radiation-induced DNA damage, especially in the local tumor regions [4,5]. Therefore, new radiosensitizing agents are still being suggested and investigated to counter this effect of radiation resistance in hypoxic regions [6].

Hypoxic cytotoxins are a kind of radiosensitizer designed to only activate under hypoxic conditions [7]. Their ability to accumulate and retain their integrity in tumor cells under low oxygen content contributes greatly to their radiosensitization property. Usually, such compounds can undergo a reduction process, converting them into reactive radicals, which are able to accumulate in hypoxic tumor cells and, subsequently, induce DNA damages [5]. Electron attachment studies with nimorazole [8], carried out in the gas phase and in the microsolvated conditions, endorse the hypothesis that the protonated radical anion of this hypoxic radiosensitizer [9] mimics the action of molecular oxygen.

One compound which has demonstrated its potential in terms of highly selective cytotoxicity is the heterocyclic–n–oxide derivative, 3–amino 1,2,4–benzotriazine 1,4–dioxide or tirapazamine (TPZ) [6,10]. Under hypoxic conditions, the antitumor prodrug TPZ is transformed into a radical intermediate $TPZ^{\bullet-}$ via a one-electron reduction process that, in a second step, becomes protonated, TPZH• [5,11]. The next reaction steps, which yield bioactive radicals considered to be the precursors for DNA damage, are still under debate. A few studies [12–14] suggested the release of OH• radicals from the protonated radical anion TPZH•, while others proposed the formation of the radical benzotriazinyl, BTZ• upon loss of neutral water from TPZH• [15,16]. Then, DNA double-strand breaks arise as the enzyme topoisomerase II gets inactivated by one of these bioactive radicals, i.e., OH• or BTZ• [11]. In the presence of oxygen, the radical intermediate, $TPZ^{\bullet -}$ is, however, back-oxidized into the nontoxic parent compound with the release of reactive oxygen species (ROS) as a byproduct [17]. Early studies suggested that TPZ can act as a substitute for molecular oxygen (O_2) in the conversion of DNA radicals into toxic strand damage [18,19].

It has been established that the radiosensitization ability of chemical sensitizers is dependent on their affine nature toward electrons [20] and the subsequent production of free radicals [21]. Low-energy electrons (LEEs), with an energy distribution peaking at about 10 eV, are released in the order 10^4/MeV upon the impact of ionizing radiation with biological tissue [21]. Water radiolysis leads to the formation of LEEs, as well as ROS in the physical and physicochemical stages of radiation damage ($<10^{-12}$ s). LEEs may become hydrated in the chemical stage ($<10^{-6}$ s) [22,23], or, if formed near a biological DNA system, they induce damage in DNA via bond cleavage, primarily driven by dissociative electron attachment (DEA) [23–25].

Low-energy electron attachment to TPZ has been studied recently [26]. In summary, this study showed the formation of the intact parent anion $(TPZ)^-$, the dehydrogenated parent anion $(TPZ–H)^-$, the fragment anion due to loss of NH_2 forming $(TPZ–NH_2)^-$ via single bond cleavage and the loss of the hydroxyl radical to form the most abundant fragment anion $(TPZ–OH)^-$ with mass of 161 u. To account for the mechanism of dissociative electron attachment, we proposed a roaming mechanism in which an OH moiety is formed in the first step and then either evaporates or roams in the vicinity of the molecule to react with various parts of the $(TPZ–OH)^-$ core. The roaming mechanism is expected to be suppressed with solvation. Single-reference quantum chemical calculations showed that the ground state potential energy surface (PES) of TPZ might cross the PES of the first excited state in TPZ^-, explaining the resonance near 0 eV. In another study using density functional calculations, Yin et al. [27] showed some possible pathways for reduced TPZ species after the N4-OH homolytic processes. It was reported that the cleavage of C-N bonds leads to ring opening by bond rupture in the triazine moiety [27,28].

Herein, we investigate additional fragmentation pathways of tirapazamine upon low-energy electron interaction in the energy range of 0–12 eV. We report ten further fragmentation channels attributed to multiple bond cleavages in the parent molecule. It is worth noting that, regardless of the dissociations observed in TPZ, the integrity of the benzene structure is preserved. All fragmentation pathways are attributed to the dissociation of the triazine moiety attached to the benzene ring.

2. Results

The DEA process may be the underlying physico-chemical mechanism for the action of some potential radiosensitizers like modified nucleobases [29–31] and for the radiation damage of molecules of biological relevance, like nucleobases, amino acids, etc. [32–34]. The mechanism of dissociation is described to proceed via Equation (1):

$$e^- + M \rightarrow M^{*-} \rightarrow (M - X) + X^- \quad (1)$$

The incoming electron is captured by the parent molecule M yielding a transient negative ion (TNI), M^{*-}, also termed resonance, which subsequently dissociates into a neutral fragment (M–X) and a fragment anion X^-. The superscript * indicates that the TNI may be formed in an excited state [33,34].

Before we discuss dissociation pathways of TPZ^-, we focus on the events that take place directly after electron attachment. As the vertical electron affinity of TPZ is about 1.3 eV, the formed TPZ^{*-} anion is prepared in an electronically excited state. Our time-dependent density functional theory (TDDFT) calculation at the TD–BMK/aug–cc–pVDZ level shows that the first excited state of $TPZ^{\bullet-}$, D_1, lies at 1.65 eV, i.e., it is energetically well-separated from the ground electronic state. The first ten excited states are relatively densely packed within 1.65–3.30 eV (see Table S1). We can thus conclude that, in any electronic state reached after electron attachment, one expects that the system switches readily to the D_1 state through nonradiative processes, due to the low energy separation between the excited states (Kasha's rule).

Figure 1 analyzes the fate of the $TPZ^{\bullet-}$ anion in the D_1 state. Note that a smaller basis is used for calculations so that the excited states are not that close to each other, as in the calculation with the aug–cc–pVDZ basis set. When we optimize the molecule in the D_1 state starting from the planar Franck–Condon (FC) structure, a minimum with a considerably distorted triazine ring is reached. This minimum almost coincides with the D_0/D_1 conical intersection (CI), through which the system might funnel to the ground electronic state; in the CI structure, the benzene ring stays almost planar, while the triazine ring shows the N-N-C-N dihedral angle of 30.3° and a pyramidalized NH_2 group. Thus, when the $TPZ^{\bullet-}$ anion enters the D_1 state (either through direct electron attachment or after passing from a higher electronic state), the D_0/D_1 conical intersection can be reached without overcoming any barrier, and we can expect the switch from D_1 to D_0 on the order of hundreds of femtoseconds. In the ground state, we expect that TPZ^- regenerates its FC structure, as the CI structure with a deformed triazine ring does not clearly hint toward any dissociation pathway.

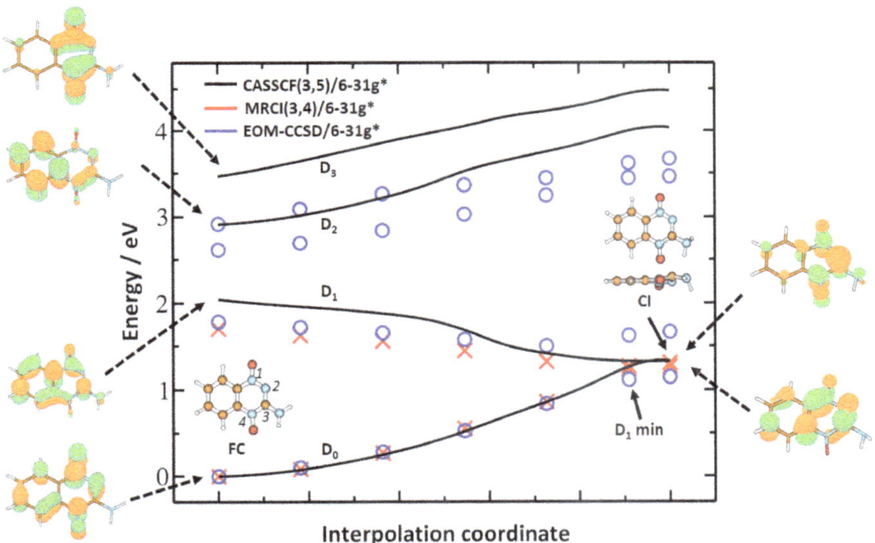

Figure 1. Interpolation from the Franck–Condon point (FC) to the minimum on the D_1 electronic state potential energy surface in TPZ^- and further to the D_0/D_1 conical intersection (CI). The FC point was optimized at the B3LYP/aug–cc–pVDZ level of theory, D_1 minimum and CI at the complete active space–self consistent field (CASSCF)(3,5)/6–31g* level. The interpolation points were recalculated at the multireference configuration interaction (MRCI)(3,4) and equation of motion-coupled clusters singles and doubles (EOM-CCSD) level with the 6–31g* basis set. For the CI, both top and side view is provided to show the triazine ring distortion. On the left-hand and right-hand sides, singly occupied orbitals, according to the configuration with the highest contribution to the respective electronic state in the configuration interaction vector, are shown as calculated in the FC point and CI at the CASSCF level.

From the methodological perspective, we can see that, when accounting for dynamical correlation through either multireference configuration interaction (MRCI) or equation of motion-coupled clusters singles and doubles (EOM-CCSD) method, the excitation energies in the FC point shift to lower values, compared to the complete active space–self consistent field (CASSCF) ones. Higher lying states are not well described by the CASSCF method, due to limited active space. The character of the conical intersection is maintained at the MRCI level, with the difference between D_0 and D_1 states of 0.04 eV on the MRCI(3,4)/6–31g*//CASSCF(3,5)/6–31g* level. Calculations at the MRCI and EOM-CCSD levels are very close to each other in the FC point and at the beginning of the interpolation pathway; when D_0 and D_1 electronic states approach each other, the single-reference EOM-CCSD method is unable to retrieve the potential energy surface of D_1 reliably. Within the single-reference TD–BMK method, a D_1 minimum of a similar structure but with a considerable D_0/D_1 gap is thus obtained [26]. Finally, the shift from the 6–31g* basis set to aug–cc–pVDZ within the TD–BMK approach has a very limited effect on the position of valence states included in Figure 1 in the FC point (with a difference in energy below 0.04 eV).

Further, we focus on the fragmentation mechanisms, resonances positions for the detected anions, experimental onsets of the various anions and quantitatively determined thermodynamic thresholds, as presented in Table 1. The mechanisms leading to the formation of the fragment anions $(TPZ–H)^-$, $(TPZ–NH_2)^-$ and $(TPZ–OH)^-$ included in Table 1 were analyzed previously [26]. The suggested reaction pathways for ions resulting from initial OH predissociation are shown in Figure 2.

Table 1. Summary of anions observed upon electron attachment to tirapazamine, along with their mass, suggested molecular structure, associated neutral fragment(s), resonance positions, experimental threshold and thermodynamically determined threshold. Energies were calculated at the B3LYP/aug–cc–pVDZ level of theory; suggested neutral fragments are based on theoretical analysis.

Mass (u)	Anion	Neutral Counterparts	Resonance Positions (eV) 1	2	3	4	5	6	7	Thresholds (eV) Exp	Theory
178	$C_7H_6N_4O_2^-$		0	0.1	0.3	-	-	-	-	~0	−1.28
177 [a]	$C_7H_5N_4O_2^-$	H	0	0.1	0.3	0.9	-	-	-	~0	−0.92
162 [a]	$C_7H_4N_3O_2^-$	NH_2	0.3	0.7	-	-	-	-	-	~0	−1.71
161 [a]	$C_7H_5N_4O^-$	OH	0	0.1	0.3	0.9	-	-	-	~0	−0.61
145	$C_7H_5N_4^-$	HO_2	0.4	1.0	1.7	4.5	5.3	-	-	~0	−0.12
133	$C_7H_5N_2O^-$	$OH + N_2$	0.0	0.1	0.3	0.8	2.4	-	-	~0	−1.92
132	$C_7H_4N_2O^-$	$H_2O + N_2$/ $OH + N_2H$	0.0	0.1	0.3	0.8	2.4	2.8	3.2	~0	−2.39/ 2.54
118	$C_7H_4NO^-$	$OH + N_3H$/ $H_2O + N_3$	0.0	0.1	2.5	3.2	3.9	-	-	~0	−0.78/ −2.01
116	$C_7H_4N_2^-$	$H_2O + N_2O$	0.0	0.1	0.9	2.6	-	-	-	~0	−1.30
90	$C_6H_4N^-$	$OH + N_2 + HOCN/H_2O + N_2 + NCO$	3.6	4.7	-	-	-	-	-	2.1	2.08/ 0.50
44	N_2O^-	$C_7H_6N_2O$	0.0	0.1	0.3	0.9	-	-	-	~0	−0.12
40	NCN^-	$C_6H_6N_2O_2$	0.0	0.1	0.3	0.7	-	-	-	~0	−0.79
26	CN^-	$C_6H_6N_3O_2$/ $C_6H_6NO_2 + N_2$	1.4	2.6	4.2	6.5	-	-	-	0.7	0.58/ −2.64
16	O^- NH_2^-	$C_7H_6N_4O$ $C_7H_4N_3O_2$	4.6	7.3	11.1	-	-	-	-	2.7	0.81 1.19

[a]—Dissociation pathways were analyzed previously [26].

Figure 3a shows the anion efficiency curve for the formation of $C_7H_5N_4^-$. We observed a first resonance region below 2 eV, with maxima at 0.4, 1.0 and 1.7 eV. A second resonance region was observed, but with low intensity, between 4 and 6 eV, with a peak maximum of around 4.5 and 5.3 eV. The formation of $C_7H_5N_4^-$ with the mass of 145 u is attributed to the loss of a hydroxyl group and oxygen at the 1–N–oxide position, see Equation (2). We expect that the anion results from the OH roaming mechanism. During the process, the roaming OH radical attaches to the O atom and dissociates as HO_2.

$$C_7H_6N_4O_2 + e^- \rightarrow (C_7H_6N_4O_2)^{-*} \rightarrow (C_7H_5N_4O \ldots OH)^{-*} \rightarrow C_7H_5N_4^- + HO_2 \quad (2)$$

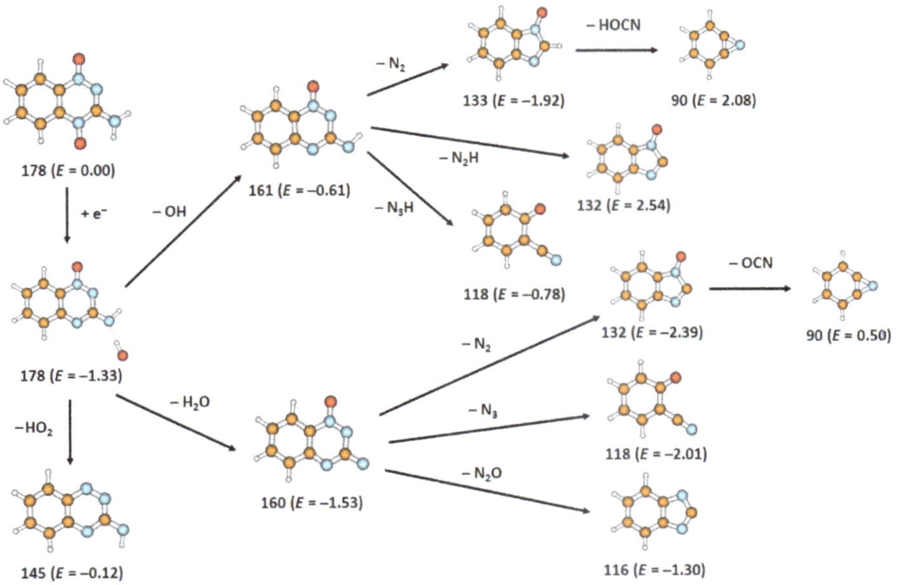

Figure 2. Suggested fragmentation pathways after electron attachment to tirapazamine starting with predissociation of the OH moiety; the mass (u) and energy E (in eV) of the respective anions are given. Calculated at the B3LYP/aug–cc–pVDZ level of theory.

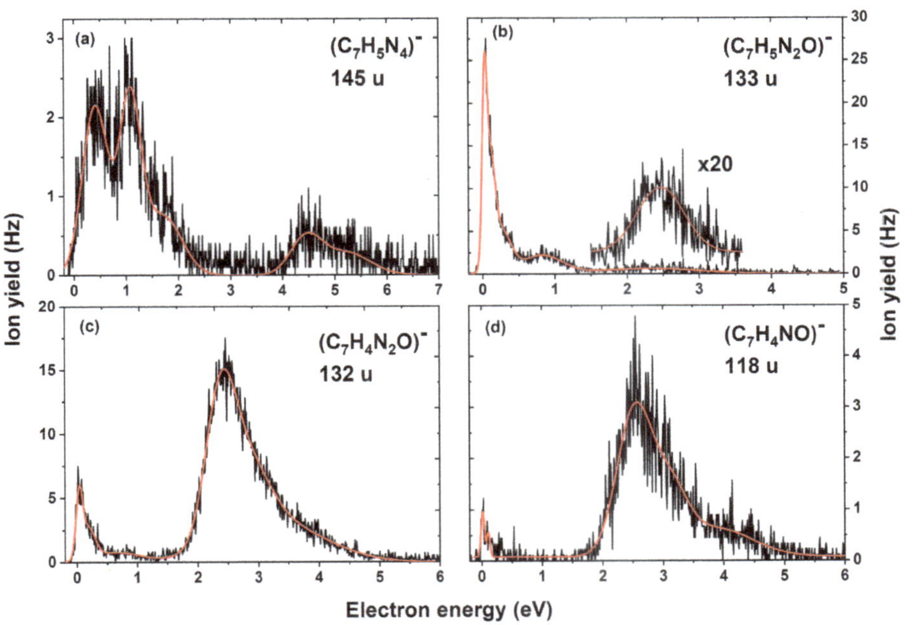

Figure 3. Anion efficiency curves for the formation of the anion with masses (**a**) 145 u, (**b**) 133 u, (**c**) 132 u and (**d**) 118 u upon electron attachment to tirapazamine. The solid red line is a cumulative sum of individual gaussian functions rendering the observed features.

The calculated thermodynamic threshold predicts an exothermic reaction with a threshold of −0.12 eV, in agreement with the observed experimental onset of 0 eV. The channel is thermodynamically less favorable than direct OH dissociation (see Figure 2), and the OH radical can evaporate before it reaches the oxygen atom, which is reflected in its low contribution to the total ion yield. Note that subsequent dissociation of OH and O would lead to a too high reaction energy. Both rings stay intact during the dissociation process.

Equations (3) and (4) show the formation of the fragment anions with masses 133 u and 132 u. These anions are assigned to the $C_7H_5N_2O^-$ and $C_7H_4N_2O^-$ structures, respectively.

$$C_7H_6N_4O_2 + e^- \rightarrow (C_7H_6N_4O_2)^{-*} \rightarrow C_7H_5N_2O^- + OH + N_2 \quad (3)$$

$$C_7H_6N_4O_2 + e^- \rightarrow (C_7H_6N_4O_2)^{-*} \rightarrow C_7H_4N_2O^- + H_2O + N_2 \quad (4)$$

From the anion efficiency curve shown in Figure 3b, we derived four main resonances for $C_7H_5N_2O^-$ (mass 133 u), with peak maxima comprising a sharp peak at 0.0 eV; low-intensity peaks at 0.1 eV, 0.3 eV, 0.8 eV and a broad resonance around 2.5 eV. Comparing Figure 3b,c, similar resonance positions are observed below 2 eV region for $C_7H_4N_2O^-$ (mass 132 u) with a lower intensity. In addition, the asymmetric shape of the main peak in the ion yield of $C_7H_4N_2O^-$ (see Figure 3c) suggests resonances at 2.8 eV and 3.2 eV. On the other hand, in the energy range of 1.5–5 eV, the intensity of $C_7H_4N_2O^-$ prevails over the one of $C_7H_5N_2O^-$ by an order of magnitude.

For both anions, we suggest that the reaction starts with predissociation of the OH moiety (Figure 2), as supported by the fact that the resonance positions below 1 eV for the formation of these anions are exactly at the same positions [26]. This implies that both result from the same TNI.

For $C_7H_5N_2O^-$, we associate the formation of the anion with the loss of a hydroxyl group, followed by a loss of N_2 upon an initial opening of the triazine ring. The final structure has two closed rings, formed through structural rearrangement, as shown in Figure 2. For $C_7H_4N_2O^-$, we expect a water molecule to be formed through the roaming mechanism, again with a subsequent N_2 dissociation (Figure 2). For electron energies above 2.54 eV (as calculated at the B3LYP/aug–cc–pVDZ level), this ion may be also formed through dissociation of OH + N_2H. This might explain the prevalence of $C_7H_4N_2O^-$ over $C_7H_5N_2O^-$ for higher electron energies.

The dissociation of N_2 or N_2H requires fundamental rearrangement of the triazine moiety; however, a considerable amount of energy is gained during the process that allows for subsequent dissociation processes (see discussion of the $C_6H_4N^-$ anion with mass 90 u below). The exothermicity of the reactions agrees with the experiment. The low yield of the subsequent dissociation reaction can be explained by energetic considerations: only if the OH radical does not take too much kinetic energy away from the system, the remaining anion can surpass a barrier for ring dissociation.

Figure 3d shows the anion efficiency curve for mass 118 u with three resonance regions, one near 0 eV, a peak with maximum around 2.5 eV followed by resonances centered at 3.2 and 3.9 eV. For this anion, two different plausible assignments are possible, $C_7H_6N_2^-$ and $C_7H_4NO^-$. In the first case, N_2O_2 would dissociate. Our calculations show that the only exothermic reaction for this dissociation would be dissociation of N_2 + O_2 with the reaction energy of −0.77 eV, which does not seem reasonable from the kinetic perspective. The $C_7H_4NO^-$ ion might be formed through dissociation of $[N_3OH_2]$, suggesting either dissociation of OH + N_3H or H_2O + N_3, Equations (5a) and (5b) and implying, again, the OH predissociation as the initial step (Figure 2).

$$C_7H_6N_4O_2 + e^- \rightarrow (C_7H_6N_4O_2)^{-*} \rightarrow C_7H_4NO^- + OH + N_3H \quad (5a)$$

$$C_7H_6N_4O_2 + e^- \rightarrow (C_7H_6N_4O_2)^{-*} \rightarrow C_7H_4NO^- + H_2O + N_3 \quad (5b)$$

Both Equations (5a) and (5b) are exothermic (−0.78 and −2.01 eV). Similarities of the peak positions in the anion efficiency curves of $C_7H_5N_2O^-$, $C_7H_4N_2O^-$ and $C_7H_4NO^-$ support our interpretation of the molecular mechanism.

We also observed a low-intensity anion with mass 116 u. The corresponding anion efficiency curve shows resonance positions that were observed previously for $(TPZ–OH)^-$ formation (0.0, 0.1, 0.9 eV); see Figure 4a. A further broad resonance with a maximum at about 2.6 eV is also present in the ion yield. We interpret this channel as formation of $C_7H_4N_2^-$ arising due to the loss of H_2O and N_2O upon multiple bond cleavages and rearrangement in the parent molecule; see Equation (6) and Figure 2. The predicted reaction energy of −1.30 eV agrees with the experimental onset.

$$C_7H_6N_4O_2 + e^- \rightarrow (C_7H_6N_4O_2)^{-*} \rightarrow C_7H_4N_2^- + H_2O + N_2O \quad (6)$$

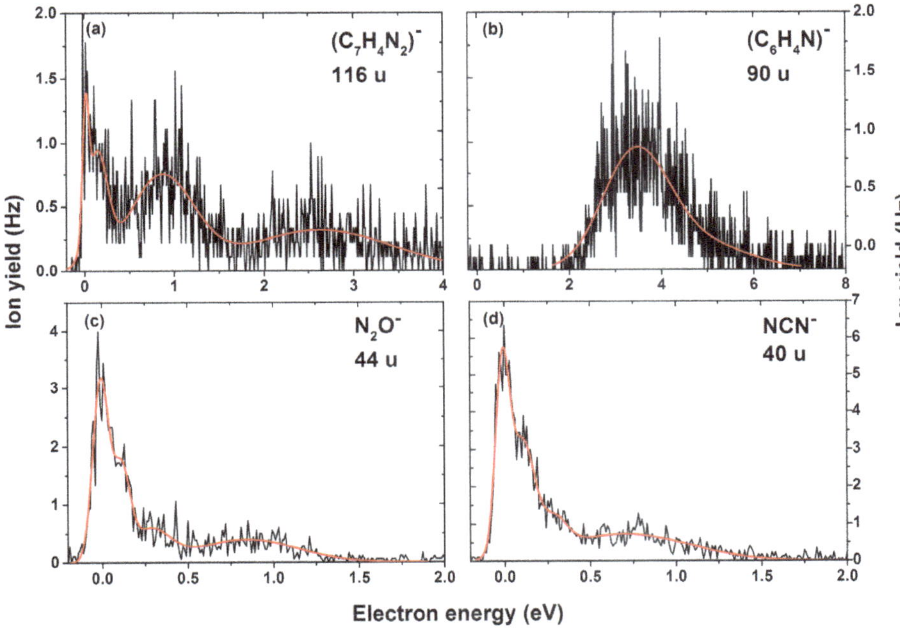

Figure 4. Anion efficiency curve for the formation of the anions with masses (**a**) 116 u, (**b**) 90 u, (**c**) 44 u and (**d**) 40 u upon electron attachment to tirapazamine.

Figure 4b shows the anion efficiency curve for the formation of the anion with mass 90 u, assigned to $C_6H_4N^-$. We observed one broad bump with a maximum at about 3.6 eV and a second resonance at 4.7 eV near the tail. We suggest that this ion emerges in a follow-up reaction of Equations (3) and/or (4):

$$C_7H_6N_4O_2 + e^- \rightarrow (C_7H_6N_4O_2)^{-*} \rightarrow C_6H_4N^- + OH + N_2 + HOCN \quad (7a)$$

$$C_7H_6N_4O_2 + e^- \rightarrow (C_7H_6N_4O_2)^{-*} \rightarrow C_6H_4N^- + H_2O + N_2 + OCN \quad (7b)$$

From the thermodynamic perspective, the first reaction is considerably less probable (2.08 eV vs. 0.50 eV). The anion is, however, observed only at higher energies and with low intensity; the experimental onset at 2.1 eV suggests that $C_6H_4N^-$ is formed upon Equation (7a), though we point out that Equation (7b) would be thermodynamically open, as well.

Our analysis shows that the formation mechanisms for the anions with masses 90–161 u have several common features: (1) The first step in their formation is predissociation of the OH group formed after proton transfer from the NH_2 group to the adjacent O atom. (2) The OH group dissociates either alone, after another proton transfer (forming H_2O) or as HO_2 (minor channel). (3) Highly stable nitrogen-containing species seem to dissociate in the subsequent step. (4) The benzene ring stays intact, while the triazine part of the molecule undergoes significant changes. It is also interesting to note that $(TPZ–H_2O)^-$ with mass 160 u, corresponding to water dissociation, is not observed in the experiment. One of the reasons could be the high energy release (−1.53 eV), enabling further dissociation reactions.

Although the energy minima presented in Figure 2 do not provide direct hints toward dissociation kinetics, one should keep in mind the high energy content in molecules (with three exceptions, all reactions depicted in Figure 2 are exothermic). Molecular dynamics runs of $(TPZ–OH)^-$ at elevated temperature show that the bonds in the triazine moiety are the weakest ones and might break preferentially.

Finally, we focus on the formation of small negative ions that can be formed through dissociation of the triazine moiety (Figure 5). Note that the calculated dissociation energies are conservative estimates as, if one allows for further dissociation of the molecule through, e.g., formation of N_2 molecules, the dissociation energy drops significantly. At the same time, it is possible to connect the dissociation pathways to the predissociation of the OH group; however, this is not needed from the thermodynamical perspective.

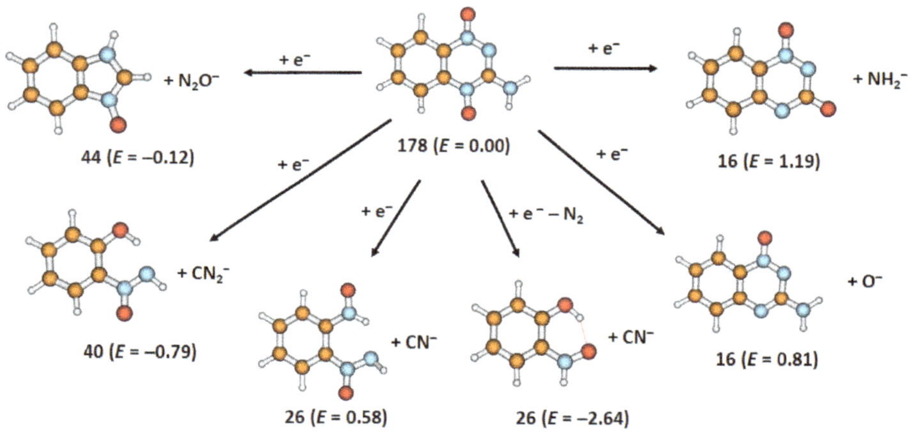

Figure 5. Suggested fragmentation pathways producing small anions after electron attachment to tirapazamine; the mass (in u) and energy E (in eV) of the respective anions are given. Calculated at the B3LYP/aug–cc–pVDZ level of theory.

Figure 4c,d show the anion efficiency curve for the formation of N_2O^- (mass 44 u) and CN_2^- (mass 40 u) due to multiple bond cleavage in the benzotriazine dissociation upon electron attachment to tirapazamine, via Equations (8) and (9), respectively. Both anions show resonance positions at 0, 0.1, 0.3 eV and a broad peak around 0.8 eV close to the tail. The calculated reaction energy is markedly exothermic for both N_2O^- (−0.12 eV) and CN_2^- (−0.79 eV) (see Figure 5).

$$C_7H_6N_4O_2 + e^- \rightarrow (C_7H_6N_4O_2)^{-*} \rightarrow N_2O^- + C_7H_6N_2O \quad (8)$$

$$C_7H_6N_4O_2 + e^- \rightarrow (C_7H_6N_4O_2)^{-*} \rightarrow CN_2^- + C_6H_6N_2O_2 \quad (9)$$

In the anion efficiency curve for the anion with mass 26 u, we observe few overlapping resonances (Figure 6a). We assign this ion yield to CN^-; the alternative explanation of $C_2H_2^-$ seems less probable, due to the need to break the benzene ring that is not observed for any other fragment. Like the formation of CN^- in nitroimidazoles [35], the anion is formed via an exothermic Equation (10) only after a rupture of the triazine side.

$$C_7H_6N_4O_2 + e^- \rightarrow (C_7H_6N_4O_2)^{-*} \rightarrow CN^- + C_6H_6N_3O_2 \quad (10a)$$

$$C_7H_6N_4O_2 + e^- \rightarrow (C_7H_6N_4O_2)^{-*} \rightarrow CN^- + C_6H_6NO_2 + N_2 \quad (10b)$$

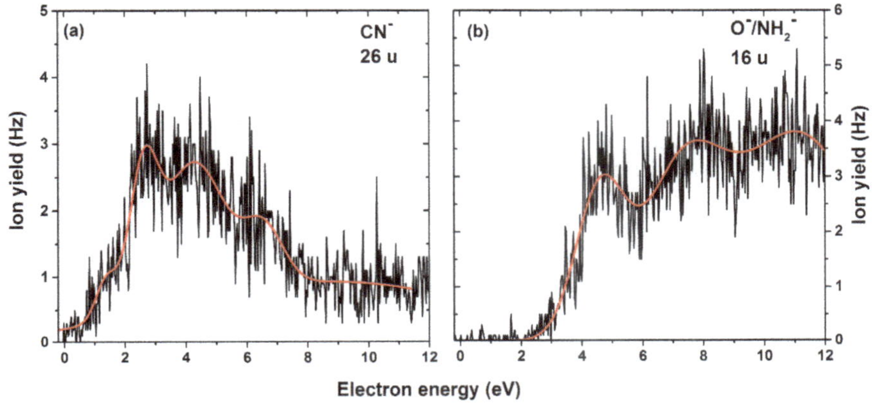

Figure 6. Anion efficiency curve for the formation of (**a**) CN^- and (**b**) O^-/NH_2^- upon electron attachment to TPZ.

While the simple dissociation of CN^- is still endothermic by 0.58 eV, the reaction energy drops to −2.64 eV if N_2 is dissociated at the same time.

We also observed, as the smallest of all fragments, the anion with mass 16 u, which could be assigned to either O^- or NH_2^-, since they have the same nominal mass. The separation of isobaric anions requires high-resolution mass spectrometers, which is a limitation in our present experimental setup, and therefore, we are not able to distinguish the two isobaric anions. Three high-energy resonances with peak maxima at 4.6, 7.3 and 11.1 eV are observed; see Figure 6b. The reactions leading to the formation of this anion yield are predicted to take place through direct dissociation (Figure 5) and are endothermic, with 0.81 eV and 1.19 eV for dissociation of O^- and NH_2^-, respectively. Therefore, at the threshold determined experimentally of 2.7 eV, the formation of both anions O^- and NH_2^- is thermodynamically possible.

3. Materials and Methods

Tirapazamine ($C_7H_6N_4O_2$) was purchased from Sigma Aldrich (Vienna, Austria) with a purity of $\geq$ 98% (HPLC) and was used as received. The experiment was performed in the gas phase under a high vacuum (10^{-8} mbar) using a crossed electron–molecule beam apparatus previously described in [36]. Briefly, the setup consists of a homemade hemispherical electron monochromator (HEM), a resistively heated copper oven, a quadrupole mass analyzer (QMA 400, Pfeiffer Vacuum, Vienna, Austria) and a detector for counting the mass-analyzed ions. The neutral beam tirapazamine in the gas phase was produced by heating the sample in the oven and expanding the vapor through a 1 mm diameter capillary. Then, the effusive molecular beam was crossed perpendicularly with the electron beam. An electron beam with energy resolution of 100 meV at FWHM (full width at half maximum) and current of 8–19 nA was used, ensuring a compromise between resulting ion beam intensity and the electron energy resolution. Before starting the measurements of negative ions, the temperature dependency of the electron ionization mass spectrum was checked to obtain a reasonable ion signal without decomposition of the sample in the oven. The compound was finally studied at the oven temperature of 395 K. The ions formed upon (dissociative) electron attachment were extracted to the QMA entrance by a weak electrostatic field, mass-analyzed in the QMA and detected by a channeltron secondary electron multiplier operated in single-pulse counting mode. The anion efficiency curves were recorded by a scan of the electron energy, while the QMA was set for transmission of a chosen anion. For calibration of the electron energy scale and determination of the energy resolution, the well-known Cl^-/CCl_4 resonance at 0 eV [37] was used. The experimental threshold for the detected DEA reactions with TPZ was determined with the method proposed in [38].

Molecular structures were optimized at the B3LYP/aug–cc–pVDZ level of theory. Wavefunction stability was checked in every point, and all reported reaction energies include zero-point corrections. To investigate excited states of the anion formed after electron attachment, we considered single-reference methods, time-dependent density functional theory (TDDFT) and equation of motion-coupled clusters singles and doubles (EOM-CCSD) [39–41], as well as multireference approaches, complete active space–self consistent field (CASSCF) and multireference configuration interaction (MRCI). For CASSCF calculations, the active space of 3 electrons in 5 orbitals (further denoted as (3,5)) was chosen for TPZ^- to cover the most important, low-lying valence electronic states. We employed averaging over four electronic states. MRCI calculations were performed in the (3,4) active space, coring 20 lowest-lying orbitals. To prevent electronic states of diffuse character from appearing, we limited ourselves to the 6–31g* basis set in multireference calculations. We tested the suitability of the approximations involved (limited active space, a small basis set) through comparison to the results of single-reference calculations in the Franck–Condon (FC) point. While our calculations rationalize the resonance at 0 eV (with D_1 as the target state), more advanced methods would be needed to reproduce positions of higher-lying resonances. For the multireference calculations, we employed the Molpro package [42], all other calculations were performed in the Gaussian program [43].

4. Conclusions

In the present publication, we analyzed processes after electron attachment to TPZ, as well as induced dissociation pathways. The ground electronic state of TPZ^- is readily reached through a D_0/D_1 conical intersection that has a distorted triazine ring and almost coincides with the D_1 minimum. We showed that the crucial step in the dissociation kinetics of all anions containing the benzene ring (anions with masses 90–161 u) is the predissociation of an OH group that roams in the anion vicinity, dissociating either directly or after forming HO_2/H_2O. When the triazine part of the molecule is destabilized by dissociation of at least one oxygen atom, degradation of the ring continues through dissociation of highly stable species with at least two nitrogen atoms. The extent of decomposition will depend on the energy carried away by the dissociating O_xH_y fragment. The benzene ring, on the other hand, plays the role of a spectator and is not altered during the dissociation process. Similarly, all small anionic species are dissociated from the triazine part of the molecule, with the benzene ring left intact. Formation of N_2O^-, CN_2^- and CN^- anions might also take place after OH group dissociation; our data, however, do not allow for any conclusion in this respect.

Supplementary Materials: The following are available online at https://www.mdpi.com/1422-0067/22/6/3159/s1, Table S1. First ten excitation energies (in eV) of TPZ^- calculated at the TD-BMK/aug-cc-pVDZ level in the structure of TPZ optimized at the B3LYP/aug-cc-pVDZ level. Cartesian coordinates (in Ångstrom) and electronic energies calculated at the B3LYP/aug-cc-pVDZ level including the zero-point energy (in Hartree).

Author Contributions: Experimental investigation and data analysis of DEA results, E.A.-B., Writing-original draft E.A.-B., J.A., M.O. and S.D.; Quantum chemical calculations, M.O.; writing—review and editing, E.A.-B., J.A., M.O. and S.D.; supervision, project administration, funding acquisition, S.D. All authors have read and agreed to the published version of the manuscript.

Funding: S.D. acknowledges support for Open Access Funding by the Austrian Science Fund (FWF), Vienna (P30332). J.A. acknowledges the Portuguese National Funding Agency FCT–MCTES through grants PD/BD/114447/2016, as well as the Radiation Biology and Biophysics Doctoral Training Programme (RaBBiT, PD/00193/2012); UIBD/04378/2020 (UCIBIO); and UIBD/00068/2020 (CEFITEC).

Institutional Review Board Statement: Not applicable.

Informed Consent Statement: Not applicable.

Data Availability Statement: The data presented in this work are available on request from the corresponding authors.

Acknowledgments: The computational results presented have been attained using the HPC infrastructure LEO of the University of Innsbruck.

Conflicts of Interest: The authors declare no conflict of interest.

References

1. Cronsky, B.T.; Knox, S.J.; Scicinski, J. Six Degrees of Separation: The Oxygen Effect in the Development of Radiosensitizers. *Transl. Oncol.* **2011**, *4*, 189–198. [CrossRef]
2. Brown, J.M.; Wilson, W.R. Exploiting tumour hypoxia in cancer treatment. *Nat. Rev. Cancer* **2004**, *4*, 437–447. [CrossRef]
3. Vaupel, P.; Mayer, A. Hypoxia in cancer: Significance and impact on clinical outcome. *Cancer Metastasis Rev.* **2007**, *26*, 225–239. [CrossRef]
4. Raviraj, J.; Bokkasam, V.K.; Kumar, V.S.; Reddy, U.S.; Suman, V. Radiosensitizers, radioprotectors, and radiation mitigators. *Indian J. Dent. Res.* **2014**, *25*, 83. [CrossRef]
5. Wang, H.; Mu, X.; He, H.; Zhang, X.D. Cancer Radiosensitizers. *Trends Pharmacol. Sci.* **2018**, *39*, 24–48. [CrossRef]
6. Seiwert, T.Y.; Salama, J.K.; Vokes, E.E. The concurrent chemoradiation paradigm—General principles. *Nat. Clin. Pract. Oncol.* **2007**, *4*, 86–100. [CrossRef]
7. Wardman, P. Chemical radiosensitizers for use in radiotherapy. *Clin. Oncol.* **2007**, *19*, 397–417. [CrossRef] [PubMed]
8. Meißner, R.; Kočišek, J.; Feketeová, L.; Fedor, J.; Fárník, M.; Limão-Vieira, P.; Illenberger, E.; Denifl, S. Low-energy electrons transform the nimorazole molecule into a radiosensitiser. *Nat. Commun.* **2019**, *10*, 2388. [CrossRef] [PubMed]
9. Overgaard, J.; Hansen, H.S.; Overgaard, M.; Bastholt, L.; Berthelsen, A.; Specht, L.; Lindeløv, B.; Jørgensen, K. A randomized double-blind phase III study of nimorazole as a hypoxic radiosensitizer of primary radiotherapy in supraglottic larynx and pharynx carcinoma. Results of the Danish Head and Neck Cancer Study (DAHANCA) Protocol 5-85. *Radiother. Oncol.* **1998**, *46*, 135–146. [CrossRef]
10. Hay, M.P.; Pruijn, F.B.; Gamage, S.A.; Liyanage, H.D.S.; Kovacs, M.S.; Patterson, A.V.; Wilson, W.R.; Brown, J.M.; Denny, W.A. DNA-Targeted 1,2,4-Benzotriazine 1,4-Dioxides: Potent Analogues of the Hypoxia-Selective Cytotoxin Tirapazamine. *J. Med. Chem.* **2004**, *47*, 475–488. [CrossRef]
11. Brown, J.M. The Hypoxic Cell: A Target for Selective Cancer Therapy. *Cancer Res.* **1999**, *59*, 5863–5870.
12. Junnotula, V.; Sarkar, U.; Sinha, S.; Gates, K.S. Initiation of DNA strand cleavage by 1, 2, 4-benzotriazine 1, 4-dioxide antitumor agents: Mechanistic insight from studies of 3-methyl-1, 2, 4-benzotriazine 1, 4-dioxide. *J. Am. Chem. Soc.* **2009**, *131*, 1015–1024. [CrossRef]
13. Li, L.C.; Zha, D.; Zhu, Y.Q.; Xu, M.H.; Wong, N.B. Theoretical study of the mechanism of hydroxyl radical release from tirapazamine's undergoing enzymatic catalysis. *Chem. Phys. Lett.* **2005**, *408*, 329–334. [CrossRef]
14. Zagorevskii, D.; Song, M.; Breneman, C.; Yuan, Y.; Fuchs, T.; Gates, K.S.; Greenlief, C.M. A mass spectrometry study of tirapazamine and its metabolites: Insights into the mechanism of metabolic transformations and the characterization of reaction intermediates. *J. Am. Soc. Mass Spectrom.* **2003**, *14*, 881–892. [CrossRef]
15. Shinde, S.S.; Maroz, A.; Hay, M.P.; Patterson, A.V.; Denny, W.A.; Anderson, R.F. Characterization of radicals formed following enzymatic reduction of 3-substituted analogues of the hypoxia-selective cytotoxin 3-amino-1, 2, 4-benzotriazine 1, 4-dioxide (tirapazamine). *J. Am. Chem. Soc.* **2010**, *132*, 2591–2599. [CrossRef]
16. Anderson, R.F.; Shinde, S.S.; Hay, M.P.; Gamage, S.A.; Denny, W.A. Activation of 3-amino-1, 2, 4-benzotriazine 1, 4-dioxide antitumor agents to oxidizing species following their one-electron reduction. *J. Am. Chem. Soc.* **2003**, *125*, 748–756. [CrossRef] [PubMed]
17. Anderson, R.F.; Shinde, S.S.; Hay, M.P.; Denny, W.A. Potentiation of the cytotoxicity of the anticancer agent tirapazamine by benzotriazine N-oxides: The role of redox equilibria. *J. Am. Chem. Soc.* **2006**, *128*, 245–249. [CrossRef] [PubMed]
18. Jones, G.D.D.; Weinfeld, M. Dual action of tirapazamine in the induction of DNA strand breaks. *Cancer Res.* **1996**, *56*, 1584–1590. [PubMed]
19. Daniels, J.S.; Gates, K.S. DNA cleavage by the antitumor agent 3-amino-1,2,4-benzotriazine 1,4-dioxide (SR4233): Evidence for involvement of hydroxyl radical. *J. Am. Chem. Soc.* **1996**, *118*, 3380–3385. [CrossRef]
20. Adams, G.E.; Dewey, D.L. Hydrated electrons and radiobiological sensitization. *Biochem. Biophys. Res. Commun.* **1963**, *12*. [CrossRef]

21. Pimblott, S.M.; LaVerne, J.A. Production of low-energy electrons by ionizing radiation. *Radiat. Phys. Chem.* **2007**, *76*, 1244–1247. [CrossRef]
22. Kumar, A.; Becker, D.; Adhikary, A.; Sevilla, M.D. Reaction of electrons with DNA: Radiation damage to radiosensitization. *Int. J. Mol. Sci.* **2019**, *20*, 3998. [CrossRef] [PubMed]
23. Alizadeh, E.; Sanche, L. Precursors of solvated electrons in radiobiological physics and chemistry. *Chem. Rev.* **2012**, *112*, 5578–5602. [CrossRef] [PubMed]
24. Boudaïffa, B.; Cloutier, P.; Hunting, D.; Huels, M.A.; Sanche, L. Resonant formation of DNA strand breaks by low-energy (3 to 20 eV) electrons. *Science* **2000**, *287*, 1658–1660. [CrossRef] [PubMed]
25. Ma, J.; Wang, F.; Denisov, S.A.; Adhikary, A.; Mostafavi, M. Reactivity of prehydrated electrons toward nucleobases and nucleotides in aqueous solution. *Sci. Adv.* **2017**, *3*, e1701669. [CrossRef] [PubMed]
26. Arthur-Baidoo, E.; Ameixa, J.; Ziegler, P.; Ferreira da Silva, F.; Ončák, M.; Denifl, S. Reactions in Tirapazamine Induced by the Attachment of Low-Energy Electrons: Dissociation Versus Roaming of OH. *Angew. Chem.-Int. Ed.* **2020**, *59*, 17177–17181. [CrossRef] [PubMed]
27. Yin, J.; Glaser, R.; Gates, K.S. On the reaction mechanism of tirapazamine reduction chemistry: Unimolecular N-OH homolysis, stepwise dehydration, or triazene ring-opening. *Chem. Res. Toxicol.* **2012**, *25*, 634–645. [CrossRef]
28. Yin, J.; Glaser, R.; Gates, K.S. Electron and spin-density analysis of tirapazamine reduction chemistry. *Chem. Res. Toxicol.* **2012**, *25*, 620–633. [CrossRef]
29. Chomicz, L.; Zdrowowicz, M.; Kasprzykowski, F.; Rak, J.; Buonaugurio, A.; Wang, Y.; Bowen, K.H. How to find out whether a 5-substituted uracil could be a potential DNA radiosensitizer. *J. Phys. Chem. Lett.* **2013**, *4*, 2853–2857. [CrossRef]
30. Ameixa, J.; Arthur-Baidoo, E.; Meißner, R.; Makurat, S.; Kozak, W.; Butowska, K.; Ferreira Da Silva, F.; Rak, J.; Denifl, S. Low-energy electron-induced decomposition of 5-trifluoromethanesulfonyl-uracil: A potential radiosensitizer. *J. Chem. Phys.* **2018**, *149*, 164307. [CrossRef]
31. Spisz, P.; Zdrowowicz, M.; Kozak, W.; Chomicz-Mańka, L.; Falkiewicz, K.; Makurat, S.; Sikorski, A.; Wyrzykowski, D.; Rak, J.; Arthur-Baidoo, E.; et al. Uracil-5-yl O-Sulfamate: An Illusive Radiosensitizer. Pitfalls in Modeling the Radiosensitizing Derivatives of Nucleobases. *J. Phys. Chem. B* **2020**, *124*, 5600–5613. [CrossRef] [PubMed]
32. Ptasińska, S.; Denifl, S.; Scheier, P.; Märk, T.D. Inelastic electron interaction (attachment/ionization) with deoxyribose. *J. Chem. Phys.* **2004**, *120*, 8505–8511. [CrossRef] [PubMed]
33. Bald, I.; Langer, J.; Tegeder, P.; Ingólfsson, O. From isolated molecules through clusters and condensates to the building blocks of life. *Int. J. Mass Spectrom.* **2008**, *277*, 4–25. [CrossRef]
34. Gorfinkiel, J.D.; Ptasinska, S. Electron scattering from molecules and molecular aggregates of biological relevance. *J. Phys. B At. Mol. Opt. Phys.* **2017**, *50*, 182001. [CrossRef]
35. Ribar, A.; Fink, K.; Probst, M.; Huber, S.E.; Feketeová, L.; Denifl, S. Isomer Selectivity in Low-Energy Electron Attachment to Nitroimidazoles. *Chem.-Eur. J.* **2017**, *23*, 12892–12899. [CrossRef]
36. Ptasinska, S.; Denifl, S.; Scheier, P.; Illenberger, E.; Märk, T.D. Bond- and site-selective loss of H atoms from nucleobases by very-low-energy electrons (<3 eV). *Angew. Chem.-Int. Ed.* **2005**, *44*, 6941–6943. [CrossRef]
37. Gallup, G.A.; Aflatooni, K.; Burrow, P.D. Dissociative electron attachment near threshold, thermal attachment rates, and vertical attachment energies of chloroalkanes. *J. Chem. Phys.* **2003**. [CrossRef]
38. Meißner, R.; Feketeová, L.; Bayer, A.; Postler, J.; Limão-Vieira, P.; Denifl, S. Positive and negative ions of the amino acid histidine formed in low-energy electron collisions. *J. Mass Spectrom.* **2019**, *54*, 802–816. [CrossRef] [PubMed]
39. Koch, H.; Jørgensen, P. Coupled cluster response functions. *J. Chem. Phys.* **1990**, *93*, 3333–3344. [CrossRef]
40. Stanton, J.F.; Bartlett, R.J. The equation of motion coupled-cluster method. A systematic biorthogonal approach to molecular excitation energies, transition probabilities, and excited state properties. *J. Chem. Phys.* **1993**, *98*, 7029–7039. [CrossRef]
41. Krylov, A.I. Equation-of-motion coupled-cluster methods for open-shell and electronically excited species: The Hitchhiker's guide to fock space. *Annu. Rev. Phys. Chem.* **2008**, *59*, 433–462. [CrossRef] [PubMed]
42. Werner, H.J.; Knowles, P.J.; Knizia, G.; Manby, F.R.; Schütz, M.; Celani, P.; Korona, T.; Lindh, R.; Mitrushenkov, A.; Rauhut, G.; et al. *MOLPRO*, version 2012.1; A package of ab initio programs; University of Cardiff Chemistry Consultants (UC3): Cardiff, UK, 2012.
43. Frisch, M.J.; Trucks, G.W.; Schlegel, H.B.; Scuseria, G.E.; Robb, M.A.; Cheeseman, J.R.; Scalmani, G.; Barone, V.; Petersson, G.A.; Nakatsuji, H.; et al. *Gaussian 16*, Revision A.03; Gaussian, Inc.: Wallingford, CT, USA, 2016.

International Journal of *Molecular Sciences*

MDPI

Article

Elastic Electron Scattering from Methane Molecule in the Energy Range from 50–300 eV

Jelena Vukalović [1,2], Jelena B. Maljković [1], Karoly Tökési [3], Branko Predojević [2] and Bratislav P. Marinković [1,*]

[1] Institute of Physics Belgrade, University of Belgrade, Pregrevica 118, 11080 Belgrade, Serbia; jelena.vukovic@pmf.unibl.org (J.V.); jelenam@ipb.ac.rs (J.B.M.)
[2] Faculty of Science, University of Banja Luka, Mladena Stojanovića 2, 78000 Banja Luka, Republic of Srpska, Bosnia and Herzegovina; bpredojevic@teol.net
[3] Institute for Nuclear Research (ATOMKI), 4026 Debrecen, Hungary; tokesi@atomki.hu
* Correspondence: bratislav.marinkovic@ipb.ac.rs; Tel.: +381-11-316-0882

Abstract: Electron interaction with methane molecule and accurate determination of its elastic cross-section is a demanding task for both experimental and theoretical standpoints and relevant for our better understanding of the processes in Earth's and Solar outer planet atmospheres, the greenhouse effect or in plasma physics applications like vapor deposition, complex plasma-wall interactions and edge plasma regions of Tokamak. Methane can serve as a test molecule for advancing novel electron-molecule collision theories. We present a combined experimental and theoretical study of the elastic electron differential cross-section from methane molecule, as well as integral and momentum transfer cross-sections in the intermediate energy range (50–300 eV). The experimental setup, based on a crossed beam technique, comprising of an electron gun, a single capillary gas needle and detection system with a channeltron is used in the measurements. The absolute values for cross-sections are obtained by relative-flow method, using argon as a reference. Theoretical results are acquired using two approximations: simple sum of individual atomic cross-sections and the other with molecular effect taken into the account.

Keywords: methane; cross-section; elastic scattering; electrons

Citation: Vukalović, J.; Maljković, J.B.; Tökési, K.; Predojević, B.; Marinković, B.P. Elastic Electron Scattering from Methane Molecule in the Energy Range from 50–300 eV. *Int. J. Mol. Sci.* **2021**, *22*, 647. https://doi.org/10.3390/ijms22020647

Received: 2 December 2020
Accepted: 1 January 2021
Published: 11 January 2021

1. Introduction

Methane (CH_4) is the simplest alkane. Its molecule has tetrahedron shape, belongs to the T_d point group symmetry, and does not possess dipole and quadrupole moments. Furthermore, methane is widely distributed in the Solar System. In general, the inner planets Mercury [1], Venus [2], and Mars [3] are methane-poor, except Earth, whereas outer planets Jupiter [4], Saturn [5], Uranus [6], and Neptune [7] have methane-rich atmospheres. Currently, the methane levels in Earth's atmosphere are around 1.6–1.8 ppmv and are considered one of the main causes of the greenhouse effect. Greenhouse effect caused by methane is about eight times that of CO_2 [8]. Emission of CH_4 in atmosphere is about 40% from natural, and about 60% from anthropogenic sources (agriculture, energy, and waste sectors [9]). In past decades methane's growth rate was changing, and in recent years, it has been increasing [10]. That is a reason why investigation of this gas again becomes consequential.

Electron collisions with methane are very important in plasma physics. In mixture with hydrogen and argon, CH_4 is used for r. f. plasma-enhanced chemical vapor deposition (R. F. PECVD) of nanocrystalline diamond films [11]. On the other hand, methane is less used than silane in design of solar cells. Its use in the preparation of amorphous SiC *p*-layers is most often emphasized in a PECVD process using high R.F. power at relatively low temperature [12]. When a Tokamak is operated at high density and high temperature methane plays a dominant role in the edge plasma region [13]. Particles and energy are expelled from the plasma and are transported to the vessel wall, which leads to complex plasma-wall interactions. These interactions create impurities in the plasma, including methane and its derivates, resulting in significant cooling of plasma, which can prevent achievement of rector relevant conditions. Therefore, understanding transport features of methane in plasma and interpretation of electron methane collisions in different physical processes plays a significant role in laboratory research in different fields of physics as well as in the investigation of properties atmospheres planets in Solar System. Finally, from a theoretical perspective and, because of its simplicity, methane can serve as a test molecule for advancing novel electron-molecule collision theories (e.g., Blanco et al. [14], Allan [15]).

Elastic electron scattering from methane molecule has been studied intensely in the past, and the most recent review of elastic differential cross-sections is given in [16]. Recommended set of data for electron/methane interactions is given by Song et al. [17]. For low incident electron energies (below 50 eV), elastic electron scattering from methane molecule has been studied experimentally and theoretically equally extensively. To cite few most recent papers, Allan [15] measured absolute differential elastic (impact electron energies 0.4–20 eV) and vibrational excitation cross-sections using an electron spectrometer with a magnetic angle changer, which allows measurements of differential cross-sections (DCSs) for backward angles. Bettega et al. [18] calculated DCSs and integral cross-sections (ICSs) for energies between 3 and 10 eV, using Schwinger multichannel method with pseudopotentials. Bundschu et al. [19] presented both experimental and theoretical results; measurements of DCSs using crossed beam apparatus and relative flow method, with He as a reference gas, and calculations using a body-fixed, single-center for close-coupled equations. Shyn and Cravens [20] reported DCSs for methane in energy range from 5 to 50 eV, using modulated crossed-beam method and He as reference gas for normalizing relative data to the absolute scale. At even lower electron energies, from 0.1 to 1.8 eV, an important study was one by Sohn et al. [21] where the references from even earlier measurements and calculations can be found. Several authors used R-matrix method to calculate cross-sections form methane [22–24].

As for papers that contain data for DCSs at intermediate-to-high energies ($E_0 \geq 50$ eV); Boesten and Tanaka [25] reported measured DCSs (electron energies between 1.5 and 100 eV), ICSs, and momentum transfer cross-sections (MTCSs) for methane. Measurements were done using crossed electron and molecular beam technique where observed DCSs were normalized point by point with the help of simultaneous measurements of DCSs of He. Vušković and Trajmar [26] obtained relative cross-sections for 20, 30, and 200 eV impact energies and normalized them to the absolute measurements of Tanaka et al. [27] for 20 and 30 eV, and to the calculations of Dhal et al. [28] for 200 eV. Cho et al. [29] published measured data for DCSs (ICSs and MTCSs as well) for electron elastic scattering from methane over scattering angles between 10° and 180° for incident electron energies from 5 to 100 eV using crossed beam spectrometer combined with a magnetic angle-changing device. Relative flow with He was exploited for normalization. They used the iterative Schwinger variational method combined with distorted-wave approximation to solve scattering equations. Sakae et al. [30] measured DCSs using crossed-beam method in angular range 5–135° for electron energies 75–700 eV. Relative DCSs were converted to the absolute values at 30° by using the ratio of elastic DCSs of the target gas to that of He. Iga et al. [31] used crossed beam apparatus to obtain scattering intensities (100–500 eV incident electron energies), which were converted to the absolute scale using relative flow method (Ne was used as a reference gas). Also, they used Schwinger variational method combined with the distorted-wave approximation to study elastic electron scattering (1–500 eV) theoretically. Jain [32] used a spherical optical complex potential model to investigate electron interaction with methane over a wide energy range from 0.1 to 500 eV. Mahato et al. [33] obtained analytical expressions for the static potentials of electron scattering from methane using Gaussian wave functions and studied elastic scattering from 10 to 500 eV incident electron energies utilizing those static potentials along with exchange and polarization potentials. Song et al. [17] presented recommended elastic DCSs and ICSs obtained by averaging other authors' data [15,19–21,25,29–31]. List of experimental and theoretical work on DCS for elastic electron scattering from methane molecule, in energy range of our interest is shown in Table 1. For the present work, the paper by Fuss et al. [34] is interesting because they provided the recommended set of data for differential and integral cross-sections for methane, including elastic electron scattering. They obtained their dataset by merging and averaging other authors' data [19,25,30,33] for lower energies and calculating ones for high energies (40–10 keV).

Table 1. List of experimental and theoretical work on differential cross-section (DCS) for elastic scattering of electrons from methane molecule, covering energy range from 50 to 300 eV.

Authors	Experiment Type with Normalization Method/Theoretical Approach	Energy Range (eV)	Angular Range (°)
Boesten and Tanaka [25]	Crossed beams, simultaneous measurements of DCS of He	1.5–100	10–130
Vušković and Trajmar [26]	Crossed beams, normalized to other authors results	20–200	8–130
Cho et al. [29]	Crossed beams, relative flow (He)/Schwinger variational method	5–100	10–180
Sakae et al. [30]	Crossed beams, relative flow (He)	75–700	5–135
Iga et al. [31]	Crossed beams, relative flow (Ne)/Schwinger variational method	100–500/ 1–500	10–135
Jain [32]	Spherical optical complex potential model	0.1–500	0–180
Mahato et al. [33]	Gaussian wave functions	10–500	0–180
Fuss et al. [34]	Optical potential method and the independent atom approximation including the screen corrected additivity rule (SCAR)	0.7–10,000	0–180

In this paper, theoretical and experimental results for elastic electron scattering from methane are shown. Obtained data include absolute differential cross-sections (DCSs) for elastic scattering for the incident electron energy range from 50 to 300 eV (with 50 eV steps) and angular range from 5° to 125° (with 5° steps), integral cross-sections, and momentum transfer cross-sections (ICSs and MTCSs, respectively) for every measured energy. The experiment was performed on a crossed-beam apparatus. As in most previous experiments [19,20,29–31], relative intensities were put on the absolute scale with help of relative flow method. The difference was in reference gas; in almost all experiments, He was used (except for Iga et al. [30] who used Ne), in our argon was reference gas. Theoretical results were obtained by calculating atomic cross-sections for molecular components, with two approximations used for molecular cross-sections simulations; simple sum (model 1) of atomic cross-sections and with molecular effect taken into the account (model 2). The existing variety of cross-section datasets for methane reflects our motivation to perform this study and, at the same time, to provide data at some of the impact energies where there are no previously measured data or where data require an independent confirmation. By exploiting different reference gas, Ar in this case, and performing new calculations that are using a coherent sum of atomic wave functions, we provide the independent and excessive datasets of cross-sections for this important molecule.

The paper is organized as follows. Theory and calculations of DCSs are explained in Section 2. Experimental setup and measurement procedure are given in Section 3. The obtained results are listed in Table 2 and presented graphically, including the comparison with the existing experimental and theoretical data, all given in Section 4. A discussion is given in Section 5. Finally, Section 6 is reserved for the conclusion.

Table 2. Experimental DCSs, integral cross-sections (ICSs), and momentum transfer cross-sections (MTCSs) for elastic scattering of electrons from methane molecule with absolute uncertainties Δ given in parentheses with 2 last digits. Values from 25° to 125° are measured and uncertainties arise from both statistical and short-term stability and include absolute uncertainties of cross-sections for reference Ar gas. Values from 0° to 20° and from 125° to 180° are extrapolated values and used to calculate ICSs and MTCSs. Uncertainties at small scattering angles are estimated to be 30%.

θ (°)	DCS (10^{-20} m^{2}sr^{-1})					
	50 (eV)	100 (eV)	150 (eV)	200 (eV)	250 (eV)	300 (eV)
0	17.5(5.3)	19.3(5.9)	25.1(7.9)	24.0(7.4)	24.5(7.0)	23.4(6.6)
5	13.8(4.1)	13.2(4.0)	15.8(4.7)	13.5(4.1)	12.9(3.9)	11.0(3.3)
10	9.8(2.9)	7.9(2.5)	7.9(2.5)	6.1(1.9)	5.0(1.4)	3.9(1.1)
15	6.3(1.9)	3.9(1.2)	3.2(1.0)	2.43(0.76)	1.50(43)	1.07(30)
20	3.7(1.1)	1.70(53)	1.09(35)	0.83(0.26)	0.43(12)	0.435(78)
25	1.96(60)	0.98(30)	0.45(14)	0.39(12)	0.325(93)	0.277(78)
30	1.30(40)	0.61(19)	0.277(87)	0.246(76)	0.200(57)	0.181(51)
35	0.73(22)	0.40(12)	0.194(61)	0.175(54)	0.136(39)	0.128(36)
40	0.50(12)	0.288(89)	0.162(51)	0.132(41)	0.106(22)	0.100(20)
45	0.384(88)	0.230(55)	0.127(31)	0.105(25)	0.072(15)	0.058(12)
50	0.302(70)	0.174(41)	0.090(22)	0.072(17)	0.053(11)	0.0423(85)
55	0.249(57)	0.128(30)	0.061(15)	0.055(13)	0.0413(85)	0.0347(70)
60	0.206(47)	0.091(22)	0.047(12)	0.044(10)	0.0324(67)	0.0264(53)
65	0.173(40)	0.068(16)	0.0398(99)	0.0362(88)	0.0271(57)	0.0214(43)
70	0.144(33)	0.055(13)	0.0340(84)	0.0306(74)	0.0229(48)	0.0186(38)
75	0.111(26)	0.0451(11)	0.0305(76)	0.0291(71)	0.0211(44)	0.0154(31)
80	0.092(21)	0.0409(98)	0.0283(70)	0.0252(61)	0.0181(38)	0.0120(25)
85	0.072(17)	0.0381(93)	0.0284(71)	0.0244(59)	0.0145(31)	0.0107(22)
90	0.067(16)	0.0403(96)	0.0263(66)	0.0212(52)	0.0121(26)	0.0096(20)
95	0.062(14)	0.043(10)	0.0274(68)	0.0195(48)	0.0120(26)	0.0088(18)
100	0.061(14)	0.047(11)	0.0256(64)	0.0183(45)	0.0118(26)	0.0088(18)
105	0.066(15)	0.049(12)	0.0252(63)	0.0185(45)	0.0106(23)	0.0085(18)
110	0.078(18)	0.051(12)	0.0250(62)	0.0160(39)	0.0099(22)	0.0078(16)
115	0.087(20)	0.057(14)	0.0250(62)	0.0154(38)	0.0094(21)	0.0078(16)
120	0.102(24)	0.061(15)	0.0260(65)	0.0155(38)	0.0092(20)	0.0073(15)
125	0.118(27)	0.063(15)	0.0264(66)	0.0141(35)	0.0092(20)	0.0075(16)
130	0.136(31)	0.066(16)	0.0258(63)	0.0139(35)	0.0092(20)	0.0072(15)
140	0.176(41)	0.071(17)	0.0249(61)	0.0140(35)	0.0091(20)	0.0066(14)
150	0.230(53)	0.074(18)	0.0244(61)	0.0142(35)	0.0089(20)	0.0061(13)
160	0.311(72)	0.077(18)	0.0243(61)	0.0144(36)	0.0086(19)	0.0059(12)
170	0.44(10)	0.079(19)	0.0243(61)	0.0144(36)	0.0084(19)	0.0058(12)
180	0.67(15)	0.079(19)	0.0243(61)	0.0145(36)	0.0083(19)	0.0057(12)
ICS's	6.1(1.8)	3.8(1.2)	2.95(0.92)	2.37(0.74)	1.89(0.55)	1.56(0.47)
MTCS's	2.09(0.53)	0.98(0.26)	0.45(0.13)	0.33(0.10)	0.222(0.054)	0.177(0.043)

2. Theory

In this work, the key to the determination of the elastic cross-section calculations of CH_4 is the atomic cross-section calculations for the components of the molecule. We assume for the determination of the effective interaction at distance **r** between a projectile electron and the target that the scattering potential can be expressed as:

$$V(r) = V_{\text{st}}(r) + V_{\text{ex}}(r) + V_{\text{cp}}(r), \tag{1}$$

where V_{st} is the electrostatic potential, V_{ex} is the electron exchange potential and V_{cp} is the correlation-polarization potential. The electrostatic potential for the interaction between an electron and the target atoms:

$$V_{\text{st}}(r) = -e[\varphi_{\text{n}}(r) + \varphi_{\text{e}}(r)], \tag{2}$$

where φ_n and φ_e are respectively the components of nucleus and the electron cloud of electrostatic potential.

The Furness-McCarthy exchange potential [35] is used for the electron exchange potential:

$$V_{ex,FM}(r) = \frac{1}{2}[E - V_{st}(r)] - \frac{1}{2}\left\{[E - V_{st}(r)]^2 + 4\pi a_0 e^4 \rho_e(r)\right\}^{1/2}. \quad (3)$$

An accurate correlation-polarization potential combines the long-range polarization potential, $V_{cp,B}(r)$, with the correlation potential $V_{cp}(r)$ obtained from the local-density approximation (LDA) and it can be expressed as [36]:

$$V_{cp,LDA}(r) \equiv \begin{cases} \max\{V_{cp}(r), V_{cp,B}(r)\}, & r < r_{cp}; \\ V_{cp,B}(r), & r \geq r_{cp}. \end{cases} \quad (4)$$

where r_{cp} is the outer radius at which $V_{cp}(r)$ and $V_{cp,B}(r)$ cross. The $V_{cp,B}(r)$ when the projectile is far from the atom can be approximated by the Buckingham potential as:

$$V_{cp,B}(r) = -\frac{\alpha_d e^2}{2(r^2 + d^2)^2}, \quad (5)$$

where α_d is dipole polarizability of the target atom and d is a phenomenological cut-off parameter that serves to prevent the polarization potential from diverging at $r = 0$. The experimental values of the atomic dipole polarizabilities from [37] are usually used in Equation (5). Perdew and Zunger [38] proposed a parameterization of the V_{cp}(r) correlation potential in the following form:

$$V_{cp}(r) = \begin{cases} -\frac{e^2}{a_0}(0.0311 \ln r_s - 0.0584 + 0.00133 r_s \ln r_s - 0.0084 r_s), & r_s < 1; \\ -\frac{e^2}{a_0}\frac{0.1423 + 0.1748 r_s^{1/2} + 0.0633 r_s}{\left(1 + 1.0529 r_s^{1/2} + 0.3334 r_s\right)^2}, & r_s \geq 1. \end{cases} \quad (6)$$

where

$$r_s = \frac{1}{a_0}\left[\frac{3}{4\pi\rho_e(r)}\right]^{1/3} \quad (7)$$

is the radius of the sphere that contains (on average) one electron of the gas, in units of the Bohr radius a_0.

For the theoretical determination of the elastic cross-sections we used the Mott's differential cross-section [39],

$$\frac{d\sigma_e}{d\Omega} = |f(\theta)|^2 + |g(\theta)|^2, \quad (8)$$

where θ is scattering angle, $f(\theta)$ and $g(\theta)$ are the spin-up and spin-down scattering amplitudes. The $f(\theta)$ and $g(\theta)$ can be expressed as:

$$f(\theta) = \sum_{l=0}^{\infty} F_\ell P_\ell(\cos\theta), \quad (9)$$

$$g(\theta) = \sum_{l=0}^{\infty} G_\ell P_\ell^1(\cos\theta), \quad (10)$$

where $P_\ell(\cos\theta)$ are the Legendre polynomials, $P_\ell{}^1(\cos\theta)$ are the associated Legendre functions. The F_ℓ and G_ℓ can be calculated according to following relations:

$$F_\ell = \frac{1}{2ik}\left\{(\ell+1)\left(e^{2i\delta_\ell^+} - 1\right) + \ell\left(e^{2i\delta_\ell^-} - 1\right)\right\}; \quad (11)$$

$$G_\ell = \frac{1}{2ik}\sum_{\ell=1}^{\infty}\left\{e^{2i\delta_\ell^-} - e^{2i\delta_\ell^+}\right\}, \quad (12)$$

where $\delta_\ell{}^+$ and $\delta_\ell{}^-$ are spin up and spin down phase shifts of the ℓth partial wave, and k is the momentum of the projectile electron, respectively.

The integration of the differential cross-section over total solid angles gives us the total elastic cross-sections in the following form:

$$\sigma_e = \int \frac{d\sigma_e}{d\Omega} d\Omega = 2\pi \int_0^\pi \sin\theta \left\{ |f(\theta)|^2 + |g(\theta)|^2 \right\} d\theta. \tag{13}$$

All calculations of elastic cross-section were performed by ELSEPA [40]. We used two approximations during the simulation of the molecular elastic cross-sections. In model 1, the calculated atomic cross-sections were simply added according to the stoichiometry numbers (additivity approximation). This approximation is frequently used approximation. However, it neglects the chemical-binding and aggregation effects. The electron distribution in molecules differs from that of an isolated atom of the same element. It was shown that this difference seems to have only a weak influence on the elastic DCS [40] at projectile energies larger than a few hundred eV. The effect of aggregation, the effect when the atoms are close together, has a stronger influence on the DCS [40]. In model 2 we have taken into account the positions of the atoms in the molecule. We used a single-scattering independent-atom approximation assuming that the interaction of the projectile with each atom of a molecule is given by the free-atom potential as for model 1. To determine the molecular DCS, the scattered wave at large distances from the molecule is approximated as the coherent sum of the wave functions scattered from all atoms in the molecule. In our calculations the carbon atom is located at the origin of the coordinate system. The positions of the hydrogen atoms are located at the following coordinates expressed in units of 10^{-10} m: $H_1(0.5541, 0.7996, 0.4965)$, $H_2(0.6833, -0.8134, -0.2536)$, $H_3(-0.7782, -0.3735, 0.6692)$, $H_4(-0.4593, 0.3874, -0.9121)$. We found the significant improvement in the cross-section calculations for model 2 compared with model 1.

3. Experiment

Experimental results presented in this paper are obtained on apparatus with crossed beams setting; incident electrons collide with a molecular beam perpendicularly. The experimental setup is placed inside the vacuum chamber, pumped with a turbomolecular pump to a typical background pressure (no gas in the chamber) about 5×10^{-7} mbar. The magnetic field in the collision region is reduced by two concentric μ-metal shields inside the chamber.

Incident electrons are derived from the hairpin tungsten filament (cathode) by thermo-electronic emission. The electron beam is then extracted and focused into the interaction volume by seven electrodes of the electron gun. Electron energy can vary in the range from 40 to 300 eV and is determined by the potential difference between the filament and the last, grounded, electrode, with energy resolution about 0.5 eV. When the current through the cathode filament was about 2.22 A, the electron current, measured with the Faradays cup without gas in the chamber, was approximately 100 nA. The electron gun can be rotated around the fixed detection system in the angular range from $-40°$ to $126°$.

Atomic/molecular beam is formed by stainless-steel gas needle. The length of the needle is l = 40 mm and its diameter, d = 0.5 mm. According to Lucas [40], beam properties can be predicted and optimized. Optimum atomic beam is obtained when $I(0)^2/(NH^2)$, where $I(0)$ is axial intensity, N throughput and H beam halfwidth, is maximum. It is shown that $I(0)^2/(NH^2) \propto l^2/d$ [41]. So, the optimum beam is acquired when the ratio of square of single tube length and its diameter is maximum, which is in our case 3200. These expressions can be employed for tubes where $\gamma = d/l$ (true dimensionless ratio of length and diameter) is below 10 ($\gamma < 10$) and gas pressure is low enough so that the mean free path of a particle is larger or equal than d ($\lambda \geq d$). In our experimental setup γ = 80 and $\lambda \approx d$.

After the interaction with the molecular beam, scattered electrons are entering the detection system. First, they are focused and slowed down to the constant pass energy of the analyzer by the four-electrode lens. Then, they are energy-analyzed by the double cylindrical mirror analyzer (DCMA). After that, elastically scattered electrons are focused by the three-electrode lens into the detector (single channel electron multiplier).

The intensity of elastically scattered electrons from methane molecule is measured as a function of scattering angle, from 25° to 125° (in 5° steps), at given incident electron energy, from 50 to 300 eV (in 50 eV steps). Experimental parameters were adjusted so that the interaction volume was kept constant. Deviations that can occur at small angles are corrected by comparing cross-sections from Ar at given energy with other authors' data [42,43]. During the measurements, working pressure in the chamber was about 4×10^{-6} mbar. For each electron energy, scattering intensities are measured at least three times. To discount background scattering contributions, a gas beam was introduced to the chamber through a side leak, away from the collision area. Scattering intensities are measured and subtracted from apparent signal for each angle.

The obtained relative DCSs are converted to the absolute DCSs using relative flow method (RFM) [44]. Briefly, the signal of scattered electrons from the target molecule is compared with elastically scattered electron intensity from reference gas, at the same scattering angle, for the same electron energy and experimental conditions. The same experimental conditions for both gasses implies that their beam profiles must be closely equal, which is acquired, according to Olander and Kruger [45], under two conditions: the mean free paths (λ) behind the gas needle for both gasses must be the same and the Knudsen number, K_L, defined as λ/l, must be in the appropriate range, $\gamma \leq K_L \leq 10$. The first condition is fulfilled when pressure ratio of test molecule and reference gas is inversely proportional to the ratio of squares of their gas kinetic diameters. In this study, Ar is used as a reference, with its diameter of approximately 3.58 Å. The gas kinetic diameter of the target gas, methane is approximately 3.8 Å, which gives ratio 1.13; almost the same gas pressure behind gas needle for both gasses must be applied. As for the second condition, gas pressures were low enough so that K_L is approximately equal to γ, although some studies have shown that even when K_L is much lower than γ, beam profiles for most gasses can still be very alike [46,47]. Besides scattering intensities of both gasses, relative flow rate is determined by measuring pressure increase in time by admitting gas into the constant volume, while a chamber outlet was closed.

The known absolute DCSs for Ar are taken from Ranković et al. [42] for incident electron energies 50–200 eV and 300 eV and from Williams and Willis [43] for electron energy of 250 eV. In both papers, the absolute values were derived by measurements of angular dependences of elastically scattered electrons using electron spectrometers, two 127° cylindrical electrostatic energy analyzer in [43], and a double cylindrical mirror analyzer in [42] but employing different normalization procedures. While in [42] the relative flow method with He as a reference gas was used, in [43] a phaseshift analysis of the relative angular distributions of electrons elastically scattered in the energy region of the resonances $^2P_{3/2,1/2}$ of Ar, i.e., between 11.0 and 11.4 eV, were used. Nevertheless, the absolute DCS values for Ar agree within mutual uncertainties as discussed in [42]. Since our normalization is based upon relative flow method, we prefer to use values from the most recent paper [42] and only for the energy of 250 eV, which is not available in [42] we used those from [43].

Since our experimental DCSs are obtained in the limited angular range, in order to obtained elastic integral (ICSs) and momentum transfer (MTCSs) cross-sections, our DCSs must be extrapolated to the smallest (0°) and the highest (180°) scattering angles. Our extrapolation takes into account the theoretically obtained shapes of DCSs (present model 2, at small (0–25°) and high (120–180°) scattering angles and present experimental values of DCSs near 25° and 120°. Thereafter, ICSs and MTCSs were obtained using extrapolated DCSs and appropriate integration defined as:

$$ICS = 2\pi \int_0^\pi DCS(\theta) \sin\theta d\theta \tag{14}$$

$$MTCS = 2\pi \int_0^\pi DCS(\theta)(1 - \cos\theta) \sin\theta d\theta \tag{15}$$

The uncertainties of the relative DCSs consisted of statistical uncertainties and short-term stability uncertainties, caused by instability of the system. This uncertainty is increased by 20% for small scattering angles, due to the potential alteration of the interaction volume. Dominant uncertainties for absolute DCSs are those from reference cross-sections for Ar [42,43], and are taken to be about 20%. The DCS uncertainties obtained in such a manner, Δ, are presented in parenthesis within Table 2. The total uncertainties for absolute DCSs are about 30% for small angles and about 20% for the rest of the angular range. The total uncertainties of ICSs and MTCs arise from the DCSs uncertainties mentioned above and uncertainties of the extrapolation of DCSs to 0° and to 180° and numerical integration (10%).

4. Results

Measured results of the absolute differential cross-sections, integral cross-sections, and momentum transfer cross-sections, together with their corresponding uncertainties, are shown in Table 2. The results cover six incident electron energies, from 50 eV to 300 eV and angular range from 25° to 125°. In Figure 1, the present theoretical and experimental DCSs for all six incident electron energies are shown graphically. For the sake of comparison, other authors' data [16,23,24,27–31], listed in Table 1, are shown in the same figure.

Theoretical results are shown in two approximations: the simple sum of individual atomic cross-sections (model 1, dashed black line) and with molecular effects taken into account (model 2, solid black line). It can be seen that those effects significantly modify DCSs. One of them, the absorption potential, plays a dominant role in this modification of cross-sections. In comparison with present experiment and other authors' results, theory with included molecular effects showed better agreement, both qualitatively and quantitatively, as expected.

Obtained ICSs, together with other authors' results [17,25,29–31,33,34], are presented in Figure 2.

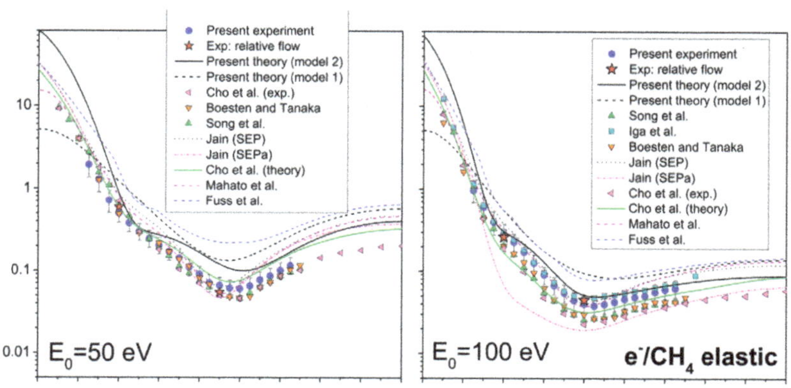

Figure 1. *Cont.*

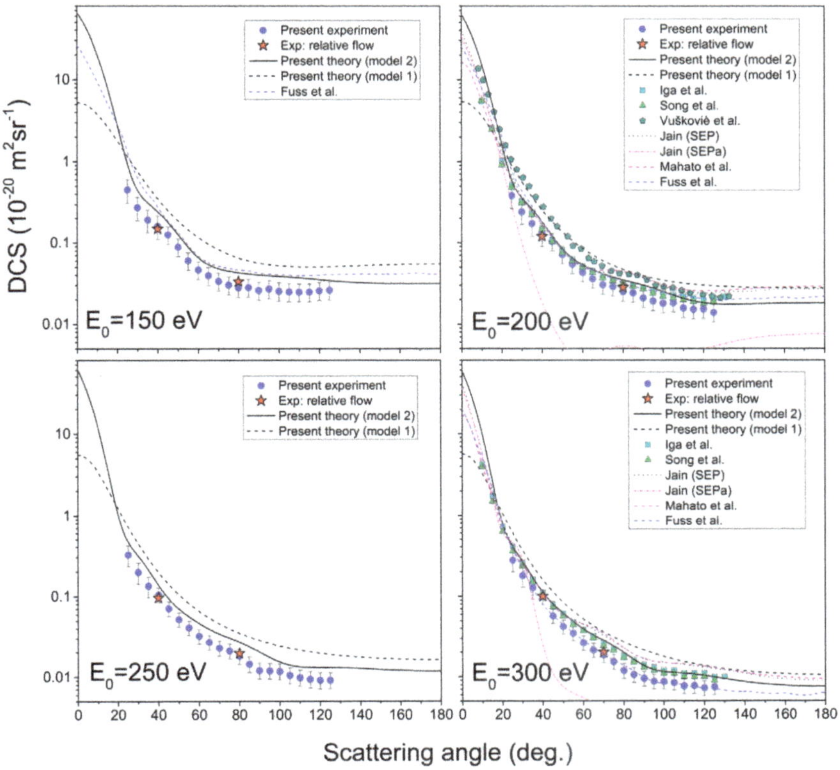

Figure 1. Angle differential cross-sections for elastic electron scattering from methane molecule, for six incident electron energies, from 50 eV to 300 eV. The present results include experiment (blue circles), relative flow absolute data (red stars), theory with molecular effects (full line) and simple sum theory (dashed line). Previous results, tabulated in Table 1, are also shown for the sake of comparison.

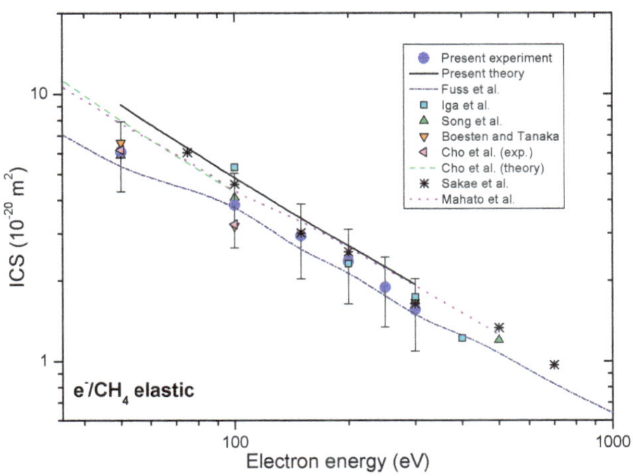

Figure 2. Integral cross-sections for elastic electron-methane collision, presented in energy range from 35 eV to 1000 eV. Present experimental (blue circles) and theoretical (solid black line) results are shown together with previous experimental, as well as the theoretical results, for comparison.

5. Discussion

Experimental DCS at 50 eV exhibits wide minimum at 100° scattering angle, for which position and depth are in good agreement with other authors' results [17,25,29,32,33], especially experimental ones [17,25,29]. Our calculations overestimate the measured DCS, but it matches with shape and is in a good agreement with other theories [29,32] at high scattering angles, above 110°. It is interesting to note that all existing experimental values at 50 eV impact energy agree among themselves within the estimated uncertainties. It is rather a problem with theories, which, in most cases, overestimate absolutes values or do so at certain angular ranges.

There are many experimental and theoretical data for DCS at 100 eV. This cross-section shows a wide and shallow local minimum at 80° scattering angle, which vanishes for incident electron energies above 150 eV. Its position is the same or similar for every shown result. Our measurement and calculation are in strong agreement with the experimental result of Iga et al. [31] and with the theories of Jain (SEP) [32] and Mahato et al. [33], except at higher scattering angles (from 105°). Other results are in good agreement with the shape, but quantitatively are underestimated in comparison with the present result. It seems that there are two classes of experimental values that differ in absolute values, one by Boesten and Tanaka [25] and Cho et al. [29] and the other by Iga et al. [31] and the present measurements. One of the significant differences between these two sets of data lie in the choice of the reference gas, the former used He while the later used Ne and Ar as a reference. Nevertheless the cross-sections for He are known with better accuracy than those of Ne and Ar, it seems plausible to conclude that more similar flow conditions between the reference gas and the target one give more reliable data. If it has been possible to obtain perfectly the same all conditions necessary for applying the relative flow method, then the choice of reference gas would be the one with the best-known cross-sections, i.e., He gas. However, since the method itself introduces additional uncertainties, in our opinion, it would be the best procedure to compare cross-sections with relatively similar flows within gas inlet system.

As for DCSs for 150 eV, there are only results by Fuss et al. [34] available for comparison. At this energy, there are no local minima, like for 50 eV and 100 eV. Instead, there is a wide plateau from 70° scattering angle. Experiment and theory are in good agreement by the shape, but the experiment is, on average, about 30% quantitatively lower.

DCSs for 200 eV (both measured and calculated) show good agreement (within experimental uncertainty) with experimental results of Iga et al. [31], recommended data by Song et al. [17] and theoretical results by Jain (SEP) [32] and Mahato et al. [33] for smaller scattering angles (below 70°). Results by Vušković and Trajm ar [26] are good by shape but higher in absolute value.

For DCS for 250 eV incident electron energy, to the best of our knowledge, there are no published experimental or theoretical results. In the graph, it can be seen that our theory is slightly higher than the measured results, but both are similar qualitatively.

At 300 eV, the shapes and values of present theory and previous results by Iga et al. [31] and Song et al. [17] are in almost perfect agreement. The experimental result is just slightly lower on the absolute scale. Same as for every other energy, calculations by Jain [32] and Mahato et al. [33] agree good for smaller scattering angles, but for higher, they little overestimate other results.

The values of DCSs span over three and four orders of magnitude in presented energy range, what is the characteristic behavior for molecular targets and noticed also in the previous targets [42,48]. The general agreement among different experimental datasets and calculations is very good, and that is what one may expect at the present level of advanced experiments and sophisticated calculations [49].

The present experimental integral cross-sections ICSs are shown in Table 1 and together with other available results in Figure 2. Our experimental ICSs are placed between recommended Fuss et al. [34] and theoretical results Mahato et al. [33]. In the electron energy range from 50 eV to 300 eV most of the other experimental ICSs, like those by

Boesten and Tanaka [25], Sakae et al. [30], Iga et al. [31], Song et al. [17], lie within the same uncertainty limits. Our experimental DCS data are obtained in rather limited angular range, and that is why the presented integral cross-sections depend strongly upon the extrapolation procedure. We have normalized our calculated DCS (model 2) to our measured absolute data to best match in shape and then used these values for integration. These values are presented in Table 2 with the uncertainty of 30% that arises from different plausible extrapolations. The absolute uncertainties of ICS and MTCS values are obtained from the difference of the corridor that is represented by maximal, DCS + $\Delta/2$, and minimal DCS − $\Delta/2$ values.

If total cross-section (TCS) we calculate as the sum of our measured ICSs, and ionization cross-sections obtained by Djurić et al. [50] plus the recommended neutral dissociation cross-sections by Fuss et al. [34], the TCS values obtained in such way agree well, within experimental uncertainties, with previously published experimental values by Zecca et al. [51] and recommended values by Song et al. [17]. Furthermore, for electron energies greater than 150 eV present TCS well agree with the semiempirical TCS by García and Manero [52] extrapolated in the interval energies of this paper. Total cross-sections for electron-methane collision in the energy range from 45 eV to 300 eV are contrasted in Figure 3a, while the corresponding Fano-Bethe plot is shown in Figure 3b. For electron energies from 150 eV to 300 eV Bethe plot of present TCS, References [17,51] and extrapolated semiempirical TCS García and Manero [52], agree within the limits of experimental uncertainties. This, under the given conditions, suggests that Born-Bethe approximation is valid in the electron energy range from 150 eV to 300 eV.

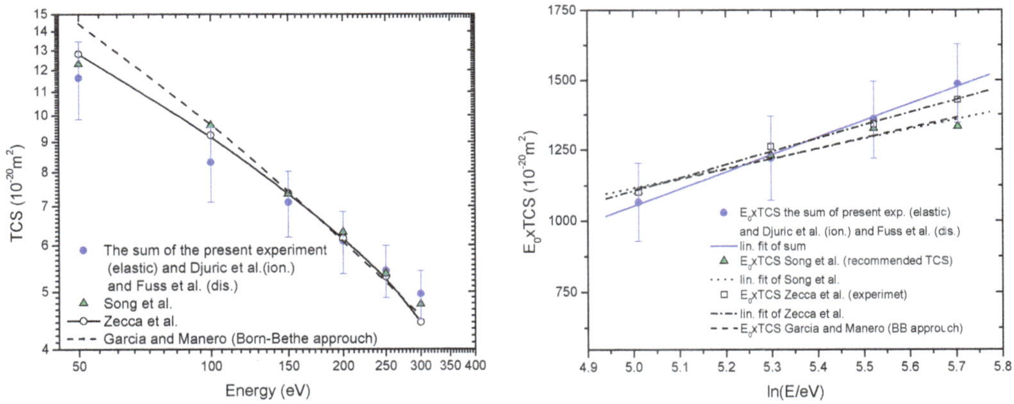

(a) (b)

Figure 3. Total cross-sections for electron-methane collision in energy range from 45 eV to 300 eV. ●, present; ▲, Song et al. (recommended values) [17]; —○—, Zecca et al. (experiment) [51]; - - -, García and Manero (Born-Bethe approach) [52]; (**a**) Absolute values; (**b**) The Fano-Bethe plot: ———, linear fit of present; • • •, linear fit of Song et al. [17]; –·–· linear fit of Zecca et al. [51].

6. Conclusions

In order to provide the insight into methane/electron interaction, measurements and calculations of elastic electron scattering from methane target were performed. To summarize, calculated and measured cross-sections for elastic CH_4-electron scattering in (50–300) eV incident electron energy range are reported and, conceivably, they will serve as a dependable standard for related investigations in the future. Good agreement between present theory and experiment and results available in the literature is noticed. Also, it is shown that molecular effects (especially absorption effects) play a crucial role on calculated cross-sections, particularly for very small scattering angles (below 20°). The experiment was performed in two independent steps. First was measuring relative DCSs for fixed electron energy in function of scattering angle. Second was obtaining pair of absolute points for every energy, by relative flow method and Ar as reference gas. These absolute points were then used for normalization of relative DCSs. These independent results agree well, which is confirmation of the reliability of the experimental method. DCSs for the incident energy 250 eV is given without previous results known to the authors.

Nevertheless there are many studies of electron elastic scattering by methane molecule, our impression is that the present study is important to pinhole the absolute cross-sections, to confirm recommended sets of data for this process given by Song et al. [17] and to bring new results at one impact energy. Last but not the least, we stress the problem of the choice of a reference gas in the relative flow measurements and the necessity of choosing gases with similar flowing conditions. That could bring methane molecule to be a new standard for cross-sections measurements of other hydrocarbons or larger organic molecules.

Author Contributions: Conceptualization, all uthors; methodology, J.B.M., B.P. and B.P.M.; software, K.T.; validation, all authors; experimental investigation, J.V. and J.B.M.; resources, B.P.M.; data curation, B.P.M.; writing—original draft preparation, J.V. and K.T.; writing—review and editing, B.P. and B.P.M.; visualization, J.V. and B.P.; supervision, B.P. and B.P.M.; project administration, B.P.M.; funding acquisition, B.P. and B.P.M. All authors have read and agreed to the published version of the manuscript.

Funding: This research was funded by the Institute of Belgrade, through the grant by the Ministry for Education, Science and Technological Development of the Republic of Serbia.

Institutional Review Board Statement: Not applicable.

Informed Consent Statement: Not applicable.

Data Availability Statement: The data presented in this study is contained within the aricle.

Acknowledgments: We are thankful to R. Srivastava for providing their data in tabular form.

Conflicts of Interest: The authors declare no conflict of interest.

Abbreviations

R. F.	radio frequency
PECVD	plasma-enhanced chemical vapor deposition
DCSs	differential cross-sections
ICSs	integral cross-sections
MTCSs	momentum transfer cross-sections
SCAR	screen corrected additivity rule
ELSEPA	Dirac partial-wave calculation of elastic scattering of electrons and positrons by atoms, positive ions and molecules
DCMA	double cylindrical mirror analyzer
RFM	relative flow method
TCS	total cross-section

References

1. Zhang, J.A.; Paige, D.A. Cold-trapped organic compounds at the poles of the Moon and Mercury: Implications for origins. *Geophys. Res. Lett.* **2009**, *36*, L16203. [CrossRef]
2. Oyama, V.I.; Carle, G.C.; Woeller, F.; Pollack, J.B.; Reynolds, R.T.; Craig, R.A. Pioneer Venus gas chromatography of the lower atmosphere of Venus. *J. Geophys. Res.* **1980**, *85*, 7891–7902. [CrossRef]
3. Krasnopolsky, V.A.; Maillard, J.P.; Owen, T.C. Detection of methane in the martian atmosphere: Evidence for life? *Icarus* **2004**, *172*, 537–547. [CrossRef]
4. Teylor, F.W.; Atreya, S.K.; Encrenaz, T.; Hunter, D.M.; Irwin, P.G.J.; Owen, T.C. The composition of the atmosphere of Jupiter. In *Jupiter: The Planet, Satellites and Magnetosphere*, 1st ed.; Bagenal, F., Dowling, T., McKinnon, W.B., Eds.; Cambridge Univ. Press: Cambridge, UK, 2004; pp. 59–78.
5. Fletcher, L.N.; Orton, G.S.; Teanby, N.A.; Irwin, P.G.J.; Bjoraken, G.I. Methane and its isotopologues on Saturn from Cassini/CIRS observations. *Icarus* **2009**, *199*, 351–367. [CrossRef]

6. Bezard, B.; Romani, P.N.; Conrath, B.J.; Maguire, W.C. Hydrocarbons in Neptune's stratosphere from Voyager infrared observations. *J. Geophys. Res.* **1991**, *96*, 18961–18975. [CrossRef]
7. Karkoschka, E.; Tomasko, M.G. The haze and methane distributions on Neptune from HST-STIS spectroscopy. *Icarus* **2011**, *211*, 780–797. [CrossRef]
8. Houghton, J. Global warming. *Rep. Prog. Phys.* **2005**, *68*, 1343–1403. [CrossRef]
9. Yusuf, R.O.; Noor, Z.; Abba, A.H.; Hassan, M.A.; Din, M.F.M. Methane emission by sectors: A comprehensive review of emission sources and mitigation methods. *Renew. Sustain. Energy Rev.* **2012**, *16*, 5059–5070. [CrossRef]
10. Dlugokencky, E.J.; Hall, B.D.; Montzka, S.A.; Dutton, G.; Muühle, J.; Elkins, J.W. Long-lived greenhouse gases. *Bull. Am. Meteor. Soc.* **2019**, *100*, S48–S50.
11. Sharma, R.; Woehrl, N.; Vrućinić, M.; Timpner, M.; Buck, V.; Barhai, P.K. Effect of microwave power and C_2 emission intensity on structural and surface properties of nanocrystalline diamond films. *Thin Solid Films* **2011**, *519*, 7632–7637. [CrossRef]
12. Fraga, M.; Rodrigo Pessoa, R. Progresses in Synthesis and Application of SiC Films: From CVD to ALD and from MEMS to NEMS. *Micromachines* **2020**, *11*, 799. [CrossRef] [PubMed]
13. Horton, L.D. Atomic and molecular data needs for fusion research. *Phys. Scr.* **1996**, *T65*, 175–178. [CrossRef]
14. Blanco, F.; Ellis-Gibbings, L.; García, G. Screening corrections for the interference contributions to the electron and positron scattering cross sections from polyatomic molecules. *Chem. Phys. Lett.* **2016**, *645*, 71–75. [CrossRef]
15. Allan, M. Improved techniques of measuring accurate electron-molecule cross sections near threshold and over a large angular range. *AIP Conf. Proc.* **2007**, *901*, 107–116. [CrossRef]
16. Marinković, B.P.; Srećković, V.A.; Vujčić, V.; Ivanović, S.; Uskoković, N.; Nešić, M.; Ignjatović, L.M.; Jevremović, D.; Dimitrijević, M.S.; Mason, N.J. BEAMDB and MOLD–Databases at the Serbian Virtual Observatory for collisional and radiative processes. *Atoms* **2019**, *7*, 11. [CrossRef]
17. Song, M.Y.; Yoon, J.S.; Cho, H.; Itikawa, Y.; Karwasz, G.P.; Kokoouline, V.; Nakamura, Y.; Tennyson, J. Cross Sections for Electron Collisions with Methane. *J. Phys. Chem. Ref. Data* **2015**, *44*, 023101. [CrossRef]
18. Bettega, M.H.F.; do Varella, M.T.N.; Lima, M.A. Polarization effects in the elastic scattering of low-energy electrons by XH_4 X=(C, Si, Ge, Sn, Pb). *Phys. Rev. A* **2003**, *68*, 012706. [CrossRef]
19. Bundschu, C.T.; Gibson, J.C.; Gulley, R.J.; Brunger, M.J.; Buckman, S.J.; Sanna, N.; Gianturco, F.A. Low-energy electron scattering from methane. *J. Phys. B* **1997**, *30*, 2239–2259. [CrossRef]
20. Shyn, T.W.; Cravens, T.E. Angular distribution of electrons elastically scattered from CH_4. *J. Phys. B* **1990**, *23*, 293–300. [CrossRef]
21. Sohn, W.; Kochem, K.-H.; Scheuerlein, K.-M.; Jung, K.; Ehrhardt, H. Elastic electron scattering from CH_4 for collision energies between 0.2 and 5 eV. *J. Phys. B* **1986**, *19*, 3625–3632. [CrossRef]
22. Varambhia, H.N.; Munro, J.J.; Tennyson, J. R-matrix calculations of low-energy electron alkane collisions. *Int. J. Mass Spectrom.* **2008**, *271*, 1–7. [CrossRef]
23. Tennyson, J. Electron–molecule collision calculations using the R-matrix method. *Phys. Rep.* **2010**, *491*, 29–76. [CrossRef]
24. Brigg, W.J.; Tennyson, J.; Plummer, M. R-matrix calculations of low-energy electron collisions with methane. *J. Phys. B* **2014**, *47*, 185203. [CrossRef]
25. Boesten, L.; Tanaka, H. Elastic DCS for e+CH_4 collisions, 1.5–100 eV. *J. Phys. B* **1991**, *24*, 821–832. [CrossRef]
26. Vušković, L.; Trajmar, S. Electron impact excitation of methane. *J. Chem. Phys.* **1983**, *78*, 4947–4951. [CrossRef]
27. Tanaka, H.; Okada, T.; Boesten, L.; Suzuki, T.; Yamamoto, T.; Kubo, M. Differential cross sections for elastic scattering of electrons by CH_4 in the energy range of 3 to 20 eV. *J. Phys. B* **1982**, *15*, 3305–3319. [CrossRef]
28. Dhal, S.S.; Srivastava, B.B.; Shingal, R. Elastic scattering of electrons by methane molecules at intermediate energies. *J. Phys. B* **1979**, *12*, 2727–2734. [CrossRef]
29. Cho, H.; Park, Y.S.; Castro, E.A.Y.; de Souza, G.L.C.; Iga, I.; Machado, L.E.; Brescansin, L.M.; Lee, M.-T. A comparative experimental–theoretical study on elastic electron scattering by methane. *J. Phys. B* **2008**, *41*, 045203. [CrossRef]
30. Sakae, T.; Sumiyoshi, S.; Murakami, E.; Matsumoto, Y.; Ishibashi, K.; Katase, A. Scattering of electrons by CH_4, CF_4 and SF_6 in the 75-700 eV range. *J. Phys. B* **1989**, *22*, 1385–1394. [CrossRef]
31. Iga, I.; Lee, M.-T.; Homem, M.G.P.; Machado, L.E.; Brescansin, L.M. Elastic cross sections for e^--CH_4 collisions at intermediate energies. *Phys. Rev. A* **2000**, *61*, 022708. [CrossRef]
32. Jain, A. Total (elastic + absorption) cross sections for e-CH_4 collisions in a spherical model at 0.10–500 eV. *Phys. Rev. A* **1986**, *34*, 3707–3722. [CrossRef]
33. Mahato, D.; Sharma, L.; Stauffer, A.D.; Srivastava, R. Electron impact elastic scattering from methane and silane molecules. *Eur. Phys. J. D* **2019**, *73*, 189. [CrossRef]
34. Fuss, M.C.; Muñoz, A.; Oller, J.C.; Blanco, F.; Hubin-Franskin, M.-J.; Almeida, D.; Limão-Vieira, P.; García, G. Electron–methane interaction model for the energy range 0.1–10,000 eV. *Chem. Phys. Lett.* **2010**, *486*, 110–115. [CrossRef]
35. Furness, J.B.; McCarthy, I.E. Semiphenomenological optical model for electron scattering on atoms. *J. Phys. B* **1973**, *6*, 2280–2291. [CrossRef]
36. Salvat, F. Optical-model potential for electron and positron elastic scattering by atoms. *Phys. Rev. A* **2003**, *68*, 012708. [CrossRef]
37. Lide, D.R. *CRC Handbook of Chemistry and Physics*, 79th ed.; CRC Press: Boca Raton, FL, USA, 1998.
38. Perdew, J.P.; Zunger, A. Self interaction correction to density functional approximations for many electron systems. *Phys. Rev. B* **1981**, *23*, 5048–5079. [CrossRef]

39. Mott, N.F. The scattering of fast electrons by atomic nuclei. *Proc. R. Soc. Lond. Ser. A* **1929**, *124*, 425–442. [CrossRef]
40. Salvat, F.; Jablonski, A.; Powell, C.J. ELSEPA Dirac partial-wave calculation of elastic scattering of electrons and positrons by atoms, positive ions and molecules. *Comput. Phys. Commun.* **2005**, *165*, 157–190. [CrossRef]
41. Lucas, B.C. *Atomic and Molecular Beams Production and Collimation*; CRC Press: Boca Raton, FL, USA, 2012; pp. 217–223.
42. Ranković, M.; Maljković, J.B.; Tökési, K.; Marinković, B.P. Elastic electron differential cross sections for argon atom in the intermediate energy range from 40 eV to 300 eV. *Eur. Phys. J. D* **2018**, *72*, 30. [CrossRef]
43. Williams, J.F.; Willis, B.A. The scattering of electrons from inert gases I. Absolute differential elastic cross sections for argon atoms. *J. Phys. B* **1975**, *8*, 1670–1682. [CrossRef]
44. Maljković, J.B.; Blanco, F.; García, G.; Marinković, B.P.; Milosavljević, A.R. Absolute cross sections for elastic electron scattering from methylformamide. *Phys. Rev. A* **2012**, *85*, 042723. [CrossRef]
45. Olander, D.R.; Kruger, V. Molecular Beam Sources Fabricated from Multichannel Arrays. III. The Exit Density Problem. *J. Appl. Phys.* **1970**, *41*, 2769–2776. [CrossRef]
46. Tanaka, H.; Ishikawa, T.; Masai, T.; Sagara, T.; Boesten, L.; Takekawa, M.; Itikawa, Y.; Kimura, M. Elastic collisions of low- to intermediate-energy electrons from carbon dioxide: Experimental and theoretical differential cross sections. *Phys. Rev. A* **1998**, *57*, 1798–1808. [CrossRef]
47. Buckman, S.J.; Gulley, R.J.; Moghbelalhossein, M.; Bennett, S.J. Spatial profiles of effusive molecular beams and their dependence on gas species. *Meas. Sci. Technol.* **1993**, *4*, 1143–1153. [CrossRef]
48. Maljković, J.B.; Vuković, J.; Tökési, K.; Predojević, B.; Marinković, B.P. Elastic electron scattering cross sections for triethyl phosphate molecule at intermediate electron energies from 50 to 250 eV. *Eur. Phys. J. D* **2019**, *73*, 27. [CrossRef]
49. Bartschat, K.; Tennyson, J.; Zatsarinny, O. Quantum-Mechanical Calculations of Cross Sections for Electron Collisions with Atoms and Molecules. *Plasma Process. Polym.* **2017**, *14*, 1600093. [CrossRef]
50. Djurić, N.; Cadež, I.; Kurepa, M. Electron impact total ionization cross-sections for methane, ethane and propane. *Int. J. Mass Spectrom. Ion Process.* **1991**, *108*, R1–R10. [CrossRef]
51. Zecca, A.; Karwasz, G.; Brusa, R.S.; Szmytkowski, C. Absolute total cross sections for electron scattering on CH_4 molecules in the 1-4000 eV energy range. *J. Phys. B* **1991**, *24*, 2747–2754. [CrossRef]
52. García, G.; Manero, F. Electron scattering by CH_4 molecules at intermediate energies (400–5000 eV). *Phys. Rev. A* **1998**, *57*, 1069–1073. [CrossRef]

International Journal of *Molecular Sciences*

MDPI

Article

Absolute Differential Cross-Sections for Elastic Electron Scattering from Sevoflurane Molecule in the Energy Range from 50–300 eV

Jelena Vukalović [1,2], Jelena B. Maljković [1], Francisco Blanco [3], Gustavo García [4,5], Branko Predojević [2] and Bratislav P. Marinković [1,*]

1 Institute of Physics Belgrade, University of Belgrade, Pregrevica 118, 11080 Belgrade, Serbia; jelena.vukovic@pmf.unibl.org (J.V.); jelenam@ipb.ac.rs (J.B.M.)
2 Faculty of Science, University of Banja Luka, Mladena Stojanovića 2, 78000 Banja Luka, Republic of Srpska, Bosnia and Herzegovina; bpredojevic@teol.net
3 Departamento de Estructura de la Materia Física Térmica y Electrónica e IPARCOS, Universidad Complutense de Madrid, Plaza de Ciencias 1, 28040 Madrid, Spain; pacobr@fis.ucm.es
4 Instituto de Física Fundamental, Consejo Superior de Investigaciones Científicas, Serrano 113-bis, 28006 Madrid, Spain; g.garcia@csic.es
5 Centre for Medical Radiation Physics, University of Wollongong, Wollongong, NSW 2522, Australia
* Correspondence: bratislav.marinkovic@ipb.ac.rs; Tel.: +381-11-316-0882

Abstract: We report the results of the measurements and calculations of the absolute differential elastic electron scattering cross-sections (DCSs) from sevoflurane molecule ($C_4H_3F_7O$). The experimental absolute DCSs for elastic electron scattering were obtained for the incident electron energies from 50 eV to 300 eV, and for scattering angles from 25° to 125° using a crossed electron/target beams setup and the relative flow technique for calibration to the absolute scale. For the calculations, we have used the IAM-SCAR+I method (independent atom model (IAM) applying the screened additivity rule (SCAR) with interference terms included (I)). The molecular cross-sections were obtained from the atomic data by using the SCAR procedure, incorporating interference term corrections, by summing all the relevant atomic amplitudes, including the phase coefficients. In this approach, we obtain the molecular differential scattering cross-section (DCS), which, integrated over the scattered electron angular range, gives the integral scattering cross-section (ICS). Calculated cross-sections agree very well with experimental results, in the whole energy and angular range.

Keywords: sevoflurane; cross-section; elastic scattering; electrons

Citation: Vukalović, J.; Maljković, J.B.; Blanco, F.; García, G.; Predojević, B.; Marinković, B.P. Absolute Differential Cross-Sections for Elastic Electron Scattering from Sevoflurane Molecule in the Energy Range from 50–300 eV *Int. J. Mol. Sci.* **2022**, *2*, 10021. https://doi.org/10.3390/ijms23010021

Academic Editor: Krzysztof Bobrowski

Received: 27 November 2021
Accepted: 15 December 2021
Published: 21 December 2021

1. Introduction

Sevoflurane (SF) is sweet-smelling non-flammable highly fluorinated methyl isopropyl ether with a boiling point at a temperature of 58.5 °C. Molar mass is 200.055 g/mol, vapor pressure 197 mmHg at 26 °C, and dipole moment 2.33 D [1]. It is one of the most commonly used inhalational anaesthetics, and it has been widely investigated. The recent review on inhaled anaesthetics [2] covered their environmental role, occupational risk, and clinical use. Gaya da Costa and co-authors [2] extensively elaborated on the case of sevoflurane molecule and its contribution to the global warming effect as a volatile anaesthetic (especially in combination with the use with N_2O), pointed out that its threshold has not yet been established in the workplace as waste anaesthetic gas, considering its side effects in a clinical context (epileptiform electroencephalogram patterns in both adults and paediatric populations; the cardio protective effect in patients with coronary artery disease undergoing vascular surgery, kidney transplantation or lung surgery). The atmospheric lifetimes of the halogenated anaesthetics, halothane, enflurane, sevoflurane, isoflurane, and desflurane to the reaction to the hydroxyl radical (OH) and UV photolysis have been determined from observations of OH reaction kinetics and UV absorption spectra [3]. Halothane, enflurane, and isoflurane showed distinct UV absorption in the range 200–350 nm, and no absorption in the wavelength range 200–350 nm was detected for sevoflurane. Tang et al. [1] analysed the effects of general anaesthetics on their potential targets by large-scale molecular simulation. The structural parameters and partial atomic charges of the anaesthetics, considered to be of great importance, were determined. Geometric optimizations using the Hartree–Fock and the B3LYP (Becke, 3-parameter, Lee–Yang–Parr) density functional theory methods with the large 6–311+G (2d,p) basis set were performed to determine the structures and charge distributions of two halogenated anaesthetics; sevoflurane and halothane. Lesarri and co-authors [4] investigated the conformational landscape of the volatile anaesthetic sevoflurane. The structure of the anaesthetic haloether sevoflurane ($CH_2F–O–CH(CF_3)_2$) has been resolved using Fourier-transform microwave (FT-MW) spectroscopy in a supersonic-jet expansion. In isolated conditions, sevoflurane adopts a single conformation characterized by a gauche fluoromethoxy group and a near-symmetric orientation of the isopropyl group with respect to the ether plane (cis H–Cipr–O–CF). A schematic drawing of sevoflurane is shown in Figure 1.

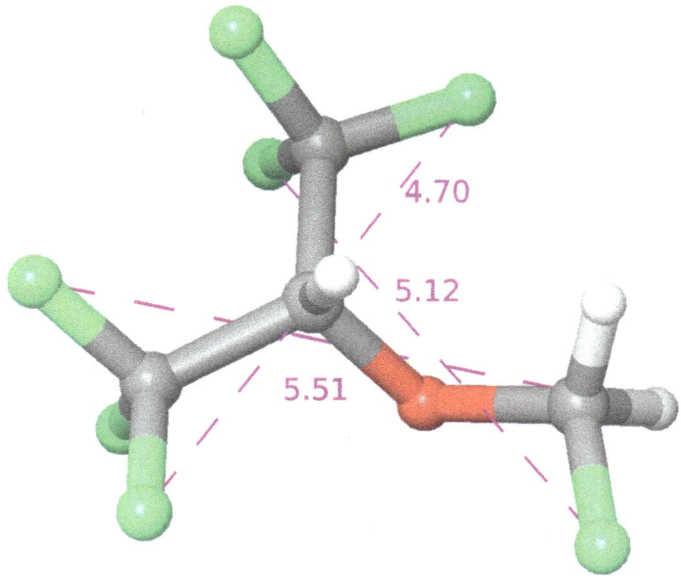

Figure 1. Schematic drawing of sevoflurane.

Don et al. [5] proposed a vibrational assignment of sevoflurane and studied its interaction with the aromatic model compound benzene, using the vibrational spectroscopy of supersonic jet expansions and of cryo solutions in liquid xenon. Sevoflurane has been investigated from the medical point of view also. Shiraishi and Ikeda [6] studied an uptake and biotransformation of sevoflurane in humans and concluded that, despite its relatively large minimum alveolar concentration (MAC), sevoflurane has a small uptake due to its low solubility, but the degradation rate is shown to be high, resulting in a higher serum fluoride concentration than for other halogenated anaesthetics. Most recently, Dong et al. [7] assessed mice and neurons treated with anaesthetics sevoflurane and desflurane, and applied nanobeam-sensor technology, an ultrasensitive method, to measure tau/p-tau amounts. They showed that the sevoflurane induces tau trafficking from neurons to microglia.

Lange et al. [8] have investigated the lowest-lying electronic states of isoflurane and sevoflurane in the 5.0–10.8 eV energy range by experimental and theoretical methods. Photoabsorption spectra of isoflurane and sevoflurane have been measured with synchrotron radiation over the photon range 5.0–10.8 eV. Low-lying excited singlet valence and Rydberg states are investigated, and the assignments supported by quantum chemical calculations, the latter also helping to identify the triplet states. The measured absolute cross-sections have been used to calculate the photolysis lifetimes of isoflurane and sevoflurane in the Earth's atmosphere.

Electron collisions with sevoflurane have also been investigated. Very recently, Lozano et al. [9] reported total scattering cross-section from sevoflurane of 1–300 eV electrons. The experimental results, obtained from an implemented magnetic beam apparatus, are compared with theoretical results from the independent atom model, with a screening corrected additivity rule including interference effects and rotational excitation.

As pointed out by Tanaka et al. [10], the absolute electron-impact excitation cross-sections could be obtained by normalization to elastic scattering cross-section (which is the subject of the present study), or in other ways to determine the binary encounter (BE) scaling curve, after measuring the DCSs at high-energy impact (>100 eV), and for small scattering angles. A comprehensive review of experimental techniques and calculation methods for determination of electron cross-sections from biomolecules, biofuels and their fragments, has been recently given by Brunger [11], covering extensive literature in the field.

In this paper, theoretical and experimental results for elastic electron scattering from sevoflurane, in the medium energy range, are shown, complementing the investigations of electron interaction with this molecule. According to our knowledge, these are the first reported results for absolute differential cross-sections in the energy range from 50–300 eV.

Experimentally obtained data include absolute differential cross-sections for elastic electron scattering for the incident electron energy range from 50 to 300 eV (with 50 eV steps) and an angular range from 25° to 125° (with 5° steps), and integral and momentum transfer cross-sections for every measured energy. The experiment was performed on a crossed-beam apparatus. Relative points were put on the absolute scale with the help of the relative flow method, using Ar as a reference gas. Theoretical results are obtained using the IAM-SCAR+I method (independent atom model (IAM) applying the screened additivity rule (SCAR) with interferences terms included (I)). A very good level of agreement has been found between the present experimental and theoretical data, within the experimental uncertainties.

The paper is organized as follows. The experimental setup and measurement procedure are given in Section 2. The theory and calculations of DCSs are explained in Section 3. Obtained results are shown graphically and in the table, and discussion is shown in Section 4. Section 5 is reserved for the conclusion.

2. Experiment

A schematic drawing of the experimental set-up is shown in Figure 2. Our beam profile has been determined in a previous set of measurements [12–14]. In our experiment, the source of molecular beam is a stainless steel gas needle, with a diameter of d = 0.5 mm and a tube length of l = 40 mm, which gives the ratio $\Gamma = d/l = 80$, while the input pressures were such that the free mean path (λ) was larger than the inner diameter of the tube. According to Lucas [15], the proposed conditions for Γ and λ are $\Gamma > 10$ and $\lambda \approx d$, so we fulfil all the requirements for the well-collimated beam, and the derived expressions can be used in our experiment. Moreover, author [15] states that the beam properties will be optimized if the ratio of the square of single tube length and its diameter is maximized, which, in our case, is 3200.

The analysing system consists of a front four-electrode lens, energy analyser, and rear three-electrode lens. The four-electrode lens is used for collecting, directing, and focusing scattered electrons to the energy analyser after slowing them down to the constant analyser pass energy. Electron energy analyser is a double cylindrical mirror analyser (DCMA), which, essentially, represents two cylindrical analysers connected in series. Pass energy is determined by the potential difference between the cylinders. The three-electrode lens is used for focusing analysed electrons into the detector (single channel electron multiplier).

Since detecting particles are electrons, detector entrance is grounded, and exit is on a high positive potential.

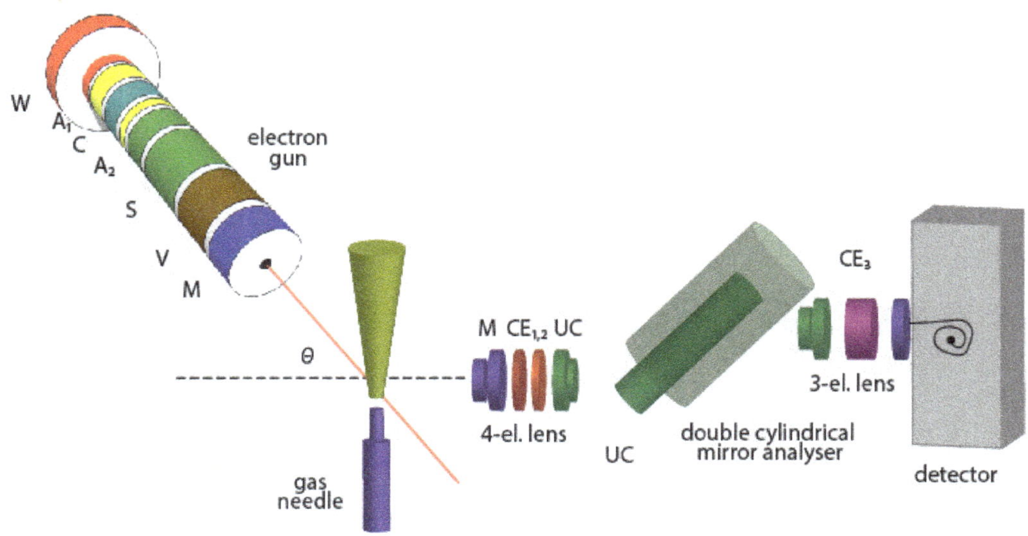

Figure 2. Schematic drawing of the experimental set-up.

The anhydrous sevoflurane ($C_4H_3F_7O$) used in this experiment had a declared purity better than 99%. Before starting the measurements a few freeze-thaw-pump cycles under vacuum have been made. Sevoflurane (SF) is a liquid at room temperature and was introduced into the scattering region from a glass container via a gas line system. SF molecule is characterized by vapor pressure (197 mm Hg at 26 °C) and relatively high dipole moment (2.33 D). Therefore, to provide stable experimental conditions, everything was heated to approximately 50 °C during the gas phase measurements in analogy with previously studied molecular targets [16,17]. In the present work, pipes and the needle were also heated providing a stable driving pressure behind the needle and also for avoiding the absorption effects [18].

In order to reduce magnetic disturbance, the experimental setup is placed in the chamber shielded by the two concentric μ-metal layers. The pre-experimental pressure in the chamber was about $5 \cdot \times 10^{-7}$ mbar and the working pressure was an order of magnitude higher.

The experimental procedure for acquiring the absolute differential cross-sections for elastic scattering of electrons from sevoflurane molecule consists of two main stages:

- Stage 1: obtaining relative DCSs by measuring the intensity of electrons elastically scattered from SF, in the function of scattering angle (from 25° to 125° in 5° steps), for a given incident electron energy (50, 100, 150, 200, 250 and 300 eV). Electron intensities are measured at least three times for every incident energy. Background contributions are suppressed by introducing gas into the chamber, away from the interaction volume, via a side leak, and by subtracting measured intensities from the apparent signal for every angle at the given energy;
- Stage 2: Obtaining absolute points (two) for every electron energy, and normalizing relative DCSs on them. Absolute points are obtained using the relative flow method [19,20].

In the present work, DCSs for elastic electron scattering from sevoflurane have been measured at selected incident electron energies, from 50 to 300 eV (in 50 eV steps), and at scattering angles from 20° to 125° (in 5° steps). At given electron energy, the relative cross-section has been derived as a function of scattering angle by measuring the elastic scattering intensity at the maximum of the elastic peak. The background contributions of the elastic electron intensities, which were around 10% at higher energies, were subtracted from the measured electron yields. It should be noted that the background contributions were generally more important at low incident energies and scattering angles (around 15%). During the measurements, potentials on electron gun electrodes and DCMA were adjusted to keep interaction volume constant. Deviations that can occur at small angles are corrected by comparing cross-sections from Ar at given energy with other authors' data [14,21].

Measured relative DCSs were normalized to the absolute scale according to the absolute points (at 40°, 80°, 90° or 100°), obtained by a Relative flow method (RFM) previously explained in detail [22]. In RFM we compare intensities of elastically scattered electrons on the target molecule (SF) and referent gas (we used Ar), under the same experimental conditions [19,20]. In order to obtain the same experimental conditions, the same beam profiles, for both target molecule and referent gas must be provided. According to Olander and Kruger [23], the same beam profiles can be obtained with two conditions: (1) the mean free paths for both gases must the same and (2) the mean free path is approximately equal to the diameter of stainless steel gas needle. $\lambda \approx d$. In the experiment, the first condition is achieved by adjusting pressures behind the gas needle inversely proportional to the ratio of the squared gas kinetic diameters, and for second condition gas pressures need to be low, which is fulfilled in our experiment [14]. Gas kinetic diameter for Ar is 3.58 Å and for sevoflurane 5.51 Å, so the pressure ratio was P_{Ar}: P_{SF} = $D^2{}_{SF}$: $D^2{}_{Ar}$ ~ 2.4. When the same beam profiles for both, target molecule (DCS_x) and referent gas (DCS_{ref}) are obtained, according to Nickel at al [19,20], absolute differential cross-sections for target molecule $DCSx$ (E, θ) can be calculated, knowing absolute differential cross-sections for the referent gas DCS_{ref} (E, θ), at the same energy and angle, using a formula:

$$DCS_x(E,\theta) = DCS_{ref}(E,\theta)\frac{N_x F_{ref}}{N_{ref} F_x}\sqrt{\frac{M_{ref}}{M_x}} \tag{1}$$

Quantities that we measure in the experiment are: Flow rate (F) and intensity of elastically scattered electrons (N). M_{ref} and M_x are molecular masses for referent gas and target molecule, respectively. The relative flow rate has been determined by closing an outlet to the chamber, admitting target gases into a closed constant volume, and then measuring the pressure increase in time. The flow is determined from the experimental curve of pressure versus time fitted by the least-squares method. The whole system was heated to minimize absorption effects [18], which can influence measuring flow rates and determining the flow.

Our experimental spectrometer UGRA is limited in performing angular dependent differential cross-sections measurements (25–125°) and to obtain integral (*ICSs*) and momentum transfer cross-sections (*MTCSs*) we need to extrapolate measured *DCSs* to 0° and 180°. For extrapolation, from 0–25°, and 125–180° we used calculated results obtained by IAM + SCAR + I procedure (see section Theory) and appropriate integration. Our experimental *DCS* data are obtained in a rather limited angular range and that is why the presented integral cross-sections depend strongly upon the extrapolation procedure. The exact formulae used for the integration are in the form:

$$ICS = 2\pi \int_0^{\pi} DCS(\theta) \sin\theta d\theta \tag{2}$$

$$MTCS = 2\pi \int_0^{\pi} DCS(\theta)(1-\cos\theta) \sin\theta d\theta \tag{3}$$

The uncertainties of the relative *DCSs* consist of statistical uncertainties and short-term stability uncertainties, caused by the instability of the system. This uncertainty is increased by 20% for small scattering angles, due to the potential alteration of the interaction volume. Dominant uncertainty for absolute *DCSs* are those from reference cross-sections for Ar [14,21], and are taken to be about 20%. The total uncertainties, derived as the square root of the sum of squares of individual independent uncertainties, for absolute *DCSs* are about 30% for small angles and about 22% for the rest of the angular range. The total uncertainties of *ICSs* and *MTCs* arise from the *DCSs* uncertainties mentioned above and uncertainties of the extrapolation of *DCSs* to 0° and 180° and numerical integration and were approximately 30%.

3. Theory

To calculate the differential elastic, integral elastic, and inelastic as well as the total cross-sections we have used the IAM-SCAR+I method (independent atom model (IAM) applying the screened additivity rule (SCAR) with interferences terms included (I)). This method has been described in detail in previous publications [9,24–28]. Briefly, the molecular target is described as an aggregate of its individual atoms (i.e., C, H, F, and O in this case). Each atomic target is represented by an ab initio interacting complex optical potential given by:

$$V_{opt}(\vec{r}) = V_R(\vec{r}) + iV_{abs}(\vec{r}) \tag{4}$$

In Equation (4), the real part accounts for elastic scattering while the imaginary part represents the inelastic processes which are considered as the 'absorption part' following the procedure of Staszewska et al. [29]. The real part is divided into three terms that include:

$$V_R(\vec{r}) = V_s(\vec{r}) + V_{ex}(\vec{r}) + V_{pol}(\vec{r}), \tag{5}$$

where vs. represents a static term derived from a Hartree-Fock calculation of the atomic charge distribution [30], V_{ex} an exchange term to account for the indistinguishability of the incident and target electrons [31], and V_{pol} a long-range polarization term [32].

The molecular cross-sections are obtained from the atomic data by using the screening corrected additivity rule (SCAR) procedure [33], incorporating interference (I) term corrections [25], by summing all the relevant atomic amplitudes, including the phase coefficients. In this approach, we obtain the molecular differential scattering cross-section (DCS), which integrated over all the scattered electron angular range gives the integral scattering cross-section (ICS). Moreover, by taking the sum of the ICS for all open channels (elastic ICS and inelastic ICS) the total cross-section (TCS) could be obtained. Note that we do not include at this stage any contribution from vibrational and rotational excitation processes.

Sevoflurane is a polar molecule with a permanent dipole moment of 2.22 D [1] and therefore rotational excitations, not accounted for by the above procedure, are relevant within the scattering scheme. In order to approximately include differential and integral rotational excitation cross-section in the present study, we used the first-Born approximation, following the procedure described in [27]. In these conditions, these cross-sections can easily be calculated by considering the molecule as a rigid rotor, with the initial rotational excited state distribution in equilibrium at 300 K, and calculating all the transitions $\Delta J = \pm 1$ (J being the rotational quantum number) assuming the Born approximation but including the correction for the larger angles given by Dickinson (see ref. [34] and references therein).

4. Results

Experimental results for DCSs, together with experimental ICSs and MTCSs, for elastic electron scattering from sevoflurane, are presented in Table 1, all with corresponding absolute uncertainties. Experimentally measured and theoretically calculated DCSs are shown graphically in Figure 3. The experiment covers six energies of impact electrons, from 50 eV to 300 eV, and the angular range from 25° to 125° (theory covers all angles, from 0° to 180°). Absolute uncertainties are about 22%, except for small angles, where uncertainties are increased to about 30% due to the potential interaction volume alterations.

Table 1. Experimental results for absolute differential cross-sections (DCSs), integral cross-sections (ICSs) and momentum transfer cross-sections (MTCSs) for elastic electron scattering from sevoflurane. In parentheses are given absolute uncertainties of the last two digits.

θ (°)	DCS (10^{-20} m^2 sr^{-1})					
	50 (eV)	**100 (eV)**	**150 (eV)**	**200 (eV)**	**250 (eV)**	**300 (eV)**
25	3.6(1.1)	2.55(76)	1.50(39)	2.25(74)	1.82(56)	1.53(45)
30	3.07(93)	1.98(59)	1.25(32)	1.44(48)	1.20(37)	1.12(33)
35	2.17(66)	1.64(49)	0.82(21)	1.01(33)	0.99(30)	0.96(28)
40	1.89(43)	1.14(34)	0.69(14)	0.80(26)	0.84(26)	0.72(21)
45	1.77(40)	0.78(23)	0.55(11)	0.71(24)	0.67(16)	0.43(13)
50	1.57(36)	0.61(14)	0.476(94)	0.61(16)	0.43(10)	0.304(66)
55	1.31(30)	0.49(11)	0.386(77)	0.41(11)	0.282(67)	0.235(51)
60	1.01(23)	0.428(98)	0.289(57)	0.270(71)	0.223(53)	0.202(44)
65	0.82(18)	0.369(84)	0.211(42)	0.214(57)	0.220(52)	0.180(39)
70	0.68(15)	0.307(70)	0.155(31)	0.194(51)	0.189(45)	0.160(35)
75	0.55(13)	0.243(56)	0.126(25)	0.195(51)	0.162(38)	0.141(31)
80	0.47(11)	0.203(47)	0.117(23)	0.179(47)	0.150(36)	0.122(27)
85	0.380(86)	0.165(38)	0.127(25)	0.165(44)	0.139(33)	0.107(24)
90	0.339(77)	0.157(36)	0.125(25)	0.160(43)	0.127(30)	0.095(21)
95	0.327(74)	0.154(35)	0.129(26)	0.162(44)	0.117(28)	0.096(21)
100	0.347(79)	0.175(40)	0.128(25)	0.167(44)	0.109(26)	0.091(20)
105	0.396(90)	0.188(43)	0.130(26)	0.160(42)	0.111(27)	0.090(21)
110	0.48(11)	0.225(51)	0.146(29)	0.160(42)	0.118(28)	0.093(20)
115	0.59(13)	0.265(61)	0.161(32)	0.165(44)	0.113(27)	0.089(20)
120	0.71(16)	0.311(71)	0.179(35)	0.173(46)	0.116(28)	0.088(20)
125	0.85(19)	0.367(84)	0.193(38)	0.170(45)	0.122(29)	0.085(19)
ICS's	37(10)	26.9(8.3)	20.5(5.6)	18.8(6.4)	16.6(5.2)	15.3(4.6)
MTCS's	13.5(4.3)	5.8(1.5)	2.94(68)	2.90(85)	2.14(58)	1.66(43)

DCSs have characteristic behavior for molecular targets, noticed previously [12,13]. At 50 eV and 100 eV DCSs exhibit a wide minimum at 100° scattering angle, which disappears at higher incident electron energies. Experiment and theory are generally in good agreement, both in shape and on the absolute scale. There are systematic discrepancies at small angles for all electron energies, probably because of the instability of interaction volume at those angles during the experiment. Also, there are obvious declinations of DCSs for 150 eV electron energies at high scattering angles.

Concerning the normalization procedure of our results, in the relative flow measurements, absolute DCSs for Ar are taken from Ranković et al. [14] for incident electron energies 50–200 eV and 300 eV and from Williams and Willis [22] for electron energy of 250 eV. In both papers, the absolute values were derived by measurements of angular dependences of elastically scattered electrons using electron spectrometers, two 127° cylindrical electrostatic energy analysers in [21], and a double cylindrical mirror analyser (DCMA) in [14]. Authors had different normalization procedures. Ranković et al. [14] used He as a reference gas and Williams and Willis [21] a phaseshift analysis of the relative angular distributions of electrons elastically scattered in the energy region of the resonances $^2P_{3/2,1/2}$ of Ar. Both absolute differential cross-sections agree within mutual uncertainties as discussed in [10]. Since our normalization is based upon RFM, we prefer to use values from the most recent paper [14] because they were obtained at the same electron spectrometer UGRA and only for the energy of 250 eV, which is not available from our apparatus [14], we used those from Williams and Willis [21].

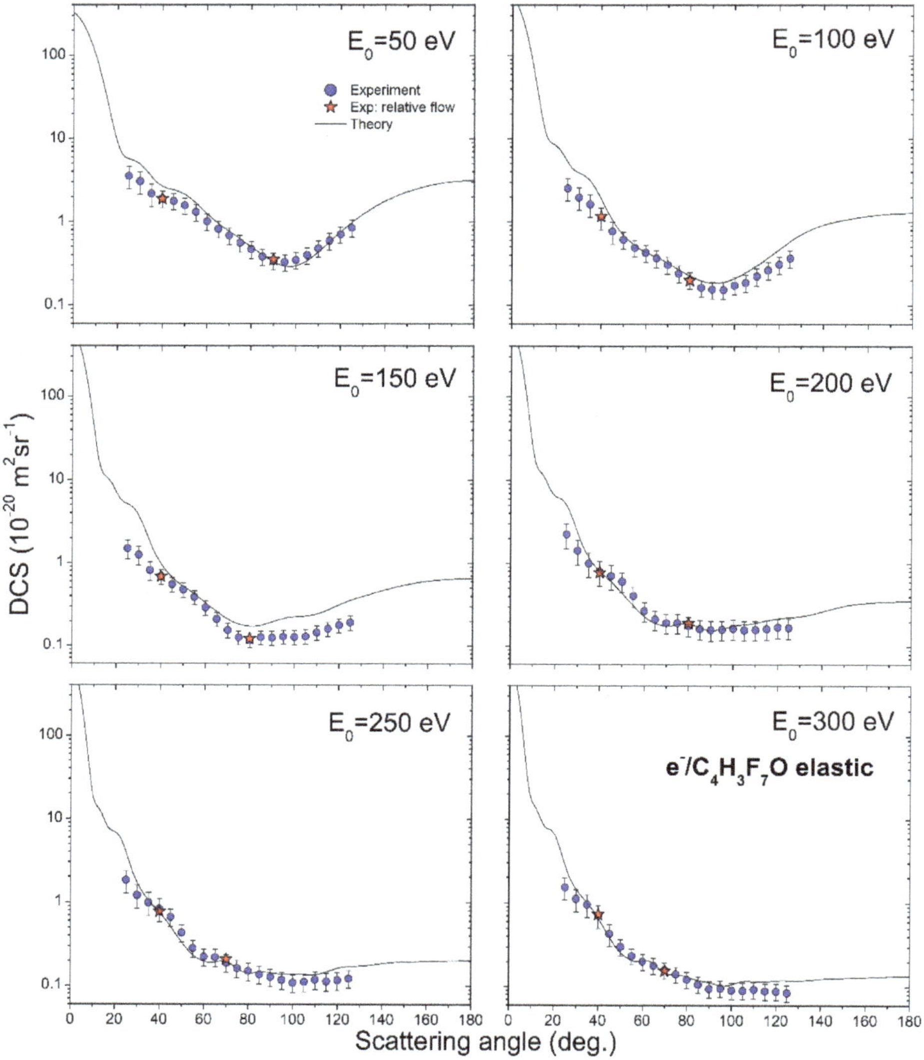

Figure 3. Angularly dependent differential cross-sections for elastic electron scattering from sevoflurane molecule, for six incident electron energies, from 50 eV to 300 eV. The present results include: experiment (blue circles), relative flow absolute data (red stars) and theory (full line).

The shapes of the experimental and theoretical cross-section plots are in excellent agreement, but measured points are 25% lower than the calculated ones. Because our experiment is angularly limited, ICS depends on the used extrapolation method, which is consisted of the normalization of our theoretical DCSs to our measured absolute ones. We have normalized calculated DCS to the measured absolute data to best match in shape and then used these values for integration. These values are presented in Table 1 with an uncertainty of 30% that arises from different plausible extrapolations. The absolute uncertainties of ICS and MTCS values are obtained from the difference of the corridor that is represented by maximal, DCS + $\Delta/2$, and minimal DCS − $\Delta/2$ values.

Measured and calculated absolute differential and integral cross-sections are graphically presented in Figures 3 and 4, and to the best of the authors knowledge, there are no other published results, neither experimental nor theoretical, for DCS or ICSs for elastic electron scattering from sevoflurane in this energy and angular range.

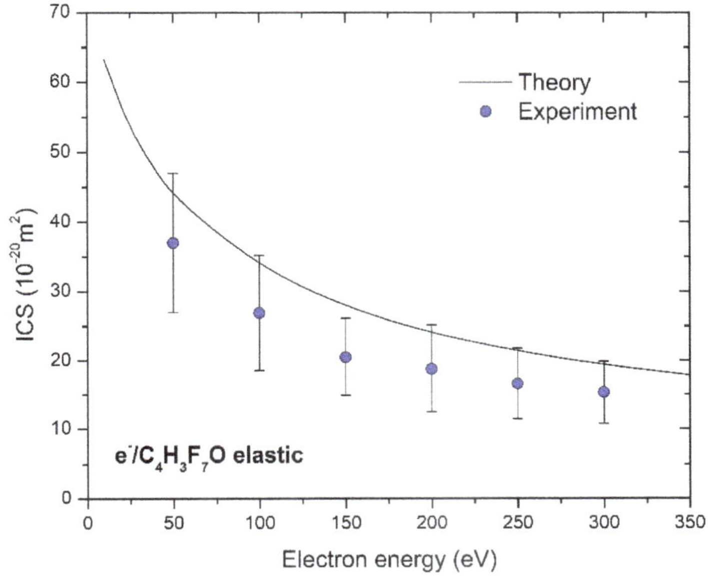

Figure 4. Integral cross-section for elastic electron collision with sevoflurane molecule in the energy range from 10 eV to 350 eV. Experiment (blue circles) and theory (solid black line) are presented.

5. Conclusions

We performed the experimental and theoretical investigation of elastic electron collisions on sevoflurane molecule at intermediate incident electron energies in order to provide insight into electron/sevoflurane interaction. Two independent sets of measurements, relative differential cross-sections at fixed energy in the function of scattering angle and absolute differential cross-sections obtained by relative flow method were performed. The last one provides us with two absolute points for every incident energy, which were used for the normalization to the absolute scale. These two independent sets of measurements showed good agreement which gives reliability to our experimental procedure. We have shown good agreement between the present experiment and theory (IAM-SCAR+I method) on the absolute scale and also in the shape. Using Ar we stress the problem of the choice of the reference gas in the relative flow measurements, since conditions should be provided for the widths (shapes) of the target beam and the reference gas beam to be approximately the same, which is easier to achieve with Ar in relation to He or Ne, since they are much lighter. The importance of this investigation is that according to our knowledge these are the first reported results for absolute differential and integral cross-sections for elastic electron scattering on sevoflurane in the energy range from (50–300) eV.

Author Contributions: Conceptualization, all authors; methodology, J.B.M. and B.P.M.; software, F.B. and G.G.; validation, all authors; experimental investigation, J.V. and J.B.M.; resources, B.P.M.; data curation, B.P.M.; writing—original draft preparation, J.V., J.B.M. and G.G.; writing—review and editing, B.P.M. and J.V.; visualization, J.V. and B.P.; supervision, B.P.M.; project administration, B.P.M.; funding acquisition, B.P.M. All authors have read and agreed to the published version of the manuscript.

Funding: This research was funded by the Institute of Physics Belgrade, through the grant by the Ministry for Education, Science and Technological Development of the Republic of Serbia. This study is partially supported by the Spanish Ministerio de Ciencia e Innovación (Project PID2019-104727RB-C21) and CSIC (Project LINKA20085).

Institutional Review Board Statement: Not applicable.

Informed Consent Statement: Not applicable.

Data Availability Statement: The data presented in this study is contained within the article. All data are possible to retrieve from the Belgrade Electron Atom/Molecule DataBase (BEAMDB) at http://servo.aob.rs/emol after the article has been published.

Acknowledgments: This article/publication is based upon work from COST Action CA18212—Molecular Dynamics in the GAS phase (MD-GAS), supported by COST (European Cooperation in Science and Technology).

Conflicts of Interest: The authors declare no conflict of interest.

References

1. Tang, P.; Zubrycki, I.; Xu, Y. Ab Initio Calculation of structures and properties of halogenated general anesthetics: Halothane and sevoflurane. *J. Comput. Chem.* **2001**, *22*, 436–444. [CrossRef]
2. Gaya da Costa, M.; Kalmar, A.F.; Struys, M.M.R.F. Inhaled Anesthetics: Environmental Role, Occupational Risk, and Clinical Use. *J. Clin. Med.* **2021**, *10*, 1306. [CrossRef] [PubMed]
3. Langbein, T.; Sonntag, H.; Trapp, D.; Hoffmann, A.; Malms, W.; Röth, E.P.; Mörs, V.; Zellner, R. Volatile anaesthetics and the atmosphere: Atmospheric lifetimes and atmospheric effects of halothane, enflurane, isoflurane, desflurane and sevoflurane. *Br. J. Anaesth.* **1999**, *82*, 66–73. [CrossRef] [PubMed]
4. Lesarri, A.; Vega-Toribio, A.; Suenram, R.D.; Brughc, D.J.; Grabowd, J.U. The conformational landscape of the volatile anesthetic sevoflurane. *Phys. Chem. Chem. Phys.* **2010**, *33*, 9624–9631. [CrossRef]
5. Dom, J.; Van der Veken, B.; Michielsen, B.; Jacobs, S.; Xue, Z.; Hesse, S.; Loritz, H.-M.; Suhmb, M.A.; Herrebout, W.A. On the weakly C–H$\cdots\pi$ hydrogen bonded complexes of sevoflurane and benzene. *Phys. Chem. Chem. Phys.* **2011**, *13*, 14142–14152. [CrossRef]
6. Shiraishi, Y.; Ikeda, K. Uptake and biotransformation of sevoflurane in humans: A comparative study with halothane, enflurane, and isoflurane. *J. Clin. Anesth.* **1990**, *2*, 381–386. [CrossRef]
7. Dong, Y.; Liang, F.; Huang, L.; Fang, F.; Yang, G.; Tanzi, R.E.; Zhang, Y.; Quan, Q.; Xie, Z. The anesthetic sevoflurane induces tau trafficking from neurons to microglia. *Commun. Biol.* **2021**, *4*, 560. [CrossRef]
8. Lange, E.; Ferreira da Silva, F.; Jones, N.C.; Hoffmann, S.V.; Duflot, D.; Limão-Vieira, P.M. The lowest-lying electronic states of isoflurane and sevoflurane in the 5.0–10.8 eV energy range investigated by experimental and theoretical methods. *Chem. Phys. Lett.* **2019**, *716*, 42–48. [CrossRef]
9. Lozano, A.I.; Ferreira da Silva, F.; Blanco, F.; Limão-Vieira, P.M.; García, G. Total electron scattering cross section from sevoflurane by 1–300 eV energy electron impact. *Chem. Phys. Lett.* **2018**, *706*, 533–537. [CrossRef]
10. Tanaka, H.; Brunger, M.J.; Campbell, L.; Kato, K.; Hoshino, M.; Rau, A.R.P. Scaled plane-wave Born cross sections for atoms and molecules. *Rev. Mod. Phys.* **2016**, *88*, 025004. [CrossRef]
11. Brunger, M.J. Electron scattering and transport in biofuels, biomolecules and biomass fragments. *Int. Rev. Phys. Chem.* **2017**, *36*, 333–376. [CrossRef]
12. Vukalović, J.; Maljković, J.B.; Tökési, K.; Predojević, B.; Marinković, B.P. Elastic Electron Scattering from Methane Molecule in the Energy Range from 50–300 eV. *Int. J. Mol. Sci.* **2021**, *22*, 647. [CrossRef] [PubMed]
13. Maljković, J.B.; Vuković, J.; Tökési, K.; Predojević, B.; Marinković, B.P. Elastic electron scattering cross sections for triethyl phosphate molecule at intermediate electron energies from 50 to 250 eV. *Eur. Phys. J. D* **2019**, *73*, 27. [CrossRef]
14. Ranković, M.; Maljković, J.B.; Tökési, K.; Marinković, B.P. Elastic electron differential cross sections for argon atom in the intermediate energy range from 40 eV to 300 eV. *Eur. Phys. J. D* **2018**, *72*, 30. [CrossRef]
15. Lucas, B.C. *Atomic and Molecular Beams Production and Collimation*; CRC Press: Boca Raton, FL, USA, 2012; pp. 217–223.
16. Milosavljević, A.R.; Giuliani, A.; Šević, D.; Hubin-Franskin, M.J.; Marinković, B.P. Elastic scattering of electrons from tetrahydrofuran molecule. *Eur. Phys. J. D* **2005**, *35*, 411–416.
17. Allan, M. Absolute angle-differential elastic and vibrational excitation cross sections for electron collisions with tetrahydrofuran. *J. Phys. B* **2007**, *40*, 3531–3544. [CrossRef]
18. Homem, M.G.P.; Iga, I.; Sugohara, R.T.; Sanches, I.P.; Lee, M. Role of adsorption effects on absolute electron-molecule cross-section calibration using the relative flow technique. *Rev. Sci. Instrum.* **2011**, *82*, 013109. [CrossRef]
19. Nickel, J.C.; Mott, C.; Kanik, I.; McCollum, D.C. Absolute elastic differential electron scattering cross sections for carbon monoxide and molecular nitrogen in the intermediate energy region. *J. Phys. B* **1988**, *21*, 1867–1877. [CrossRef]

20. Nickel, J.C.; Zetner, P.V.; Shen, G.; Trajmar, S. Principles and procedures for determining absolute differential electron-molecule (atom) scattering cross sections. *J. Phys. E* **1989**, *22*, 730–738. [CrossRef]
21. Williams, J.F.; Willis, B.A. The scattering of electrons from inert gases I. Absolute differential elastic cross sections for argon atoms. *J. Phys. B* **1975**, *8*, 1670–1682. [CrossRef]
22. Maljković, J.B.; Blanco, F.; García, G.; Marinković, B.P.; Milosavljević, A.R. Absolute cross sections for elastic electron scattering from methylformamide. *Phys. Rev. A* **2012**, *85*, 042723. [CrossRef]
23. Olander, D.R.; Kruger, V. Molecular Beam Sources Fabricated from Multichannel Arrays. III. The Exit Density Problem. *J. Appl. Phys.* **1970**, *41*, 2769–2776. [CrossRef]
24. Blanco, F.; Rosado, J.; Illana, A.; García, G. Comparison of two screening corrections to the additivity rule for the calculation of electron scattering from polyatomic molecules. *Phys. Lett. A* **2010**, *374*, 4420–4424. [CrossRef]
25. Blanco, F.; Ellis-Gibbings, L.; García, G. Screening corrections for the interference contributions to the electron and positron scattering cross sections from polyatomic molecules. *Chem. Phys. Lett.* **2016**, *645*, 71–75. [CrossRef]
26. Traore Dubuis, A.; Verkhovtsev, A.; Ellis-Gibbings, L.; Krupa, K.; Blanco, F.; Jones, D.B.; Brunger, M.J.; García, G. Total cross section of furfural by electron impact: Experiment and theory. *J. Chem. Phys.* **2017**, *147*, 054301. [CrossRef]
27. Fuss, M.C.; Sanz, A.G.; Blanco, F.; Oller, J.C.; Limão-Vieira, P.; Brunger, M.J.; García, G. Total electron-scattering cross sections from pyrimidine as measured using a magnetically confined experimental system. *Phys. Rev. A* **2013**, *88*, 042702. [CrossRef]
28. Lozano, A.I.; Oller, J.C.; Jones, D.B.; da Costa, R.F.; do, N.; Varella, M.T.; Bettega, M.H.F.; Ferreira da Silva, F.; Limão-Vieira, P.; Lima, M.A.P.; et al. Total electron scattering cross sections from para-benzoquinone in the energy range 1–200 eV. *Phys. Chem. Chem. Phys.* **2018**, *20*, 22368–22378. [CrossRef] [PubMed]
29. Staszewska, G.; Schwenke, D.W.; Thirumalai, D.; Truhlar, D.G. Quasifree-scattering model for the imaginary part of the optical potential for electron scattering. *Phys. Rev. A* **1983**, *28*, 2740–2751. [CrossRef]
30. Cowan, R.D. *The Theory of Atomic Structure and Spectra*; University of California Press: Berkeley, CA, USA, 1981.
31. Riley, M.E.; Truhlar, D.G. Approximations for the exchange potential in electron scattering. *J. Chem. Phys.* **1975**, *63*, 2182–2191. [CrossRef]
32. Zhang, X.; Sun, J.; Liu, Y. A new approach to the correlation polarization potential-low-energy electron elastic scattering by He atoms. *J. Phys. B At. Mol. Opt. Phys.* **1992**, *25*, 1893–1897. [CrossRef]
33. Blanco, F.; García, G. Screening corrections for calculation of electron scattering differential cross sections from polyatomic molecules. *Phys. Lett. A* **2004**, *330*, 230–237. [CrossRef]
34. Sanz, A.G.; Fuss, M.C.; Blanco, F.; Sebastianelli, F.; Gianturco, F.A.; García, G. Electron scattering cross sections from HCN over a broad energy range (0.1–10000 eV): Influence of the permanent dipole moment on the scattering process. *J. Chem. Phys.* **2012**, *137*, 124103. [CrossRef] [PubMed]

International Journal of *Molecular Sciences*

MDPI

Article

Total Electron Detachment and Induced Cationic Fragmentation Cross Sections for Superoxide Anion (O_2^-) Collisions with Benzene (C_6H_6) Molecules

Carlos Guerra [1], Sarvesh Kumar [2], Fernando Aguilar-Galindo [3], Sergio Díaz-Tendero [4,5,6,*], Ana I. Lozano [2], Mónica Mendes [2], Juan C. Oller [7], Paulo Limão-Vieira [2] and Gustavo García [1,8,*]

1 Instituto de Física Fundamental, Consejo Superior de Investigaciones Científicas, Serrano 113-bis, 28006 Madrid, Spain; carlosguerra@iff.csic.es
2 Laboratório de Colisões Atómicas e Moleculares, CEFITEC, Departamento de Física, Faculdade de Ciências e Tecnologia, Universidade NOVA de Lisboa, 2829-516 Caparica, Portugal; s.kumar@campus.fct.unl.pt (S.K.); anita_ilm@iff.csic.es (A.I.L.); mf.mendes@fct.unl.pt (M.M.); plimaovieira@fct.unl.pt (P.L.-V.)
3 Donostia International Physics Center (DIPC), Paseo Manuel de Lardizabal 4, 20018 Donostia-San Sebastián, Spain; fernando.aguilar@dipc.org
4 Departamento de Química, Módulo 13, Universidad Autónoma de Madrid, 28049 Madrid, Spain
5 Condensed Matter Physics Center (IFIMAC), Universidad Autónoma de Madrid, 28049 Madrid, Spain
6 Institute for Advanced Research in Chemical Science (IAdChem), Universidad Autónoma de Madrid, 28049 Madrid, Spain
7 Centro de Investigaciones Energéticas Medioambientales y Tecnológicas (CIEMAT), Avenida Complutense 22, 28040 Madrid, Spain; jc.oller@ciemat.es
8 Centre for Medical Radiation Physics, University of Wollongong, Wollomgong, NSW 2522, Australia
* Correspondence: sergio.diaztendero@uam.es (S.D.-T.); g.garcia@csic.es (G.G.)

Citation: Guerra, C.; Kumar, S.; Aguilar-Galindo, F.; Díaz-Tendero, S.; Lozano, A.I.; Mendes, M.; Oller, J.C.; Limão-Vieira, P.; García, G. Total Electron Detachment and Induced Cationic Fragmentation Cross Sections for Superoxide Anion (O_2^-) Collisions with Benzene (C_6H_6) Molecules. *Int. J. Mol. Sci.* **2022**, *23*, 1266. https://doi.org/10.3390/ijms23031266

Academic Editor: Mihai V. Putz

Received: 31 December 2021
Accepted: 21 January 2022
Published: 23 January 2022

Abstract: In this study, novel experimental total electron detachment cross sections for O_2^- collisions with benzene molecules are reported for the impact energy range (10–1000 eV), as measured with a transmission beam apparatus. By analysing the positively charged species produced during the collision events, relative total ionisation cross sections were derived in the incident energy range of 160–900 eV. Relative partial ionisation cross sections for fragments with $m/z \leq 78$ u were also given in this energy range. We also confirmed that heavier compounds ($m/z > 78$ u) formed for impact energies between 550 and 800 eV. In order to further our knowledge about the collision dynamics governing the fragmentation of such heavier molecular compounds, we performed molecular dynamics calculations within the framework of the Density Functional Theory (DFT). These results demonstrated that the fragmentation of these heavier compounds strongly supports the experimental evidence of m/z = 39–42, 50, 60 (u) cations formation, which contributed to the broad local maximum in the total ionisation observed from 550 to 800 eV. This work reveals the reactivity induced by molecular anions colliding with hydrocarbons at high energies, processes that can take place in the interstellar medium under various local conditions.

Keywords: anion–molecule collisions; electron detachment cross sections; positive ion-induced fragmentation; molecular dynamics

1. Introduction

Benzene is one of the simplest and more stable aromatic ring molecules. It has been considered a prototype for the study of chemical reactions involving biomolecules, such as pyridine [1] and carbamates [2], a building block for the synthesis of carbon nano-cages [3] and a precursor of hydrocarbon structures (e.g., phenol, toluene, aniline, anthracene) with important technological applications [4]. The functionalization of fullerene-based nanostructures using the Density Functional Theory (DFT) has opened new investigations on their transport properties in biological media and their capability to be used as molecular sensors or drug carriers [5–7]. Additionally, oxygen superoxide anion, O_2^-, is one of the reactive oxygen species (ROS) which are responsible for numerous biochemical processes leading to oxidative damage in living organisms [8] as well as in materials [9]. In particular, 8-oxoguanine is frequently formed by the interaction of ROS with the guanine base in DNA under conditions of oxidative stress, yielding an efficient way of damaging DNA [10].

Due to benzene's relevance for some of the applications listed above where charge transfer (electron transfer) is a pivotal mechanism in such reactions, electron scattering cross section data have been the subject of numerous theoretical and experimental studies. We recently reported an experimental and theoretical analysis of the total electron scattering cross section (TCS) with the aim of providing some reference data for modelling procedures connected with these applications [11]. Apart from TCS measurements and calculations, elastic scattering, vibrational excitation and electron attachment cross sections have also been published (see Ref. [11] and references therein). However, surprisingly, there is a lack of data on electron impact ionization and electronic excitation cross sections. Bull et al. in 2014 [12] evidenced this scarcity of data for aromatic molecules and provided new experimental data, which showed important discrepancies with the old measurements [13]. More recently, the double ionization of benzene by electron impact has been studied by Wolf et al. [14] and Sigaud and Montenegro [15], both presenting the remarkable probability of double ionization of benzene for increasing impact energies above 20 eV. Yet, being the projectile a superoxide anion, O_2^-, collision-induced ionization of a target molecule is certainly more complex to describe from the dynamical point of view of the interaction. To our knowledge, except for our recent study on the anomalous oxidation of benzene in collisions with superoxide anions [16], no experimental or theoretical data for O_2^- collision with benzene are available in the literature. Notwithstanding, we report on absolute total electron detachment cross sections for such collision processes in the impact energy range of 200–900 eV. Measurements were obtained with a beam attenuation technique which has been previously described [16,17]. The total uncertainty limits of these results are within 6 and 10%. Additional experimental information on the induced cationic fragmentation, i.e., positively charged fragments with a mass-to-charge ratio (m/z) less than 78 u, is also given in terms of relative cross sections normalized to that of the parent ion formation with uncertainty limits within 10 and 20%. These relative fragmentation probabilities were also obtained by means of molecular dynamics calculations carried out in the framework of the density functional (DFT) theory. The remainder of this paper has been divided into the following sections: in Section 2, we present our experimental and theoretical data together with details of their associated uncertainties. These results are discussed and compared in Section 3. A detailed description of the theoretical and experimental methods used in this study are presented in Section 4. Finally, some conclusions are drawn in Section 5.

2. Results

The theoretical and experimental results of this study are presented in this section and discussed in Section 3, and a full description of the Material and Methods utilized to obtain these results is provided in Section 4.

2.1. Total Electron Detachment Cross Sections

The experimental absolute total electron detachment cross sections (TEDCS) are presented in Table 1. Total uncertainty limits, according to the error estimation procedure described in Section 4, are also included in this table. These results are also plotted in Figure 1 together with their corresponding error bars.

As to the authors' knowledge, no previous experimental or theoretical TEDCS data have been reported in the literature. If we compare these results with our recent measurements for O_2^- collisions with N_2 [17], we see that the average magnitude of the TEDCS in the common energy range (50–1000 eV) is in concordance with their respective molecular masses, i.e., 3–4 times higher for benzene than for nitrogen. In addition, the energy dependence of the TEDCSs shows some similarities for both molecular targets. In both cases, the magnitude of the cross section has some local maxima/minima superimposed to a relatively flat tendency. For N_2, a prominent local maximum is shown around 200 eV (see Ref. [17]), while for benzene, two local maxima at 40 and 400 eV and a minimum around 320 eV are discernible. In contrast with nitrogen, these features are not well defined for benzene, and their respective magnitudes follow the flat tendency of the TEDCS if we consider their total uncertainty limits. These features should be connected with the excitation of specific energy levels of the target molecule. Figure 2 shows the time-of-flight spectra once calibrated in kinetic energy of the incident anion beam with and without benzene in the scattering region. As derived from this figure, the average energy loss by the primary beam after the collision with 1.2 mTorr of benzene was about 16 eV. This energy transferred was enough to induce single and double electronic excitations and/or ionisations to the benzene target molecule.

Table 1. Experimental total electron detachment cross sections in SI units (10^{-20} m^2). Total uncertainty limits are also included.

Energy (eV)	Total Electron Detachment Cross Section (10^{-20} m^2)	Uncertainty Limit (%)
10	33.93	7.8
20	34.46	6.7
40	36.20	6.0
60	34.46	6.2
80	34.51	6.7
100	35.02	8.4
120	33.91	7.6
140	34.11	9.7
160	32.51	6.7
180	32.40	5.2
200	32.39	7.1
220	31.89	8.9
240	32.12	6.6
260	31.92	9.6
280	30.65	7.9
300	30.22	7.9
320	29.42	6.6
340	30.03	5.6
360	30.27	6.8
380	31.72	8.1
400	33.60	6.7
420	32.10	5.5
440	32.43	7.9
460	32.47	5.6
480	33.93	8.0
500	34.84	6.1
600	34.97	6.1
700	35.40	6.0
750	35.65	7.3
800	35.49	6.2
900	35.85	7.6
950	35.75	5.8
1000	36.78	5.9

2.2. Induced Fragmentation of Benzene by O_2^- Impact

The relative fragmentation yields of benzene cations formed after O_2^- collisions in the energy range 160–900 eV were obtained by measuring the intensity of the TOF features, corresponding to identified fragments with $m/z < 78$ u, with respect to that of the parent ion ($m/z = 78$ u). A typical mass spectrum for an 850 eV impact energy is shown in Figure 3.

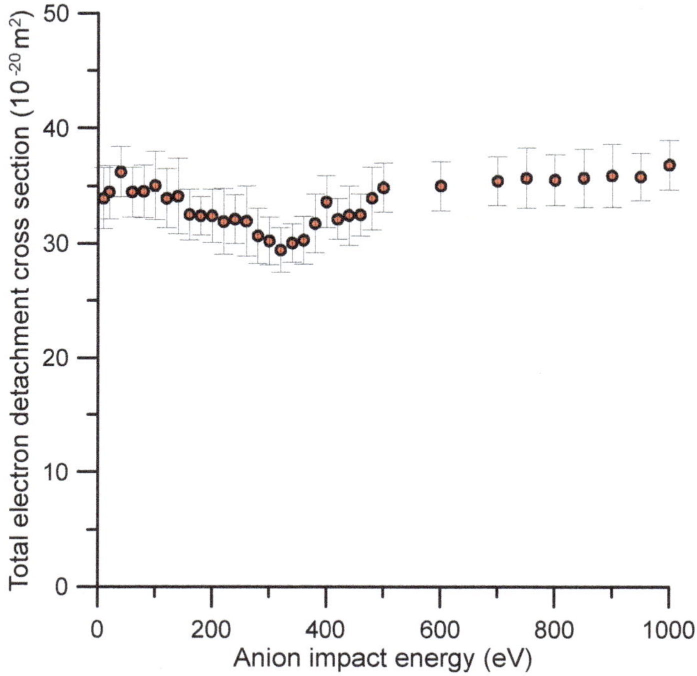

Figure 1. Total electron detachment cross section (TEDCS) for O_2^- collisions with benzene (C_6H_6) as measured with the described transmission beam experimental setup (see Section 4 for details).

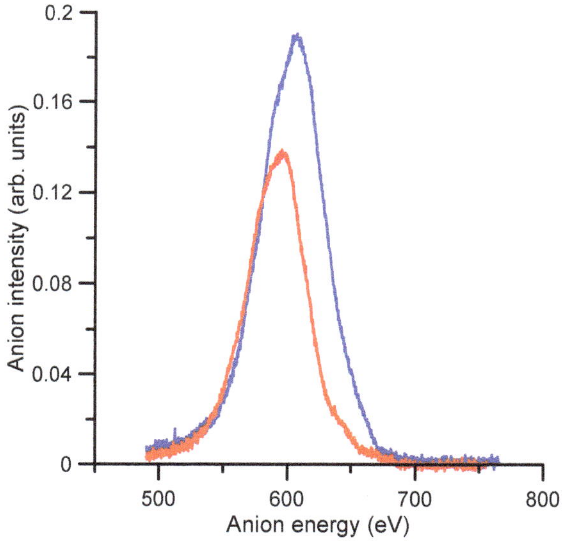

Figure 2. Intensity and energy of the transmitted anion beam with: —, no benzene gas (0 mTorr) in the gas cell; —, 1.2 mTorr of benzene in the gas cell.

As described in Section 4, due to the specific anion formation process in the hollow cathode discharge, the time (mass) resolution was not enough to fully resolve each single fragment ion produced, but allowed distinguishing different fragment series clearly characterised by the number of carbon atoms constituting each group of cations. Nonetheless, using a gaussian fitting analysis procedure, we could identify the mass of the fragments that mainly contributed to each fragment group. For the anion impact energy shown in Figure 3 (850 eV), the most intense feature corresponds to the parent ion ($C_6H_6^+$), and other significative ion fragment peaks were identified as $C_4H_2^+$, $C_5H_5^+$, $C_2H_6^+$, $C_3H_6^+$, $C_2H_2^+$, C_5^+, CH_4^+, C^+ and H/H_2^+ (in decreasing order of their relative intensities). Note that some features corresponding to ions heavier than the parent ion ($m/z > 78$ u) also appear in the mass spectrum. The formation of these heavy complexes can only be explained by the aggregation of oxygen molecules to some of the generated fragments, in particular [C_6H_6 + O] (m = 94 u) and [C_6H_6 + O_2] (m = 110 u). The mechanisms by which these abducts are generated were investigated in a recent study, and the experimental evidence together with its theoretical justification can be found in Ref. [16]. We focused the present study on the formation of cationic fragments that were lighter than the parent ion, i.e., $m/z < 78$ u.

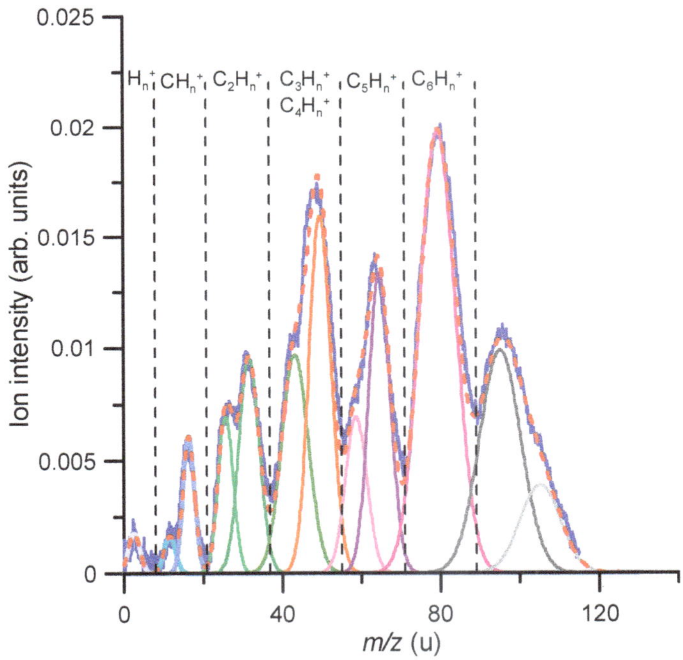

Figure 3. TOF mass spectrum of the positive ion-induced fragmentation of benzene in collisions with 850 eV oxygen anions (O_2^-). Gaussian fitting analysis: $m/z(u)$ = 2 (—); 12 (—); 16 (—); 26 (—); 32 (—); 42 (—); 50 (—); 60 (—); 65 (—); 78 (—); 94 (—); 110 (—); fit sum (---).

It is interesting to compare the fragmentation induced by the anion beam to that produced by an electron beam in similar collision conditions. As described in Section 4, the experimental arrangement incorporated an electron gun normal to the anion beam, just in front of the TOF spectrometer used to analyse ions species of the generated fragments. Figure 4 shows the cationic fragmentation spectrum induced by 850 eV anion collisions (same as that shown in Figure 3) together with the induced fragmentation mass spectrum generated by an 850 eV electron beam. Both spectra are normalised to the parent ion ($m/z = 78$) intensity.

A close inspection of Figure 4 shows that superoxide anion projectiles produced a higher fragmentation pattern relative to electrons as projectiles. At 850 eV, 60% of the electron impact ionization corresponded to the formation of the parent ion, while in the case of O_2^-, it represented only 32.6% of the total ion yield. Thus, a clear enhancement of the target's fragmentation induced by the anion beam was noted. In general, the relative intensity of the observed peaks obtained with the superoxide anion beam increased with respect to those obtained with the electron beam, but in particular, those with m/z = 65, 50 and 42 were especially enhanced, showing also a double-peak structure. This seemed to indicate that losses of CH, C_2H_4 and C_3 radicals from the benzene ring were favoured in anion collisions.

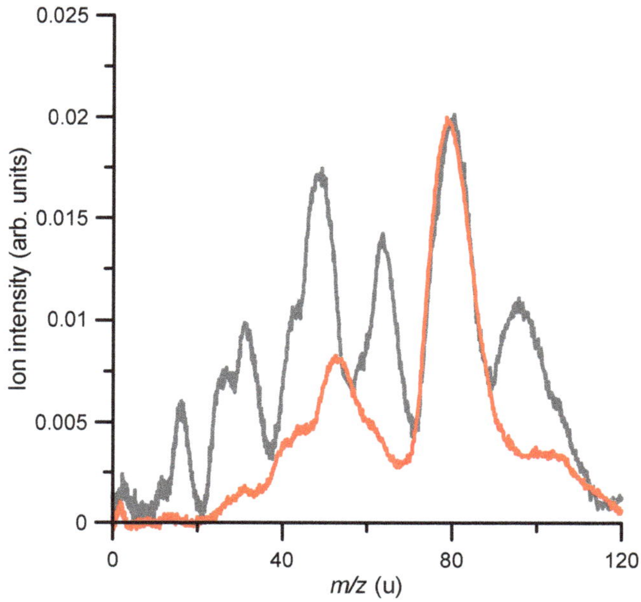

Figure 4. Positive ion-induced fragmentation by the superoxide anion beam (—) and the electron beam (—), respectively, both at 850 eV impact energy.

In order to further our knowledge on the collision dynamics governing such cation formation, molecular dynamics calculations using the Density Functional Theory (DFT) were performed (see Section 4 for further details). In our previous work [16], we proposed a collision model with the formation of a stable complex in which O_2 molecules remained bonded to the ionized benzene, $[C_6H_6–O_2]^+$. Non-associative collisions, where the O_2 molecule did not remain bonded to benzene, might also occur. Therefore, we considered both situations and we ran two sets of trajectory calculations. The first one assumed chemical interaction and thus formation of a stable $[C_6H_6–O_2]^+$ molecule; therefore, these simulations started with the structure shown in Figure 5a. The starting point of the dynamics in the second one was a weakly bonded van der Waals complex $[C_6H_6 \cdots O_2]^+$, shown in Figure 5b. Thus, both situations kept the same number of atoms but with a quite different bonding strength. Both initial structures were optimized and, thus, they were minima of the potential energy surfaces. The final results were obtained by computing statistics over the molecular dynamics simulations: 100 trajectories for each structure and each excitation energy value. Note that the excitation energy was randomly redistributed among the nuclear degrees of freedom in each trajectory. We considered that a fragment was created when the atoms forming this species were separated by at least 3 Å from other fragments in the last point of the trajectory.

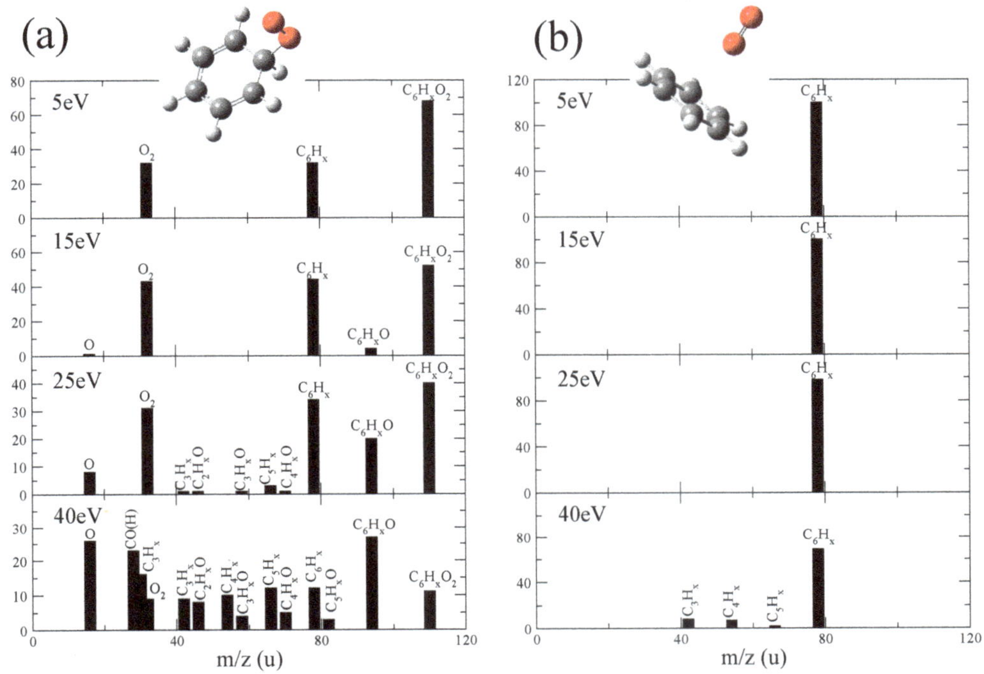

Figure 5. Simulated mass spectra for (**a**) $[C_6H_6–O_2]^+$ and (**b**) $[C_6H_6 \cdots O_2]^+$ at different excitation energies derived from the molecular dynamics simulations.

In the simulated mass spectra in Figure 5, we included the percentage of trajectories where a given fragment appeared. In these spectra, we grouped all fragments with a different number of H atoms in a single peak, e.g., C_6H_X included trajectories in which C_6H_6, C_6H_5, C_6H_4, etc. molecules were produced. The spectra corresponding to the weakly bonded $[C_6H_6 \cdots O_2]^+$ were dominated by a single peak that corresponded to $C_6H_X{}^+$ molecules. Further fragmentation was observed only for E_{exc} = 40 eV, in which a slightly amount of $C_5H_X{}^+$, $C_4H_X{}^+$ and $C_3H_X{}^+$ could be noted. In the case of $[C_6H_6–O_2]^+$, a much richer spectrum was obtained, where oxygenated species were of particular importance: $C_6H_XO_2{}^+$ and $C_6H_XO^+$, together with $C_6H_X{}^+$, were the most prominent peaks at low excitation energies. At the highest excitation energy here considered, mono-oxygenated fragments, $C_5H_XO^+$, $C_4H_XO^+$, $C_3H_XO^+$, gained in relative intensity. It is worth stressing that for $[C_6H_6–O_2]^+$, the relative intensity of non-oxygenated species, i.e., $C_5H_X{}^+$, $C_4H_X{}^+$, and $C_3H_X{}^+$, was much higher than for the weakly bonded case.

In the experimental mass spectrum, we could also discern double-peak structures, which we assigned to oxygenated and non-oxygenated species, e.g., the double-peak structure shown in the region 40–50 u, corresponded to C_3H_X and C_2H_XO. Therefore, by only considering the fragmentation of $[C_6H_6–O_2]^+$, these double-peak structures can be explained. A comparison between the experimental measurements and the theoretical simulations is shown in Figure 6. For this purpose, we added both theoretical spectra obtained at 40 eV, i.e., we considered that both possibilities might occur, i e., ionization and excitation of benzene (i) in a non-reactive collision and (ii) with formation of stable oxygenated species. A qualitative agreement was obtained, showing the validity of the model and corroborating the associative mechanism proposed in Ref. [16].

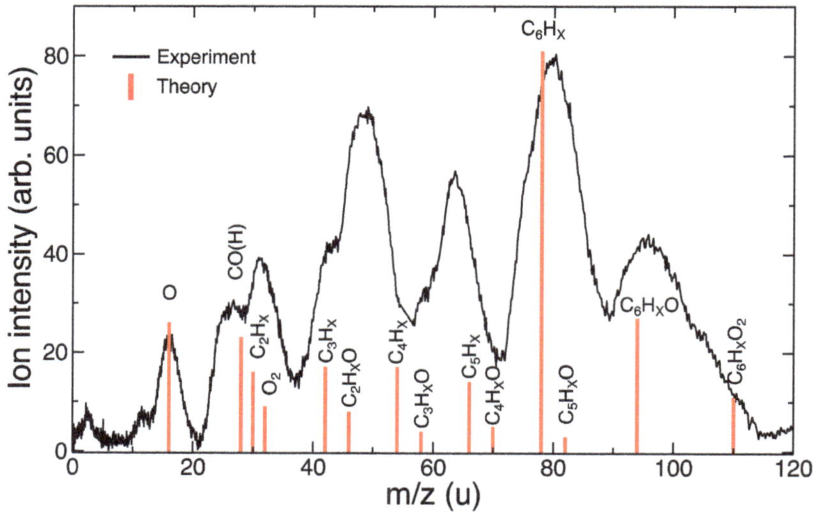

Figure 6. Comparison of experimental (collision energy = 850 eV) and theoretical (internal excitation energy = 40 eV) fragmentation spectra.

2.3. Relative Partial and Total Ionization Cross Sections of Benzene by O_2^- Impact

The relative intensities were normalised to the primary beam intensity and, according to Equation (2) (see Section 4, they accounted for the target gas pressure and geometry of the ion extraction region. However, as the ratio between the ion extracted beam and the actual primary beam intensity in the interaction region was not known, we could only provide relative values of the fragment yields. The energy dependence of these yields for the identified cation fragments is shown in Figure 7 for the whole impact energy range considered in this study (160–900 eV).

The fragment cation intensity distributions shown in Figure 3 were analysed as a function of the incident energy. For a given m/z fragment, its production cross section was given by the ratio between its detected ion intensity and the primary anion intensity divided by the target molecular density and the length of the ion extraction region (see Section 4 for technical details).

Following this procedure, we could in principle derive the absolute values of those partial ionisation cross sections. However, as already mentioned, we did not accurately know the ratio between the cation and the primary anion currents in the interaction region and therefore we could only report here on their relative cross section values. The error bars shown in Figure 7 correspond to the combination in quadrature of all the known uncertainty sources (ion intensity statistical uncertainties and those derived from the gas pressure and temperature determination, see Section 4 for details).

The total ionisation cross sections (TICS) derived by adding the intensities of all detected cationic species generated by the anion–molecule collisions considered in this study are shown in Figure 8. In such figure, the TICS present a prominent feature from 550 eV to 800 eV. Note that these species included the heavier adducts (m/z = 94–95, m/z = 110), whose formation process is described in Ref. [16]. According to the fragmentation pattern shown in Figure 7 and the above theoretical justification, we deduced that this prominent peak was mainly due to the fragmentation of these heavier adducts which mainly formed for impact energies from 550 to 800 eV. The dashed red line in Figure 6 represents the contribution to the TICS of the fragments coming from the dissociation of the parent ion, i.e., the TICS values we would obtain if these adducts did not form.

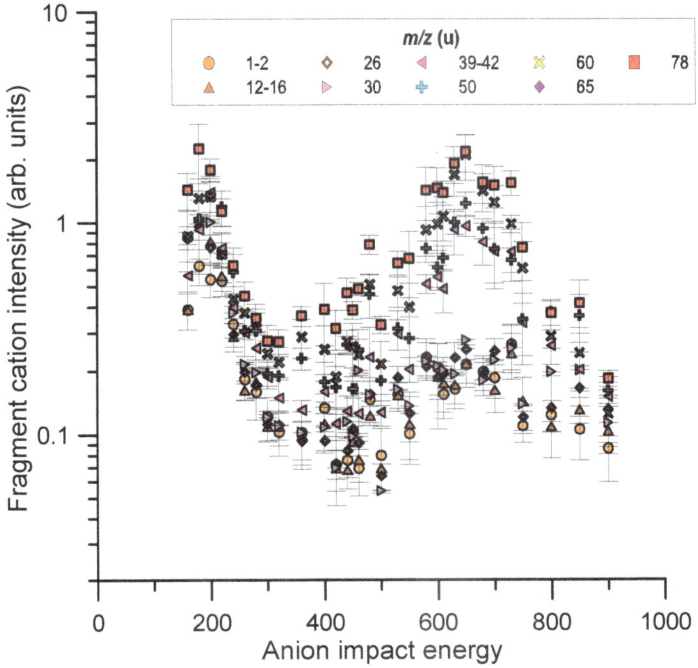

Figure 7. Relative cation intensities for m/z 78 u fragments induced by collisions with oxygen anions in the impact energy range (160–900 eV). See legend for the different cation species.

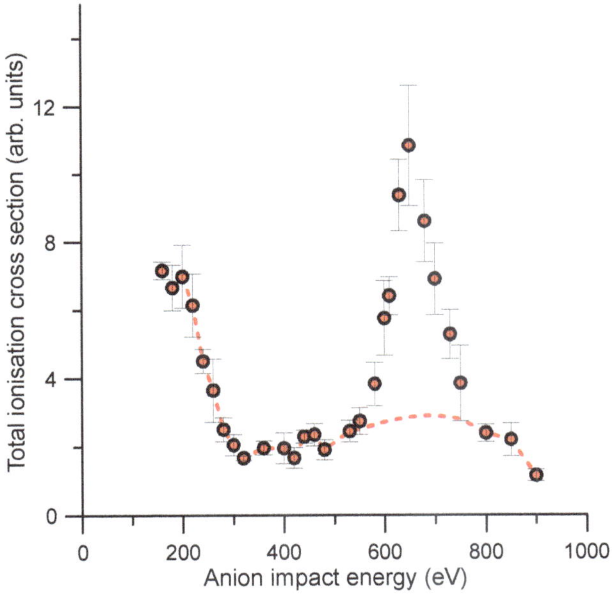

Figure 8. Relative total ionisation cross sections (TICS) accounting for all the cationic species generated by single O_2^{-}–C_6H_6 collisions for impact energies ranging from 160 up to 900 eV.

3. Discussion

The energy dependence of the TEDCS showed a quite flat behaviour except for two relative maxima and a local minimum, cross section values which were not clearly defined within the quoted uncertainty limits. These could be due to specific electronic excitations of the target molecule, but no theoretical evidence has been found to support such assumption. However, the relative TICS energy dependence showed a prominent local maximum in the cross section values for impact energies between 550 and 800 eV. According to the present theoretically predicted ionization and further fragmentation patterns, this local maximum mainly formed by the fragmentation of the heavier adducts formed by the aggregation model we previously proposed in Ref. [16]: a sudden double ionization of benzene and the subsequent electrostatic attraction between the dication and the anion formed a stable covalently bonded $C_6H_6O_2^+$ molecule, which evolved towards the formation of benzene–diol conformers. Here, we demonstrated that the fragmentation of these heavier molecules led to the formation of those species that mainly contributed to the local maximum in the total ionisation cross section. Thus, these results strongly support our previous adduct formation model, verifying the validity of the mechanism proposed in Ref. [16].

4. Materials and Methods

4.1. Experimental Setup and Procedure

The experimental setup used in this study was already described in detail in refs. [16,17]. Here, just a brief description with some improvements and modifications will be given. The experimental apparatus consisted of a sequence of high-vacuum chambers (8.4×10^{-7} mbar background pressure) connected to each other: a projectile chamber, a collision chamber and a transmission chamber. A schematic diagram of the apparatus is shown in Figure 1 of Ref. [16].

The oxygen anionic beam was produced inside the projectile chamber through a plasma discharge process. The plasma pulse was generated within a hollow cathode discharge mechanism by applying −500 V to the cathode while the anode was grounded. The negative ions formed in the afterglow region, once the plasma-generated species de-excited, and secondary electron attachment and/or charge exchange processes occurred. A commercial Parker pulsed valve (VAC1250) was used to introduce the projectile (O_2) into the vacuum chamber. It operated at 350 μs pulse width in an 80 ms duty cycle and at a gas pressure of typically 500 mbar. The background pressure in the projectile chamber during the pulsed valve operation was maintained below 10^{-5} mbar. Afterwards, the projectile beam left the discharge chamber, being focused and deflected towards the collision chamber through a combination of an Einzel lens (L1), placed just after the anode, and XY deflecting plates (D1 and D2), placed at the entrance of the second chamber (see Figure 1 of Ref. [13]). In the same region of the deflection plates, two magnets were placed outside the chamber in order to avoid stray electrons passing to the collision region. The oxygen anionic beam projectile reached the collision chamber entering through a 2 mm diameter hole after being focused by an Einzel lens (L2) and travelling into the gas cell (GC). The GC was a 36 mm-diameter and 27 mm-length cylindrical chamber, where a negative or positive voltage was applied to accelerate or decelerate the anion beam to define the kinetic energy of the O_2^- anions and, therefore, the collision energy. During all measurements, a negative variable voltage of 200–280 V was applied to the central electrode of L2 to increase the beam's intensity in the collision region.

The molecular interactions occurred in the GC when the molecular target was introduced at low pressure (<2 mTorr) through a 15 mm diameter aperture controlled by a sapphire leak valve. The positive ions formed in the GC were extracted and accelerated by means of an extractive plate system, through a perpendicular 1.40 m-long time-of-flight (TOF) mass spectrometer, placed over the collision region, where ions were mass-analysed and detected by a microchannel plate (MCP1) operating in single-pulse counting mode. The extractive parallel plates (E1, E2) were connected to the positive (0 to +900) and negative (0 to −900) V1 and V2 pulsed voltages, respectively. Pulse voltages and durations were optimized to ensure a total and uniform ion extraction but maintaining a reasonable mass resolution. The mass resolution used in this study was limited by the length of the hollow cathode. The uncertainty in the position where anions formed along the cathode was converted into a time of flight, and therefore mass, an uncertainty which determined our inherent mass resolution limit ($\Delta m/m$ = 0.05).

Under the GC, a homemade electron gun provided an energy-controlled electron beam (0–500 eV electron incident kinetic energy) entering the interaction region normal to the anion beam and opposite to the TOF mass analyser. The electron gun was not strictly necessary, but it was useful to analyse the molecular composition of the background and the gas target, as well as to heat the chamber to facilitate its evacuation when needed. It also provided a reference electron-induced fragmentation spectrum to compare with that produced by anion collisions.

A second microchannel plate (MCP2) was placed in the transmission chamber (see Figure 1 of Ref. [16]), 10 cm above the normal plane of the beam and 0.47 m from the hollow cathode source. The anionic beam was there repelled by a continuous −250 V voltage applied to E3 (E4 is grounded) towards the MCP2 detector which monitored the primary anion beam intensity (Figure 1). The MCP2 detector signal was also used to determine the primary beam energy distribution by using a set of three grids (G2) as a retarding field analyser placed at the entrance of the transmission chamber. A typical TOF mass spectrum of the projectile anionic beam is shown in Figure 2, comprising three features assigned to O^-, O_2^- and O_3^-, where O_3^- intensity was about 50% of O_2^- intensity. By tuning the extractive pulses, we could select the area of the primary beam which was really generating the positive ion fragments being mass-analysed (see refs. [16,17] for details).

In this study, we obtained the total electron detachment cross sections (σ_t) for O_2^- collisions with O_2 by measuring the primary anion beam attenuations as a function of the target O_2 pressure from 0 to 2 mTorr. The absolute pressure value was measured by two capacitance manometers (MKS Baratron 627B) placed at opposite sides of the GC (PGC1 and PGC2, respectively), normal to the primary beam direction, so ensuring that pressure gradients were not affecting the present measurements. Under these conditions, the total electron detachment cross-section value was derived from the Beer–Lambert attenuation law:

$$I = I_o \exp\left(-\frac{Pl\sigma_t}{kT}\right) \quad (1)$$

where I is the transmitted anion intensity trough the GC, P is the O_2 pressure [$P_{GC} = (P_{GC1} + P_{GC2})/2$], I_o is the initial intensity (no gas in the GC), l is the effective path length of the collision chamber (36 mm), k is the Boltzmann constant, and T is the absolute temperature. The O_2^- transmitted intensity was measured with a single-ion counting system (see Ref. [16] for details) as a function of the O_2 gas pressure. For each measurement, pressure varied from 0 to a maximum value (P_{max}) which was chosen to ensure we were measuring the transmitted intensity under single-collision conditions [16]. According to the above expression (Equation (1)), for different impact energies, transmitted intensities, as a function of pressure, follow an exponential law with a slope directly providing the σ_t value.

Moreover, the total ionisation cross sections (σ_+) can be derived from the ratio between the total positive ion detected intensity and the primary anion intensity, according to the expression:

$$\sigma_+ = \frac{I_+}{I_-}\frac{1}{nl} \quad (2)$$

where I_+ is the total positive ion intensity as recorded by the MCP1 detector, I_- is the O_2^- intensity measured by the MCP2 detector (transmitted O_2^- intensity), n is the molecular gas density derived from the gas pressure (P_{GC}) and the temperature (T) by assuming an ideal gas behaviour, and l is the effective ion extraction length which is supposed to be the diameter of the ion extraction aperture (2 mm). Under these conditions, Equation (2) provides absolute values of the cross sections. In the conditions of this experiment, the counting efficiency of both MCP1 and MCP2 was about 100%. However, as we could not ensure that the primary beam intensity detected by MCP2 corresponded to the actual intensity of this beam in the gas cell, only relative values of the ionization cross sections, as a function of the anion impact energy, could be obtained.

4.2. Uncertainty Analysis

Concerning the total electron detachment cross-section, statistical uncertainties were constrained below 5% by repeating at least 5 times the anion attenuation measurements and their corresponding data acquisition and analysis procedures for each impact energy considered in this study. The accuracy of the pressure determination was given by the uncertainty limit of the MKS-Baratron gauge (1%, according to the manufacturer's data). However, due to pressure gradients in the gas cell, we considered the gas pressure as the average of those measured at both sides of the cell, so introducing an additional error contribution to the experimental cross sections of about 8%. By adding in quadrature all the known error sources and statistical uncertainties, a total uncertainty limit within 6–10% was determined for the absolute electron detachment cross sections. With respect to the relative ionisation cross section, instabilities in the hollow cathode discharge led to statistical uncertainties within 8 and 20%, depending on the impact energy.

4.3. Computational Details

Calculations were performed in the framework of the Density Functional Theory (DFT). The Gaussian16 package [18] was used to calculate stationary points of the potential energy surface and to perform molecular dynamics (MD) simulations. In order to carry out the MD, we used the Atom-Centered Density Matrix Propagation (ADMP) method [19–21]. In this method, nuclei move classically in the potential computed at a quantum level, in our case using the DFT. We chose the M06-2X functional [22] because it is adequate for systems with main group elements, including weak interaction forces, in combination with the 6-31++G(d,p) basis set [23–25], which includes diffuse functions, thus allowing a proper description of the electron density far from the nuclei. We considered singly ionized complexes with doublet spin multiplicity. Trajectories with the ADMP method were propagated up to 200 fs, with a time step of 0.1 fs and using a fictitious mass of $\mu = 0.1$ amu, thus ensuring the adiabaticity in the dynamics. We ran trajectories with four different values of internal excitation energy, $E_{exc} = 5, 15, 25$ and 40 eV. In each trajectory, E_{exc} was randomly redistributed among the nuclear degrees of freedom. This method has been used with success in the past to study the fragmentation dynamics of ionized molecules and clusters of different nature [26–33].

5. Conclusions

Novel experimental total electron detachment cross sections for O_2^- collisions with benzene molecules were reported for impact energies ranging from 10 to 1000 eV as measured with a transmission beam technique. By analysing the positively charged species produced during the collisions, relative total ionisation cross sections were derived in the incident energy range of 160–900 eV. Relative partial ionisation cross sections for fragments with masses $m \leq 78$ were also given in this energy range. From the mass analysis of the fragments, we confirmed that the heavier compounds ($m > 78$) discussed in Ref. [13] formed for impact energies between 550 and 800 eV. By performing molecular dynamics calculations within the framework of the Density Functional Theory (DFT), we demonstrated that the fragmentation of these heavier compounds reinforced the creation of $m/z = 39$–42, 50, 60 u cations, which contributed to the formation of a local maximum in the total ionisation cross section within this energy range (550–800 eV). We did not find any previously published data to compare with the present results.

Author Contributions: Methodology, G.G., P.L.-V. and S.D.-T.; measurements, C.G., M.M., A.I.L. and S.K.; software and experimental analysis, J.C.O.; calculations, F.A.-G. and S.D.-T.; writing original draft—review and editing, G.G., S.D.-T. and P.L.-V. All authors have read and agreed to the published version of the manuscript.

Funding: This research was funded by the Spanish Ministry of Science and Innovation (MCIN; MCIN/AEI/10.13039/501100011033), grants number PID2019-104727RB-C21 and PID2019-110091GB-I00, the 'María de Maeztu' (CEX2018-000805-M) Program for Centres of Excellence in R&D and CSIC (Project LINKA20085).

Institutional Review Board Statement: Not applicable.

Informed Consent Statement: Not applicable.

Data Availability Statement: Not applicable.

Acknowledgments: This article is based upon work from COST action CA18212—Molecular Dynamics in the GAS phase (MD-GAS), supported by COST (European Cooperation in Science and Technology). The authors acknowledge the generous allocation of computer time at the Centro de Computación Científica at the Universidad Autónoma de Madrid (CCC-UAM). CG acknowledges funding from the IFHARU-SENACYT Program of Panamá through the Grant No. 270-2018-364. SK acknowledges the Portuguese National Funding Agency FCT through the Grant PD/BD/142831/2018 and ERASMUS+ 29191(533)107/2019/SMT together with PLV the Research Grants CEFITEC (UIDB/00068/2020) and PTDC/FIS-AQM/31281/2017. This work was also supported by the Radiation Biology and Biophysics Doctoral Training Programme (RaBBiT; Grant PD/00193/2012) and UCIBIO (Grant UIDB/04378/2020).

Conflicts of Interest: No conflict of interest to declare.

References

1. Costa, F.; Traoré-Dubuis, A.; Álvarez, L.; Lozano, A.I.; Ren, X.; Dorn, A.; Limão-Vieira, P.; Blanco, F.; Oller, J.C.; Muñoz, A.; et al. A Complete Cross Section Data Set for Electron Scattering by Pyridine: Modelling Electron Transport in the Energy Range 0–100 EV. *Int. J. Mol. Sci.* **2020**, *21*, 6947. [CrossRef]
2. Bak, A.; Kozik, V.; Kozakiewicz, D.; Gajcy, K.; Strub, D.; Swietlicka, A.; Stepankova, S.; Imramovsky, A.; Polanski, J.; Smolinski, A.; et al. Novel Benzene-Based Carbamates for AChE/BChE Inhibition: Synthesis and Ligand/Structure-Oriented SAR Study. *Int. J. Mol. Sci.* **2019**, *20*, 1524. [CrossRef]
3. Matsui, K.; Segawa, Y.; Itami, K. All-Benzene Carbon Nanocages: Size-Selective Synthesis, Photophysical Properties, and Crystal Structure. *J. Am. Chem. Soc.* **2014**, *136*, 16452–16458. [CrossRef]
4. Silveira, J.F.R.V.; Muniz, A.R. Functionalized Diamond Nanothreads from Benzene Derivatives. *Phys. Chem. Chem. Phys.* **2017**, *19*, 7132–7137. [CrossRef]
5. Chigo-Anota, E.; Escobedo-Morales, A.; Hernández-Cocoletzi, H.; López y López, J.G. Nitric oxide adsorption on non-stoichiometric boron nitride fullerene: Structural stability, physicochemistry and drug delivery perspectives. *Phys. E* **2015**, *74*, 538–543. [CrossRef]
6. Chigo-Anota, E.; Salazar Villanueva, M.; Valdez, S.; Castro, M. In silico studies of the magnetic octahedral B_6^- cluster-nitric oxide and $[B_6^-NO]^-O_2$ interactions. *Struct. Chem.* **2017**, *28*, 1757–1764. [CrossRef]
7. Chigo Anota, E.; Cortes Arriagada, D.; Bautista Hernández, A.; Castro, M. In silico characterization of nitric oxide adsorption on a magnetic $[B_{24}N_{36}$ fullerene$/(TiO_2)_2]^-$ nanocomposite. *Appl. Surf. Sci.* **2017**, *400*, 283–292. [CrossRef]
8. Marx, J. Oxygen Free Radicals Linked to Many Diseases. *Science* **1987**, *235*, 529–531. [CrossRef] [PubMed]
9. Hayyan, M.; Hashim, M.A.; AlNashef, I.M. Superoxide Ion: Generation and Chemical Implications. *Chem. Rev.* **2016**, *116*, 3029–3085. [CrossRef] [PubMed]
10. Shigdel, U.K.; Ovchinnikov, V.; Lee, S.-J.; Shih, J.A.; Karplus, M.; Nam, K.; Verdine, G.L. The Trajectory of Intrahelical Lesion Recognition and Extrusion by the Human 8-Oxoguanine DNA Glycosylase. *Nat. Commun.* **2020**, *11*, 4437. [CrossRef]
11. Costa, F.; Álvarez, L.; Lozano, A.I.; Blanco, F.; Oller, J.C.; Muñoz, A.; Souza Barbosa, A.; Bettega, M.H.F.; da Silva, F.F.; Limão-Vieira, P.; et al. Experimental and theoretical analysis for total electron scattering cross sections of benzene. *J. Chem. Phys.* **2019**, *151*, 084310. [CrossRef] [PubMed]
12. Bull, J.N.; Lee, J.W.L.; Vallance, C. Absolute electron total ionization cross-sections: Molecular analogues of DNA and RNA nucleobase and sugar constituents. *Phys. Chem. Chem. Phys.* **2014**, *16*, 10743–10752. [CrossRef] [PubMed]
13. Schram, M.J.; van der Wiel, M.J.; de Heer, F.J.; Moustafa, H.R. Absolute gross ionization cross sections for electrons (0.6–12 keV) in hydrocarbons. *J. Chem. Phys.* **1966**, *44*, 49–54. [CrossRef]
14. Wolff, W.; Perlin, A.; Oliveira, R.R.; Fantuzzi, F.; Coutinho, L.H.; de A Ribeiro, F.; Hilgers, G. Production of Long-Lived Benzene Dications from Electron Impact in the 20–2000 EV Energy Range Combined with the Search for Global Minimum Structures. *J. Phys. Chem. A* **2020**, *124*, 9261–9271. [CrossRef]
15. Sigaud, L.; Montenegro, E.C. Absolute cross sections for production of molecular dications by electron impact. *J. Phys. Conf. Ser.* **2020**, *1412*, 052007. [CrossRef]
16. Guerra, C.; Kumar, S.; Aguilar-Galindo, F.; Díaz-Tendero, S.; Lozano, A.I.; Mendes, M.; Limão-Vieira, P.; García, G. Unexpected benzene oxidation in collisions with superoxide anions. *Sci. Rep.* **2021**, *11*, 23125. [CrossRef]

17. Mendes, M.; Guerra, C.; Lozano, A.I.; Rojo, D.; Oller, J.C.; Limão-Vieira, P.; García, G. Experimental Electron-Detachment Cross Sections for Collisions of O2$^-$ with N_2 Molecules in the Energy Range 50–7000 EV. *Phys. Rev. A* **2019**, *99*, 062709. [CrossRef]
18. Frisch, M.J.; Trucks, G.W.; Schlegel, H.B.; Scuseria, G.E.; Robb, M.A.; Cheeseman, J.R.; Scalmani, G.; Barone, V.; Petersson, G.A.; Nakatsuji, H.; et al. *Gaussian 16, Revision C.01*; Gaussian, Inc.: Wallingford, CT, USA, 2019; Available online: https://gaussian.com (accessed on 31 December 2021).
19. Iyengar, S.; Schlegel, H.B.; Millam, J.M.; Voth, G.A.; Scuseria, G.E.; Frisch, M.J. Ab initio molecular dynamics: Propagating the density matrix with Gaussian orbitals. II. Generalizations based on mass-weighting, idempotency, energy conservation and choice of initial conditions. *J. Chem. Phys.* **2001**, *115*, 10291–10302. [CrossRef]
20. Schlegel, H.B.; Millam, J.M.; Iyengar, S.; Voth, G.A.; Daniels, A.D.; Scuseria, G.E.; Frisch, M.J. Ab initio molecular dynamics: Propagating the density matrix with Gaussian orbitals. *J. Chem. Phys.* **2001**, *114*, 9758–9763. [CrossRef]
21. Schlegel, H.B.; Iyengar, S.; Li, X.; Millam, J.M.; Voth, G.A.; Scuseria, G.E.; Frisch, M.J. Ab initio molecular dynamics: Propagating the density matrix with Gaussian orbitals. III. Comparison with Born-Oppenheimer dynamics. *J. Chem. Phys.* **2002**, *117*, 8694–8704. [CrossRef]
22. Zhao, Y.; Truhlar, D.G. The M06 suite of density functionals for main group thermochemistry, thermochemical kinetics, noncovalent interactions, excited states, and transition elements: Two new functionals and systematic testing of four M06-class functionals and 12 other functionals. *Theor. Chem. Account* **2018**, *120*, 215–241. [CrossRef]
23. Clark, T.; Chandrasekhar, J.; Spitznagel, G.W.; Schleyer, P.V.R. Efficient diffuse function-augmented basis sets for anion calculations. III. The 3-21+G basis set for first-row elements, Li-F. *J. Comput. Chem.* **1983**, *4*, 294–301. [CrossRef]
24. Hariharan, P.C.; Pople, J.A. The influence of polarization functions on molecular orbital hydrogenation energies. *Theor. Chim. Acta* **1973**, *28*, 213–222. [CrossRef]
25. Hehre, W.J.; Ditchfield, R.; Pople, J.A. Self-Consistent Molecular Orbital Methods. XII. Further Extensions of Gaussian-Type Basis Sets for Use in Molecular Orbital Studies of Organic Molecules. *J. Chem. Phys.* **1972**, *56*, 2257–2261. [CrossRef]
26. Maclot, S.; Delaunay, R.; Piekarski, D.G.; Domaracka, A.; Huber, B.H.; Adoui, L.; Martín, F.; Alcamí, M.; Avaldi, L.; Bolognesi, P.; et al. Determination of Energy-Transfer Distributions in Ionizing Ion-Molecule Collisions. *Phys. Rev. Lett.* **2016**, *117*, 073201. [CrossRef] [PubMed]
27. Capron, M.; Diaz-Tendero, S.; MacLot, S.; Domaracka, A.; Lattouf, E.; Lawicki, A.; Maisonny, R.; Chesnel, J.Y.; Mery, A.; Poully, J.C.; et al. A Multicoincidence Study of Fragmentation Dynamics in Collision of γ-Aminobutyric Acid with Low-Energy Ions. *Chem. Eur. J.* **2012**, *18*, 9321–9332. [CrossRef]
28. Maclot, S.; Piekarski, D.G.; Domaracka, A.; Mery, A.; Vizcaino, V.; Adoui, L.; Martín, F.; Alcamí, M.; Huber, B.A.; Rousseau, P.; et al. Dynamics of Glycine Dications in the Gas Phase: Ultrafast Intramolecular Hydrogen Migration versus Coulomb Repulsion. *J. Phys. Chem. Lett.* **2013**, *4*, 3903–3909. [CrossRef]
29. Piekarski, D.G.; Delaunay, R.; Maclot, S.; Adoui, L.; Martín, F.; Alcamí, M.; Huber, B.A.; Rousseau, P.; Domaracka, A.; Díaz-Tendero, S. Unusual hydroxyl migration in the fragmentation of β-alanine dication in the gas phase. *Phys. Chem. Chem. Phys.* **2015**, *17*, 16767–16778. [CrossRef]
30. Kling, N.G.; Díaz-Tendero, S.; Obaid, R.; Disla, M.R.; Xiong, H.; Sundberg, M.; Khosravi, S.D.; Davino, M.; Drach, P.; Carroll, A.M.; et al. Time-resolved molecular dynamics of single and double hydrogen migration in ethanol. *Nat. Commun.* **2019**, *10*, 2813. [CrossRef]
31. Oostenrijk, B.; Barreiro, D.; Walsh, N.; Sankari, A.; Månsson, E.P.; Maclot, S.; Sorensen, S.L.; Díaz-Tendero, S.; Gisselbrecht, M. Fission of charged nano-hydrated ammonia clusters—Microscopic insights into the nucleation processes. *Phys. Chem. Chem. Phys.* **2019**, *21*, 25749–25762. [CrossRef]
32. McDonnell, M.; LaForge, A.C.; Reino-González, J.; Disla, M.; Kling, N.G.; Mishra, D.; Obaid, R.; Sundberg, M.; Svoboda, V.; Díaz-Tendero, S.; et al. Ultrafast Laser-Induced Isomerization Dynamics in Acetonitrile. *J. Phys. Chem. Lett.* **2020**, *11*, 6724–6729. [CrossRef] [PubMed]
33. Rousseau, P.; Piekarski, D.G.; Capron, M.; Domaracka, A.; Adoui, L.; Martín, F.; Alcamí, M.; Díaz-Tendero, S.; Huber, B.A. Polypeptide formation in clusters of β-alanine amino acids by single ion impact. *Nat. Commun.* **2020**, *11*, 3818. [CrossRef] [PubMed]

International Journal of
Molecular Sciences

MDPI

Article

Double and Triple Differential Cross Sections for Single Ionization of Benzene by Electron Impact

Ana I. Lozano [1,2,*], Filipe Costa [1], Xueguang Ren [3], Alexander Dorn [3], Lidia Álvarez [1], Francisco Blanco [4], Paulo Limão-Vieira [2] and Gustavo García [1,*]

1 Instituto de Física Fundamental, Consejo Superior de Investigaciones Científicas, Serrano 113-bis, 28006 Madrid, Spain; filipe2626@gmail.com (F.C.); lid.alvarez@iff.csic.es (L.Á.)
2 Atomic and Molecular Collisions Laboratory, CEFITEC, Department of Physics, Universidade NOVA de Lisboa, 2829-516 Caparica, Portugal; plimaovieira@fct.unl.pt
3 Max Planck Institute for Nuclear Physics, 69117 Heidelberg, Germany; ren@mpi-hd.mpg.de (X.R.); A.Dorn@mpi-k.de (A.D.)
4 Departamento de Estructura de la Materia Física Térmica y Electrónica e IPARCOS, Universidad Complutense de Madrid, Plaza de Ciencias 1, 28040 Madrid, Spain; pacobr@fis.ucm.es
* Correspondence: ai.lozano@fct.unl.pt (A.I.L.); g.garcia@csic.es (G.G.)

Abstract: Experimental results for the electron impact ionization of benzene, providing double (DDCS) and triple differential cross sections (TDCS) at the incident energy of 90 eV, measured with a multi-particle momentum spectrometer, are reported in this paper. The most intense ionization channel is assigned to the parent ion ($C_6H_6^+$) formation. The DDCS values are presented for three different transferred energies, namely 30, 40 and 50 eV. The present TDCS are given for two fixed values of the ejected electron energy (E_2), at 5 and 10 eV, and an electron scattering angle (θ_1) of 10°. Different features related to the molecular orbitals of benzene from where the electron is extracted are observed. In addition, a semi-empirical formula to be used as the inelastic angular distribution function in electron transport simulations has been derived from the present DDCS result and compared with other expressions available in the literature.

Keywords: electron scattering cross sections; benzene ionization

Citation: Lozano, A.I.; Costa, F.; Ren, X.; Dorn, A.; Álvarez, L.; Blanco, F.; Limão-Vieira, P.; García, G. Double and Triple Differential Cross Sections for Single Ionization of Benzene by Electron Impact. *Int. J. Mol. Sci.* **2021**, 22, 4601. https://doi.org/10.3390/ijms22094601

Academic Editor: Linda A. Peteanu

Received: 8 April 2021
Accepted: 23 April 2021
Published: 27 April 2021

1. Introduction

Ionizing radiation transfers energy along its path within a biological system being closely related to linear energy transfer (LET). As a consequence of local photoelectric and Compton effects, a considerable number of low-energy electrons (LEE) are produced as final products of the radiation-matter interactions [1]. Nowadays, it is well-established that those electrons may significantly alter the local chemistry and potentiate strong biochemical changes in the medium, leading to biological damage, yielding single and double DNA strand breaks via dissociative electron attachment resonances [2,3] and non-resonant processes as ionization [4], among others. The characterization of the damage induced by LEE at the molecular level requires detailed knowledge of the interaction probabilities [5] (i.e., the cross sections), which provide relevant input information for the Monte Carlo simulation procedures used in radiotherapy [6]. Despite considerable progress in this area in recent decades from both experimental and theoretical methodologies, the study of several molecules becomes harder, or even impossible, as the size of the molecule increases. For this reason, electron interactions with model molecules are being extensively investigated, to serve as a benchmark for testing the validity of electron scattering of more complex molecules.

Benzene (C_6H_6) is the simplest aromatic hydrocarbon, and therefore acts as a reference model for understanding the physico-chemical properties of a vast set of biologically relevant molecules. Nevertheless, although one finds in the literature [7–11] (and references therein) several studies on electron interactions with benzene, detailed investigations of the electron-impact ionization dynamics have not been performed to date, as far as the authors are aware. Nonetheless, different studies on the ionization and fragmentation dynamics of benzene produced by intense laser fields [12–14] or excited metastable atoms [15] can be found in the literature. On the other hand, different studies on the electron-impact ionization dynamics for other complex biologically relevant molecules such as DNA bases or pyrimidine are also reported [16–20].

The main goals of this study are twofold. Firstly, we aim to provide an experimental characterization of the electron-impact ionization dynamics of benzene at relatively low-electron energy (90 eV). For that, kinematically complete experiments, (*e,e*) and (*e,2e*) studies, were performed with a reaction microscope apparatus [21,22]. In this experiment, one can simultaneous probe energy and angular distributions for all the particles involved in the ionization process (scattered electron, secondary ejected electron and recoil ion), providing double and triple differential cross section (DDCS and TDCS) values. Secondly, through the obtained experimental results, we seek to check the accuracy of a semi-empirical formula [23] to reproduce the inelastic angular distribution of scattered electrons, as a function of that for the elastic scattering and the energy transferred during the collision.

The remainder of this paper comprises three parts. The obtained experimental results are presented and discussed in the next part. In addition, the accuracy of the semi-empirical formula is checked, and some improvements are proposed. In the following section, the experimental setup is described. Finally, the main findings are summarized, and some conclusions are drawn.

2. Results and Discussion

The outermost molecular orbitals (MOs) of benzene in terms of D_{6h} symmetry in its ground state ($^1A_{1g}$) are tabulated in Table 1 [24], together with their binding energies [24], i.e., ionization energy (IE), as well as with their respective ionization cross sections at 90 eV electron impact energy obtained by means of the binary-encounter-Bethe (BEB) method [25].

Table 1. Ionization cross sections for the outermost MOs of benzene as calculated with the BEB method. The input is available online (https://physics.nist.gov).

MO	IE (eV)	σ_{BEB} (10^{-20} m^2)
$1e_{1g}$	9.25	3.93
$3e_{2g}$	11.50	2.49
$1a_{2u}$	12.15	1.34
$3e_{1u}$	13.85	1.96
$1b_{2u}$	14.63	0.89
$2b_{1u}$	15.40	0.83
$3a_{1g}$	16.85	0.76
$2e_{2g}$	18.60	1.10
$2e_{1u}$	22.10	0.56
$2a_{1g}$	28.70	0.22

In the present experiment, the most intense ionization channel was found to be the parent ion ($C_6H_6^+$), whereas the signal for H-loss ($C_6H_5^+$) was difficult to discern from the contribution of other $C_6H_n^+$ fragments, so we focused on the investigation of $C_6H_6^+$ only.

2.1. Double Differential Ionization Cross Sections (DDCS)

The single ionization of a target by an incoming electron produces two outgoing electrons, together with the residual ion. DDCS, ($d^3\sigma/d\Omega_1 dE_1$), would be obtained by measuring the energy and angular distribution of one outgoing electron. Therefore, DDCS are differential both in energy and angle.

The experimental results for DDCS to ionize benzene at an impact energy of 90 eV (Figure 1) are summarized in Table 2, together with their estimated uncertainties (δ). The data were obtained by analyzing the angular distribution of the scattered electrons as a function of their energies, i.e., the energy transferred to the target from the recorded 2D surfaces. Here, we present cuts through the two-dimensional (2D) DDCS surfaces for three transferred energies (ΔE), namely 30, 40 and 50 eV. The absolute values were assigned from normalization with those calculated for the ionization of He atoms [26], by applying the well-established convergent close-coupling method. As the He to benzene density ratios, which are necessary to normalize the results, were determined using the total ion yields as well as the total ionization cross sections, we conservatively consider an uncertainty associated with the absolute values of about 20%.

A close inspection of the DDCS as a function of the transferred energy in Figure 1 reveals that, below 40°, the cross section increases, with the decreasing scattering angle being more pronounced as the transferred energy decreases. In contrast, above 40°, an almost isotropic behavior is observed, regardless of the transferred energy. A similar behavior was recently observed in our experimental results of DDCS for the production of the pyridine parent ion ($C_6H_5N^+$) [20]. Yet, the cross-section magnitude to produce the parent ion is slightly higher for the case of the benzene molecule.

2.2. *Triple Differential Ionization Cross Sections (TDCS)*

As mentioned above, an ionizing collision produces an electron pair of one scattered and one ejected electron in addition to the residual ion. The ionization kinematics is completely determined if in addition to the known projectile momentum, e.g., the emission angles of both outgoing electrons and the energy of one of the electrons are measured. The respective TDCS, ($d^3\sigma/d\Omega_1 d\Omega_2 dE_2$), therefore, is fully differential. Figure 2 shows the scattering geometry considered in the present work which includes the incoming projectile with momentum $\vec{p}_0$ and the two outgoing electrons with momenta $\vec{p}_1$ and $\vec{p}_2$ as well as emission angles θ_1 and θ_2 with respect to the projectile forward direction. It is most likely that the energy sharing among the electrons is asymmetric, with one faster electron ($\vec{p}_1$) which is preferentially emitted to the forward direction and which can be well identified with the scattered projectile. Then, the momentum transferred by the projectile to the target is given by $\vec{q} = \vec{p}_0 - \vec{p}_1$. Generally, the TDCS can be obtained for the emission of the ionized electron into the full solid angle and is represented, e.g., as a 3D surface [27,28]. In this study, we only consider a cut of the 3D surface, in particular through the projectile scattering plane (xz) (see ref. [29] for further details). In the following, the TDCS is presented for the fixed projectile scattering angle θ_1 and particular energies E_2 of the ionized electron, as a function of its emission angle θ_2 within the plane.

Figure 3 shows the experimental TDCS in the xz-scattering plane where the values are not on an absolute scale. The panels (a) and (b) are for ejected electron energies E_2 = 5 and 10 eV, respectively, and the scattered electron angle is $\theta_1 = -10°$. The direction of the momentum $\vec{q}$ transferred by the projectile to the molecular target and its opposite direction $-\vec{q}$ are indicated by arrows. In the same way as the DDCS, these data were recorded in coincidence with the parent ion and, therefore, for the ionization of the highest three valence orbitals $1e_{1g}$, $3e_{2g}$ and $1a_{2u}$. Normally, atomic and molecular TDCS show two maxima, one for electron emission roughly along $\vec{q}$ and one for $-\vec{q}$. The first can be assigned to a binary collision of the projectile and the bound target electron and is named binary peak. The second involves additional backscattering of the ionized electron in the potential of the residual ion and is called recoil peak. In Figure 2, we see pronounced and broad binary peaks approximately along $\vec{q}$ extending to $\theta_2 = 0°$ and beyond. Interestingly, the regularly observed angular shift away from $\vec{q}$ to larger angles due to the so-called post collision interaction (PCI), which is the repulsion of the two outgoing electrons, is not observed. In contrast, the binary peaks are slightly shifted towards the forward direction. The binary peak width and shape are determined by the collision dynamics, but also by the bound orbital momentum profile. This is because for ionization in a binary collision, the ejected electron carries the initial bound state momentum to which the momentum transferred in the collision is added. The uppermost three valence orbitals of benzene have π-character with amplitude minima at zero momentum [30]. As a result, a minimum in the cross section along $\vec{q}$ is also expected for the so-called Bethe ridge condition where the absolute value of the ejected electron momentum p_2 is close to the magnitude of the momentum transfer q. Here, this is the case for $E_2 = 5$ eV, where $p_2 = 0.61$ a.u., which must be compared to $q = 0.5$ a.u. At $E_2 = 10$ eV, the respective electron momentum is $p_2 = 0.86$ a.u. and, thus, more different from q. As a consequence, no minimum is found in the binary peak. In the $-\vec{q}$ direction, no clear recoil peak can be identified. Nevertheless, there is a finite cross section in this angular region which is lower for $E_2 = 10$ eV. For this energy, there is an indication of a small recoil peak at about 180° according to the wings observed around 150° and 210°, while its center is in the angular region, which is not covered by the spectrometer.

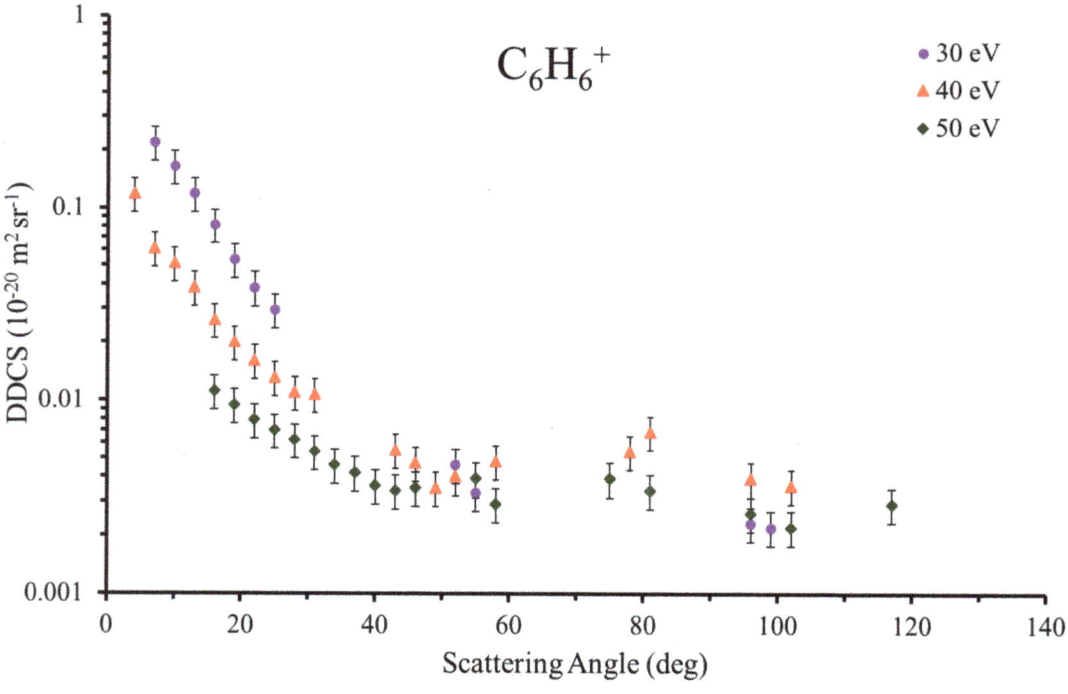

Figure 1. Experimental absolute double differential cross sections (DDCS) for producing the ion parent of benzene ($C_6H_6^+$) at 90 eV of impact energy for different transferred energies (30, 40, and 50 eV).

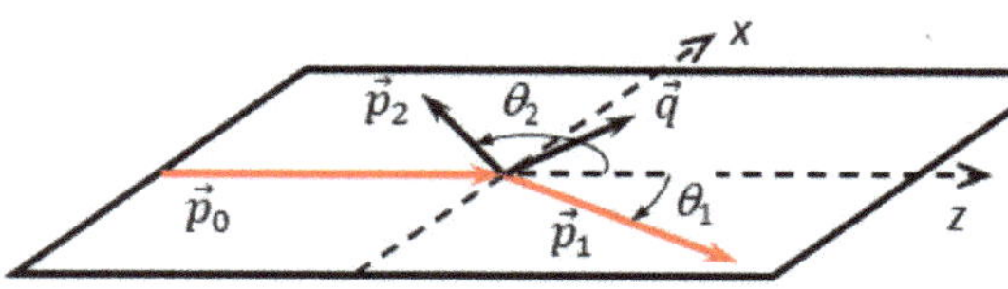

Figure 2. The coplanar scattering geometry. For details see text.

Table 2. Absolute DDCS for the production of the benzene parent ion $C_6H_6^+$ at the impact energy of 90 eV for different transferred electron energies together with their estimated uncertainties (units are 10^{-20} m^2 sr^{-1}).

	30 eV		40 eV		50 eV	
Angle (deg)	**DDCS**	**δ**	**DDCS**	**δ**	**DDCS**	**δ**
4	-	-	0.1191	0.0238	-	-
7	0.2193	0.0438	0.0614	0.0123	-	-
10	0.1647	0.0329	0.0513	0.0102	-	-
13	0.1188	0.0237	0.0386	0.0077	-	-
16	0.0814	0.0163	0.0262	0.0052	0.0112	0.0022
19	0.0534	0.0107	0.02	0.0040	0.0095	0.0019
22	0.0383	0.0077	0.0161	0.0032	0.0079	0.0016
25	0.0294	0.0059	0.0131	0.0026	0.007	0.0014
28	-	-	0.011	0.0022	0.0062	0.0012
31	-	-	0.0107	0.0021	0.0054	0.0011
34	-	-	-	-	0.0046	0.0009
37	-	-	-	-	0.0042	0.0008
40	-	-	-	-	0.0036	0.0007
43	-	-	0.0055	0.0011	0.0034	0.0007
46	-	-	0.0047	0.0009	0.0035	0.0007
49	-	-	0.0035	0.0007	-	-
52	0.0046	0.0009	0.004	0.0008	-	-
55	0.0033	0.0007	-	-	0.0039	0.0008
58	-	-	0.0048	0.0010	0.0029	0.0006
75	-	-	-	-	0.0039	0.0008
78	-	-	0.0054	0.0011	-	-
81	-	-	0.0068	0.0014	0.0034	0.0007
96	0.0023	0.0005	0.0039	0.0008	0.0026	0.0005
99	0.0022	0.0004	-	-	-	-
102	-	-	0.0036	0.0007	0.0022	0.0004
117	-	-	-	-	0.0029	0.0006

It is interesting to compare these data with results from other cyclic molecules. Recent results from tetrahydrofuran (THF) at similar kinematics [31] show a much more isotropic and less structured TDCS, which is typical and regularly observed for larger and relatively complex molecules [32,33]. In these cases, this was tentatively assigned to strong scattering of the ionized electron in the multi-center potential of the residual ion. Here, aromatic rings seem to behave differently; in addition to the present results for benzene, for phenol at a higher impact energy, a rather distinct binary peak was found accompanied with rather small intensity in the recoil region [34]. This could be explained by the strong delocalization of the highest π-orbitals in the benzene aromatic ring, which results in a comparatively narrow momentum distribution of bound orbitals, and possibly also by a reduced scattering of the ejected electron in the ionic potential. In this case, further, more systematic and detailed experimental studies including a theoretical treatment are required.

2.3. Semi-Empirical Formula

Input information required for event-by-event Monte Carlo procedures include inelastic angular distribution functions. Apart from the growing efforts to obtain accurate inelastic differential cross sections either from theoretical or experimental methods, Monte Carlo simulations often rely on approximate equations. With the main objective being to find an accurate approximation to comprehensively describe the interaction dynamics, a semi-empirical formula (Equation (1)) was derived from differential energy loss spectra, as thoroughly explained in Ref. [30]. This equation approximates the inelastic scattering angular distribution to that of the elastic contribution by including a correction exponent which depends on the incident and transferred energies [30]. The efficiency of this empirical equation has been proven in different electron transport simulations for an extensive set of biological relevant molecules [5,20,23,35].

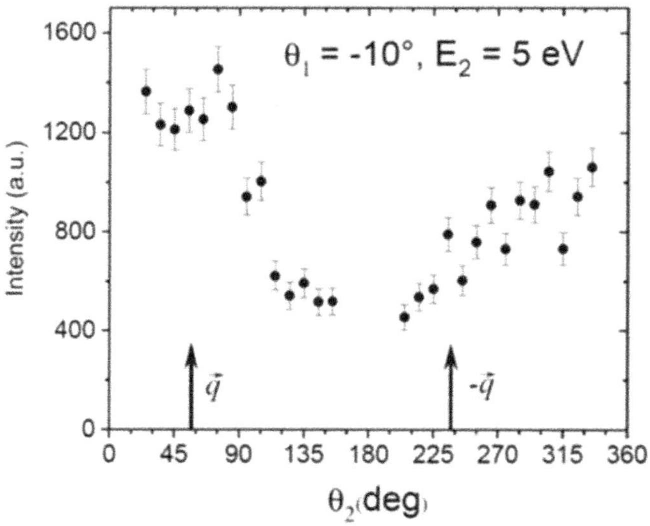

(a)

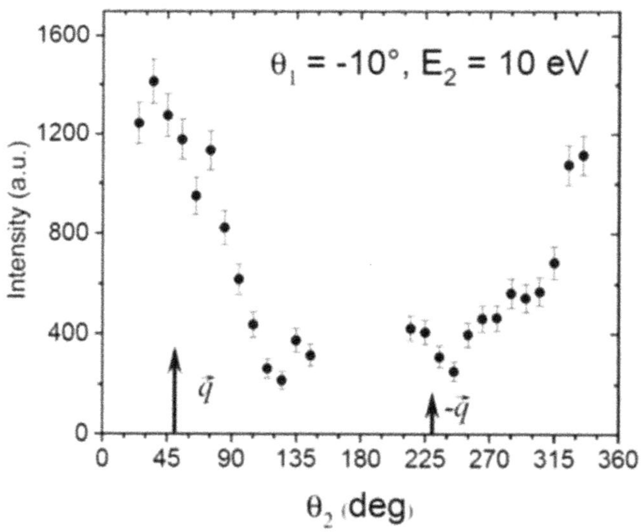

(b)

Figure 3. Experimental relative TDCS (in arbitrary units) for producing the parent ion of benzene ($C_6H_6^+$) at 90 eV electron impact energy as a function of the ejected electron emission angle θ_2. The scattered projectile angle is fixed to $\theta_1 = -10°$ while the ejected electron energy E_2 is chosen to be 5 eV (**a**) and 10 eV (**b**). The experimental results are for the scattering plane (xz). The angles of the momentum transfer vector $\vec{q}$ and its opposite $-\vec{q}$ are indicated by arrows.

Figure 4 shows the experimental DDCS results together with the angular distribution given by the elastic scattering, as calculated with our IAM-SCAR+I [36,37] procedure, and that derived from the empirical Equation (1). As shown in this figure, Equation (1) reasonably accurately reproduces the experimental data for transferred energies of about 30 eV. However, for $\Delta E > 30$ eV, especially for scattering angles below 40°, Equation (1) does not give a good description of the scattering dynamics, overestimating the experimental findings. A similar behavior has been observed for the case of pyridine [20].

$$\frac{d^2\sigma(E)}{d\Omega d\Delta E} \propto \left(\frac{d\sigma(E)}{d\Omega}\right)_{el}^{1-\Delta E/E} \quad (1)$$

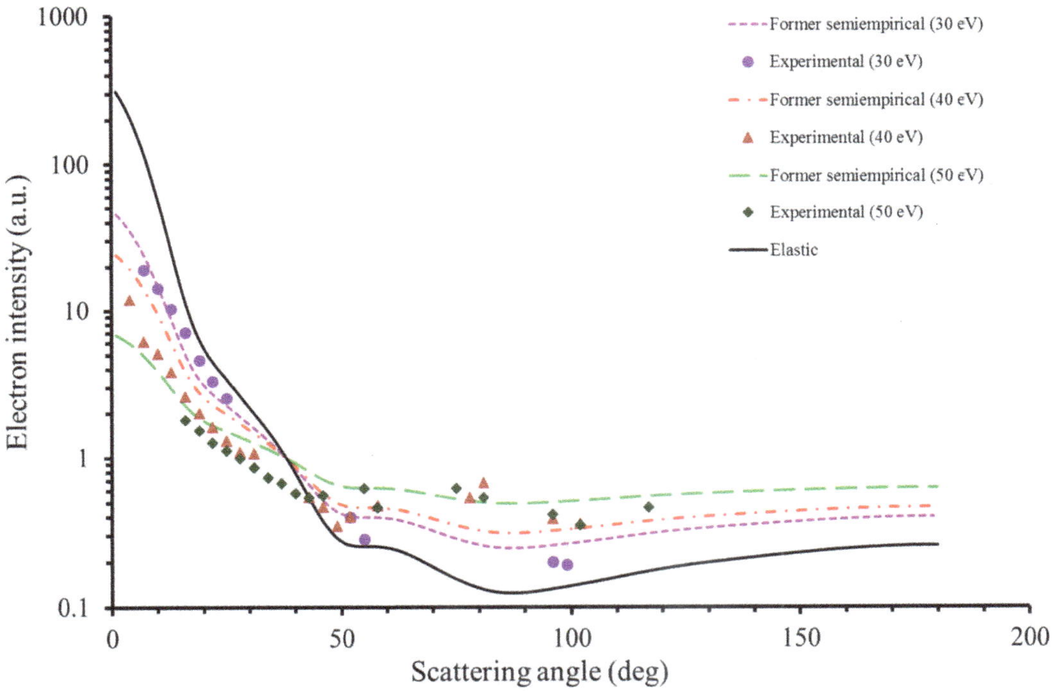

Figure 4. Inelastic angular distribution obtained from our former semi-empirical formula (Equation (1)) together with the experimental DDCS for transferred energies of 30, 40 and 50 eV and impact energy of 90 eV. Note that the elastic distribution is also included.

In order to accurately reproduce the experimental results for transferred energies above 30 eV, we propose an alternative equation which is based on the present experimental DDCS results and is shown in (Equation (2)).

$$\frac{d^2\sigma(E)}{d\Omega d\Delta E} \propto \left(\frac{d\sigma(E)}{d\Omega}\right)_{el}^{(1-\Delta E/E)^{\alpha}} \quad (2)$$

This new approximation results from applying an exponent α to the $(1 - \Delta E/E)$ term, where α takes the value of 1.3, according to the present experimental data, as well as to those previously obtained for pyridine [20]. As seen in Figure 5, the resulting angular distribution dependence obtained with Equation (2) more accurately reproduces the experimental data for all the transferred energies considered in this study.

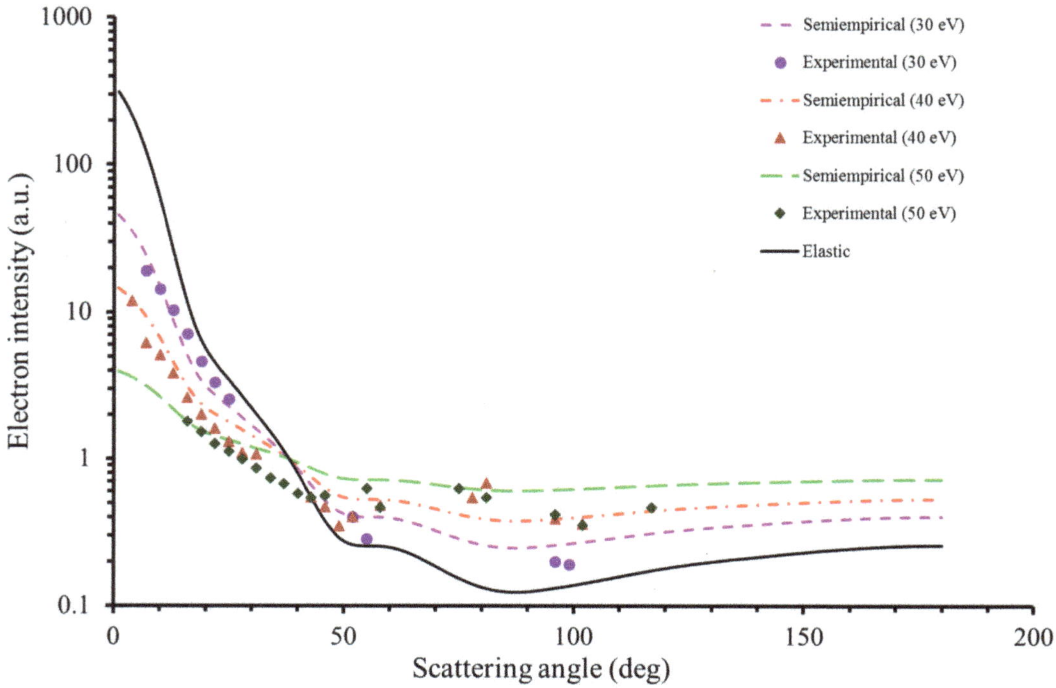

Figure 5. Same as in Figure 3 but now using the new adopted Equation (2). See text for details.

3. Experimental Setup

The experimental setup used to investigate DDCS and TDCS is a spectrometer of multicoincidence multielectron recoil-ion momentum (reaction microscope), which was specifically designed for electron-impact ionization studies (Figure 6). This apparatus has been described extensively elsewhere [21,22]. In brief, a well-focused pulsed electron beam of about 0.5 ns duration, 0.5 eV energy spread, and 40 kHz pulse repetition frequency is crossed with a cold target beam created by supersonic expansion. The benzene expands through different stages and is admitted into the main scattering chamber where the background pressure is about 10^{-8} mbar. The scattered and the ejected electrons and the positive recoil ion, produced in the ionizing collisions, are extracted by parallel electric and magnetic fields, which are then projected into two position- and time-sensitive detectors in opposite directions. Triple-coincidences of both outgoing electrons and the benzene cation were recorded (*e*, *2e* + *ion*), therefore providing the necessary information to derive DDCS and TDCS. From the positions of the hits on the detector and the corresponding time-of-flight measurements, the momentum vectors of the final state particles can be determined. The full solid angle is almost entirely covered with this configuration, in particular 100% of the recoil ions and about 80% of secondary electrons with energies below 15 eV are detected. The projectile beam axis is adjusted to be exactly parallel to the electric and magnetic extraction field directions. Because of this alignment, after crossing the target gas jet, the projectile beam reaches the center of the electron detector where a central hole in the micro-channel plates allows the beam to pass undetected. This hole causes a blind angular range for electrons with small forward and backward emission angles which depends on the electron energy. Some secondary electrons, which undergo an integer number of cyclotron revolutions in the magnetic field, can also hit the mentioned hole. These events require several experimental runs, with different extraction conditions, in order to maximize the accessible angular range. For this purpose, two different electric extraction fields were applied during the measurements. Data for coincidence analysis were accumulated simultaneously for all covered scattering angles (θ_1) and the respective scattered and ejected electron energies (E_1 for DDCS and E_2 for TDCS). With this procedure, the obtained DDCS and TDCS values were internormalized. The electron energy was calibrated using the ionization energy of neon, which was measured in the same experimental setup. The angular and energy resolution of this experimental system for the impact energy studied lie at around 2° and 2 eV, respectively.

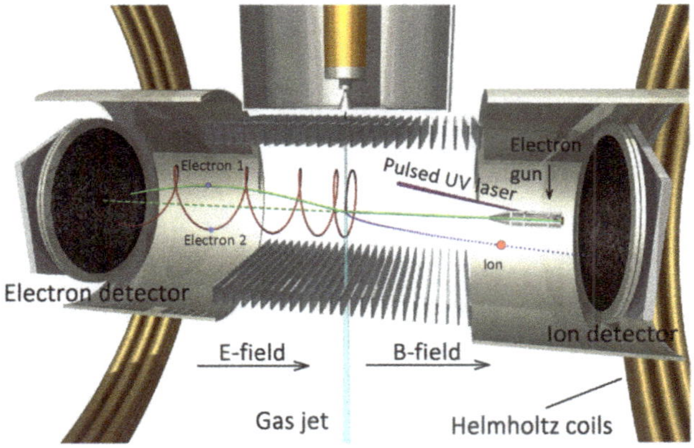

Figure 6. Schematic diagram of the reaction microscope used to analyze the angular and energy resolved double and triple differential cross sections, for electron induced ion fragmentation to benzene molecules.

4. Conclusions

The present study constitutes the first kinematically complete experiment on the electron-impact ionization dynamics of benzene. Novel results of DDCS and TDCS for this highly relevant molecule have been reported for the parent ion ($C_6H_6^+$) at 90 eV impact energy. The DDCS have been obtained for three different transferred energies, showing that below 40°, the cross section continues to get more pronounced as the transferred energy decreases. As far as TDCS are concerned, experimental results in the coplanar xz-plane have been shown for two fixed values of the ejected electron energy E_2 and the scattered electron angle $\theta_1 = -10°$. The TDCS show pronounced binary peaks with indications of a central minimum at Bethe ridge kinematics. The binary peak angular positions are shifted away from the momentum transfer and towards the forward direction and, therefore, are not in accordance with PCI, which, for ionization of atoms and smaller molecules, usually shifts electron emission to the backwards direction. This might originate in the interaction of the outgoing electrons with the comparatively complex residual ion, which should be confirmed by future calculations. This set of experimental data will help to comprehensively evaluate the accuracy of theoretical models that account for the quantum few-body dynamics. Additionally, the absolute values of DDCS provide new input parameters for electron transport models based on Monte Carlo simulations. Finally, the semi-empirical formula we commonly used to obtain the inelastic angular distribution function for such models has been improved by extending its accuracy to higher energy transferred values, where the agreement with experimental data is highly satisfactory.

Author Contributions: Methodology, P.L.-V., F.B., A.D., X.R. and G.G.; experimental data collection, analysis and interpretation, X.R., A.I.L., F.C. and L.Á.; writing—original draft preparation, G.G., A.I.L. and F.C. All authors have read and agreed to the published version of the manuscript.

Funding: This research was partially funded by the Spanish Ministerio de Ciencia, Innovación y Universidades-MICIU (Project PID2019-104727RB-C21) and CSIC (Project LINKA 20085).

Institutional Review Board Statement: Not applicable.

Informed Consent Statement: Not applicable.

Data Availability Statement: Not applicable.

Acknowledgments: F.C. acknowledges the Radiation Biology and Biophysics Doctoral Training Programme (RaBBiT, PD/00193/2012); UCIBIO (UIDB/04378/2020) and together with P.L.V. and A.I.L. the Portuguese National Funding Agency FCT through the Research Grants CEFITEC (UIDB/00068/2020) and PTDC/FISAQM/31281/2017. L.Á. thanks the local CAM government and MICIU, respectively, for their corresponding grants within the "Garantía Juvenil" programmes.

Conflicts of Interest: The authors declare no conflict of interest. The funders had no role in the design of the study; in the collection, analyses, or interpretation of data; in the writing of the manuscript, or in the decision to publish the results.

Abbreviations

DDCS	Double differential cross section
TDCS	Triple differential cross section
MO	Molecular orbital
HOMO	Highest occupied molecular orbital
IAM-SCAR+I	Independent atom model with screening corrected additivity rule and interference effects
BEB	Binary-encounter-Bethe method
PCI	Post collision interaction

References

1. García Gómez-Tejedor, G.; Fuss, M.C. Biological and Medical Physics, Biomedical Engineering. In *Radiation Damage in Biomolecular Systems*; García Gómez-Tejedor, G., Fuss, M.C., Eds.; Springer: Dordrecht, The Netherlands, 2012; ISBN 978-94-007-2563-8.
2. Boudaïffa, B.; Cloutier, P.; Hunting, D.; Huels, M.A.; Sanche, L. Resonant formation of DNA strand breaks by low-energy (3 to 20 eV) electrons. *Science* **2000**, *287*, 1658–1660. [CrossRef]
3. Sanche, L. Low energy electron-driven damage in biomolecules. *Eur. Phys. J. D* **2005**, *35*, 367–390. [CrossRef]
4. Pimblott, S.M.; LaVerne, J.A. Production of low-energy electrons by ionizing radiation. *Radiat. Phys. Chem.* **2007**, *76*, 1244–1247. [CrossRef]
5. Blanco, F.; Muñoz, A.; Almeida, D.; Da Silva, F.F.; Limão-Vieira, P.; Fuss, M.C.; Sanz, A.G.; García, G. Modelling low energy electron and positron tracks in biologically relevant media. *Eur. Phys. J. D* **2013**, *67*. [CrossRef]
6. Sanz, A.G.; Fuss, M.C.; Muñoz, A.; Blanco, F.; Limão-Vieira, P.; Brunger, M.J.; Buckman, S.J.; García, G. Modelling low energy electron and positron tracks for biomedical applications. *Int. J. Radiat. Biol.* **2012**, *88*, 71–76. [CrossRef]
7. Costa, F.; Álvarez, L.; Lozano, A.I.; Blanco, F.; Oller, J.C.; Muñoz, A.; Barbosa, A.S.; Bettega, M.H.F.; Ferreira da Silva, F.; Limão-Vieira, P.; et al. Experimental and theoretical analysis for total electron scattering cross sections of benzene. *J. Chem. Phys.* **2019**, *151*, 084310. [CrossRef]
8. De Souza, G.L.C.; Dos Santos, A.S.; Lucchese, R.R.; Machado, L.E.; Brescansin, L.M.; Manini, H.V.; Iga, I.; Lee, M.T. Theoretical investigation on electron scattering by benzene in the intermediate-energy range. *Chem. Phys.* **2012**, *393*, 19–24. [CrossRef]
9. Barbosa, A.S.; Bettega, M.H.F. Shape resonances, virtual state, and Ramsauer-Townsend minimum in the low-energy electron collisions with benzene. *J. Chem. Phys.* **2017**, *146*. [CrossRef]
10. Makochekanwa, C.; Sueoka, O.; Kimura, M. Comparative study of electron and positron scattering from benzene (C6H6) and hexafluorobenzene (C6F6) molecules. *Phys. Rev. A* **2003**, *68*, 8. [CrossRef]
11. Gianturco, F.A.; Lucchese, R.R. One-electron resonances and computed cross sections in electron scattering from the benzene molecule One-electron resonances and computed cross sections in electron scattering from the benzene molecule. *J. Chem. Phys.* **2017**, *108*, 5144–6159.
12. Itakura, R.; Watanabe, J.; Hishikawa, A.; Yamanouchi, K. Ionization and fragmentation dynamics of benzene in intense laser fields by tandem mass spectroscopy. *J. Chem. Phys.* **2001**, *114*, 5598–5606. [CrossRef]
13. Kjeldsen, T.K.; Bisgaard, C.Z.; Madsen, L.B.; Stapelfeldt, H. Influence of molecular symmetry on strong-field ionization: Studies on ethylene, benzene, fluorobenzene, and chlorofluorobenzene. *Phys. Rev. A* **2005**, *71*, 1–12. [CrossRef]
14. Winney, A.H.; Lee, S.K.; Lin, Y.F.; Liao, Q.; Adhikari, P.; Basnayake, G.; Schlegel, H.B.; Li, W. Attosecond Electron Correlation Dynamics in Double Ionization of Benzene Probed with Two-Electron Angular Streaking. *Phys. Rev. Lett.* **2017**, *119*, 1–6. [CrossRef] [PubMed]
15. Horio, T.; Maruyama, R.; Kishimoto, N.; Ohno, K. A crossed-molecular beam study on collisional ionization dynamics of acetonitrile and benzene molecules with He*(23S) metastable atoms. *Chem. Phys. Lett.* **2004**, *384*, 73–79. [CrossRef]

16. Colyer, C.J.; Bellm, S.M.; Hanne, G.F.; Al-Hagan, O.; Madison, D.; Ning, C.G.; Lohmann, B. Dynamical (e, 2e) studies using tetrahydrofuran as a DNA analogue. *J. Phys. Conf. Ser.* **2011**, *288*. [CrossRef]
17. Builth-Williams, J.D.; Bellm, S.M.; Jones, D.B.; Chaluvadi, H.; Madison, D.H.; Ning, C.G.; Lohmann, B.; Brunger, M.J. Experimental and theoretical investigation of the triple differential cross section for electron impact ionization of pyrimidine molecules. *J. Chem. Phys.* **2012**, *136*, 1–8. [CrossRef] [PubMed]
18. Dal Cappello, C.; Charpentier, I.; Houamer, S.; Hervieux, P.A.; Ruiz-Lopez, M.F.; Mansouri, A.; Roy, A.C. Triple-differential cross sections for the ionization of thymine by electrons and positrons. *J. Phys. B At. Mol. Opt. Phys.* **2012**, *45*. [CrossRef]
19. Khelladi, M.F.; Mansouri, A.; Cappello, C.D.; Charpentier, I.; Hervieux, P.A.; Ruiz-Lopez, M.F.; Roy, A.C. Angular distributions in the double ionization of DNA bases by electron impact. *J. Phys. B* **2016**, *49*, 1–12. [CrossRef]
20. Costa, F.; Traoré-Dubuis, A.; Álvarez, L.; Lozano, A.I.; Ren, X.; Dorn, A.; Limão-Vieira, P.; Blanco, F.; Oller, J.C.; Muñoz, A.; et al. A Complete Cross Section Data Set for Electron Scattering by Pyridine: Modelling Electron Transport in the Energy Range 0–100 eV. *Int. J. Mol. Sci.* **2020**, *21*, 6947. [CrossRef]
21. Dürr, M.; Dimopoulou, C.; Dorn, A.; Najjari, B.; Bray, I.; Fursa, D.V.; Chen, Z.; Madison, D.H.; Bartschat, K.; Ullrich, J. Single ionization of helium by 102 eV electron impact: Three-dimensional images for electron emission. *J. Phys. B* **2006**, *39*, 4097–4111. [CrossRef]
22. Ullrich, J.; Moshammer, R.; Dorn, A.; Dörner, R.; Schmidt, L.P.H.; Schmidt-Böcking, H. Recoil-ion and electron momentum spectroscopy: Reaction-microscopes. *Rep. Prog. Phys.* **2003**, *66*, 1463–1545. [CrossRef]
23. Fuss, M.C.; Sanz, A.G.; Muñoz, A.; Do, T.P.D.; Nixon, K.; Brunger, M.J.; Hubin-Franskin, M.-J.; Oller, J.C.; Blanco, F.; García, G. Interaction model for electron scattering from ethylene in the energy range 1–10000 eV. *Chem. Phys. Lett.* **2013**, *560*, 22–28. [CrossRef]
24. Potts, A.W.; Price, W.C.; Streets, D.G.; Williams, T.A. Photoelectron spectra of benzene and some fluorobenzenes. *Gen. Discuss. Faraday Soc.* **1972**, *54*, 168–181. [CrossRef]
25. Stone, P.M.; Kim, Y.-K. An overview of the BEB method for electron-impact ionization of atoms and molecules. *Surf. Interface Anal.* **2005**, *37*, 966–968. [CrossRef]
26. Karlsson, L.; Mattsson, L.; Jadrny, R.; Bergmark, T.; Siegbahn, K. Valence electron spectra of benzene and the hexafluorides of sulphur, molybdenum, tungsten and uranium: An application of multichannel detector technique to UV-valence electron spectroscopy. *Phys. Scr.* **1976**, *14*, 230–241. [CrossRef]
27. Ren, X.; Amami, S.; Hossen, K.; Ali, E.; Ning, C.; Colgan, J.; Madison, D.; Dorn, A. Electron-impact ionization of H2O at low projectile energy: Internirmalized triple-differential cross sections in three-dimensional kinematics. *Phys. Rev. A* **2017**, *95*, 022701. [CrossRef]
28. Hossen, K.; Ren, X.; Wang, E.; Gong, M.; Li, X.; Zhang, S.B.; Chen, X.; Dorn, A. Triple-differential cross sections for single ionization of CO 2 by 100 eV electron impact. *J. Phys. B* **2018**, *51*, 215201. [CrossRef]
29. Ali, E.; Ren, X.; Dorn, A.; Ning, C.; Colgan, J.; Madison, D. Experimental and theoretical triple-differential cross sections for tetrahydrofuran ionized by low-energy 26-eV-electron impact. *Phys. Rev. A* **2016**, *93*, 062705. [CrossRef]
30. Samardzic, O.; Brunger, M.J.; Grisogono, A.-M.; Weigold, E. Electron momentum spectroscopy studies on ring compounds. I. Benzene. *J. Phys. B* **1993**, *26*, 3921–3935. [CrossRef]
31. Wang, E.; Ren, X.; Gong, M.; Ali, E.; Wang, Z.; Ma, C.; Madison, D.; Chen, X.; Dorn, A. Triple-differential cross sections for (e, 2e) electron-impact ionization dynamics of tetrahydrofuran at low projectile energy. *Phys. Rev. A* **2020**, *102*, 062813. [CrossRef]
32. Builth-Williams, J.D.; da Silva, G.B.; Chiari, L.; Jones, D.B.; Chaluvadi, H.; Madison, D.H.; Brunger, M.J. Dynamical (e,2e) studies of tetrahydropyran and 1,4-dioxane. *J. Chem. Phys.* **2014**, *140*, 214312. [CrossRef] [PubMed]
33. Jones, D.B.; Ali, E.; Ning, C.G.; Ferreira da Silva, F.; Ingólfsson, O.; Lopes, M.C.A.; Chakraborty, H.S.; Madison, D.H.; Brunger, M.J. A dynamical (e,2e) investigation into the ionization of the outermost orbitals of R -carvone. *J. Chem. Phys.* **2019**, *151*, 124306. [CrossRef] [PubMed]
34. Da Silva, G.B.; Neves, R.F.C.; Chiari, L.; Jones, D.B.; Ali, E.; Madison, D.H.; Ning, C.G.; Nixon, K.L.; Lopes, M.C.A.; Brunger, M.J. Triply differential (e,2e) studies of phenol. *J. Chem. Phys.* **2014**, *141*, 124307. [CrossRef] [PubMed]
35. Lozano, A.I.; Oller, J.C.; Jones, D.B.; Bettega, M.H.F.; Lima, P.; Lima, M.A.P.; White, R.D.; Brunger, M.J.; Blanco, F.; Mun, A. Total electron scattering cross sections from para -benzoquinone in the energy range 1–200 eV. *Phys. Chem. Chem. Phys.* **2018**. [CrossRef] [PubMed]
36. Blanco, F.; Rosado, J.; Illana, A.; García, G. Comparison of two screening corrections to the additivity rule for the calculation of electron scattering from polyatomic molecules. *Phys. Lett. Sect. A Gen. At. Solid State Phys.* **2010**, *374*, 4420–4424. [CrossRef]
37. Blanco, F.; García, G. Screening corrections for calculation of electron scattering differential cross sections from polyatomic molecules. *Phys. Lett.* **2004**, *330*, 230–237. [CrossRef]

International Journal of *Molecular Sciences*

MDPI

Article

Excited States of Bromopyrimidines Probed by VUV Photoabsorption Spectroscopy and Theoretical Calculations

Mónica Mendes [1,*], Fábris Kossoski [2,*], Ana I. Lozano [1], João Pereira-da-Silva [1], Rodrigo Rodrigues [1], João Ameixa [1], Nykola C. Jones [3], Søren V. Hoffmann [3] and Filipe Ferreira da Silva [1]

1 CEFITEC, Departamento de Física, NOVA School of Science and Technology, FCT NOVA, Universidade NOVA de Lisboa, 2829-516 Caparica, Portugal; ai.lozano@fct.unl.pt (A.I.L.); jvp.silva@campus.fct.unl.pt (J.P.-d.-S.); rar09843@campus.fct.unl.pt (R.R.); j.ameixa@campus.fct.unl.pt (J.A.); f.ferreiradasilva@fct.unl.pt (F.F.d.S.)

2 Laboratoire de Chimie et Physique Quantiques (UMR 5626), Université de Toulouse, CNRS, UPS, CEDEX 09, 31062 Toulouse, France

3 ISA, Department of Physics and Astronomy, Aarhus University, Ny Munkegade 120, 8000 Aarhus C, Denmark; nykj@phys.au.dk (N.C.J.); vronning@phys.au.dk (S.V.H.)

* Correspondence: mf.mendes@fct.unl.pt (M.M.); fkossoski@irsamc.ups-tlse.fr (F.K.)

Abstract: We report absolute photoabsorption cross sections for gas-phase 2- and 5-bromopyrimidine in the 3.7–10.8 eV energy range, in a joint theoretical and experimental study. The measurements were carried out using high-resolution vacuum ultraviolet synchrotron radiation, with quantum chemical calculations performed through the nuclear ensemble approach in combination with time-dependent density functional theory, along with additional Franck–Condon Herzberg–Teller calculations for the first absorption band (3.7–4.6 eV). The cross sections of both bromopyrimidines are very similar below 7.3 eV, deviating more substantially from each other at higher energies. In the 7.3–9.0 eV range where the maximum cross-section is found, a single and broad band is observed for 5-bromopyrimidine, while more discernible features appear in the case of 2-bromopyrimidine. Several $\pi^* \leftarrow \pi$ transitions account for the most intense bands, while weaker ones are assigned to transitions involving the nitrogen and bromine lone pairs, the antibonding σ^*_{Br} orbital, and the lower-lying Rydberg states. A detailed comparison with the available photo-absorption data of bromobenzene is also reported. We have found significant differences regarding the main absorption band, which is more peaked in bromobenzene, becoming broader and shifting to higher energies in both bromopyrimidines. In addition, there is a significant suppression of vibrational structures and of Rydberg states in the pair of isomers, most noticeably for 2-bromopyrimidine.

Keywords: VUV photoabsorption; halopyrimidines; valence and Rydberg states; time-dependent density functional theory; radiosensitizers

Citation: Mendes, M.; Kossoski, F.; Lozano, A.I.; Pereira-da-Silva, J.; Rodrigues, R.; Ameixa, J.; Jones, N.C.; Hoffmann, S.V.; Ferreira da Silva, F. Excited States of Bromopyrimidines Probed by VUV Photoabsorption Spectroscopy and Theoretical Calculations. *Int. J. Mol. Sci.* **2021**, *22*, 6460. https://doi.org/10.3390/ijms22126460

Academic Editor: Paschalis Alexandridis

Received: 21 May 2021
Accepted: 12 June 2021
Published: 16 June 2021

1. Introduction

The absorption of ionizing radiation by living cells has long been recognized as a significant cause of long-term biological damage, leading to cellular death, mutations and/or cancer. Most biological effects of radiation arise from permanent modifications of the DNA structure, which can result in apoptosis [1]. It is well known that such consequences are mainly related to the effects of secondary low energy electrons and free radicals, which are produced by the interaction of the primary ionization radiation with the cellular molecular constituents [2,3]. Along the ionization track, secondary electrons are also able to excite biomolecular targets through inelastic collisions, thereby populating electronically excited states. These, in turn, might decay into further reactive species.

The ability of radiation in inducing modifications of the cellular components is particularly important and useful for treatment of diseases, especially in cancer therapy. Nowadays, the investigation and development of more efficient therapies based on the combination of radiation and chemotherapeutic agents represent an important and growing line of research. In chemoradiation, radiosensitizers are administered to the patient in order to enhance the radiosensitivity of tumor cells. In the subsequent radiotherapy stage, the damage is thus expected to be more localized in regions rich with radiosensitizers drugs. Through this combination, it is possible to increase tumor cell killing, with less damage to healthy tissues [4,5].

Pyrimidine is part of an important class of organic molecules, since it can be regarded as a building block of three nucleic acids, namely thymine, cytosine, and uracil. For this reason, its physicochemical properties have been studied through several experimental and theoretical techniques [6–15]. Chemical modification of pyrimidines by single halogenation, for example, has brought out their role as anti-cancer drugs. Indeed, it has been demonstrated that modified pyrimidines enhance the sensitivity of cells to ionizing radiation when incorporated into DNA as a substitute for thymine [16–19]. Halogenated analogues of thymidine, such as bromodeoxyuridine and iododeoxyuridine have shown great impact on the enhancement of the radiotherapeutic effects via the Auger electron cascade produced by the decay of the ionized heavy halogen atom [20–22]. Recently, some fluorine analogues, especially 5-fluorouracil, 5-fluorodeoxyuridine, have also demonstrated an important role in radiosensitization [5,23].

A full characterization of the electronic structure and electronic spectroscopy of pyrimidine and its derivatives is particularly relevant, to provide insights into photo-induced processes and their role in DNA damage. In particular, 2-bromopyrimidine (2BrPyr) and 5-bromopyrimidine (5BrPyr) are isomers generated by replacing a hydrogen atom by a bromine atom at positions C_2 and C_5 of pyrimidine, respectively. The effect of such substitutions in pyrimidine and pyrimidine-related compounds is important in order to better understand the site-specific chemical bonding in DNA/RNA molecules. The vibrational spectra of halogenated pyrimidines, including 2BrPyr and 5BrPyr, have been studied, with vibrational modes of different 2- and 5-monosubstituted pyrimidines [24,25]. Additionally, the vibrations of 2BrPyr and 5BrPyr were also investigated through infrared and Raman spectroscopy experiments [26–28].

Bolognesi and co-workers have performed an extensive experimental and theoretical analysis of halogen-substituted pyrimidines by X-ray photoemission spectroscopy [29], near edge X-ray absorption fine structure spectroscopy [30], photofragmentation by absorption of vacuum ultraviolet (VUV) and soft X-ray radiation [31,32], and Auger spectroscopy [33]. Castrovilli et al. [34] have measured the photoionization mass spectra of bromopyrimidines in the 9–15 eV energy range. The valence photoelectron spectra (PES) of halogenated pyrimidines, including 2BrPyr and 5BrPyr, were investigated by O'Keeffe et al. [35]. The authors observed that the halogen atom induces changes in the electronic structure of pyrimidine through two main mechanisms, the inductive and the resonant effect, which involve the π orbitals localized on the ring and the lone pair orbitals of the halogen atom. More recently, a higher resolution PES spectrum of 2BrPyr was reported by Śmiałek et al. [36]. Shape and core-excited resonances of halopyrimidines were investigated by electron transmission and dissociative electron attachment spectroscopies [37], as well as by electron scattering [38] and quantum chemical calculations [39]. Furthermore, the effect of bromination of thymine and uracil was the subject of investigations of electronic and vibrational excitation [40,41] and the formation of negative ions [41–45].

There is only one experimental investigation on the electronic excited states of 2BrPyr, by Joshi et al. [27]. They reported the VUV absorption spectrum (in arbitrary units of intensity), for photon energies ranging between 3.81 eV and 4.13 eV. VUV spectra were recorded in the gas-phase as well as in different solutions, with some vibrational assignments tentatively reported for the gas phase spectrum. This previous study was limited to lower energies and nothing is known about the higher-lying excited states of 2BrPyr. To the best of our knowledge, the excited states of 5BrPyr have never been studied.

In the present work, we investigate the electronically excited states of 2BrPyr and 5BrPyr, by means of experimental and theoretical methodologies. Specifically, we report high-resolution VUV absolute photoabsorption cross sections in the energy range of 3.7–10.8 eV, time-dependent density functional theory (TDDFT) calculations in combination with the nuclear ensemble approach (NEA), and Franck–Condon Herzberg–Teller (FCHT) calculations, for both bromopyrimidine isomers. Experimental and theoretical methodologies are explained in detail in Section 3. Our results bring forth a comprehensive picture of their excited states, and could shed further light on the subsequent production of radicals, which strongly impact the local site chemistry. Considering the radio-sensitizing potential of bromopyrimidines, the present data may also provide new insights on the effects of radiation damage at a nanoscale level and their role in chemoradiation applications.

2. Results and Discussion

Both 2BrPyr (Figure 1a) and 5BrPyr (Figure 1b) belong to the C_{2v} point group. In the electronic ground state (A_1 symmetry), the outermost molecular orbitals are … $(18a_1)^2$ $(2a_2)^2$ $(5b_1)^2$ $(19a_1)^2$ $(10b_2)^2$ $(11b_2)^2$ $(6b_1)^2$ for 2BrPyr, and … $(18a_1)^2$ $(5b_1)^2$ $(2a_2)^2$ $(19a_1)^2$ $(10b_2)^2$ $(11b_2)^2$ $(6b_1)^2$ for 5BrPyr, as obtained at the density functional theory/CAM-B3LYP/aug-cc-pVDZ + 2s2p2d level of theory. The frontier canonical orbitals are shown in the Supporting Information, in Figures S2 and S3. However, most electronic excitations involve more than one dominant pair of occupied and unoccupied canonical orbitals, making the interpretation of the transitions less straightforward. For this reason, we have also computed natural transition orbitals, which provide a compact representation of the dominant hole and particle orbitals that define the character of the transition [46]. While such orbitals are state-specific, they presented the same qualitative character along most of the excited states discussed here, and therefore our assignments are based on the set of typical natural transition orbitals displayed in Figures 2 and 3. Most can be immediately associated with a corresponding canonical orbital: $\pi^*(a_2)$ with the lowest unoccupied molecular orbital (LUMO), $\pi^*(b_1)$ with the LUMO + 1, $\pi(b_1)$ with the highest occupied molecular orbital (HOMO), $n_+(a_1)$ with the HOMO-3, $\pi(a_2)$ and $\pi Br(b_1)$ with either HOMO-4 or HOMO-5, depending on the isomer. In 5BrPyr, the $n_-(b_2)$ natural transition orbital corresponds to the HOMO-1 and in 2BrPyr it arises as the combination of the HOMO-1 and the HOMO-2. The HOMO-2 of the former and the orthogonal combination of the latter provide the $n_{Br}(b_2)$ orbital. Finally, the $\sigma^*_{Br}(a_1)$ orbital also appears in some excitations, and corresponds to higher-lying LUMOs.

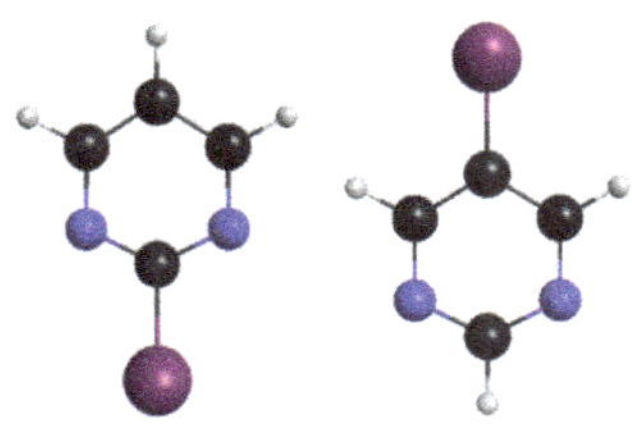

Figure 1. Molecular structure of (**a**) 2-bromopyrimidine, and (**b**) 5-bromopyrimidine. Carbon atoms in black, hydrogen in white, nitrogen in blue, and bromine in purple.

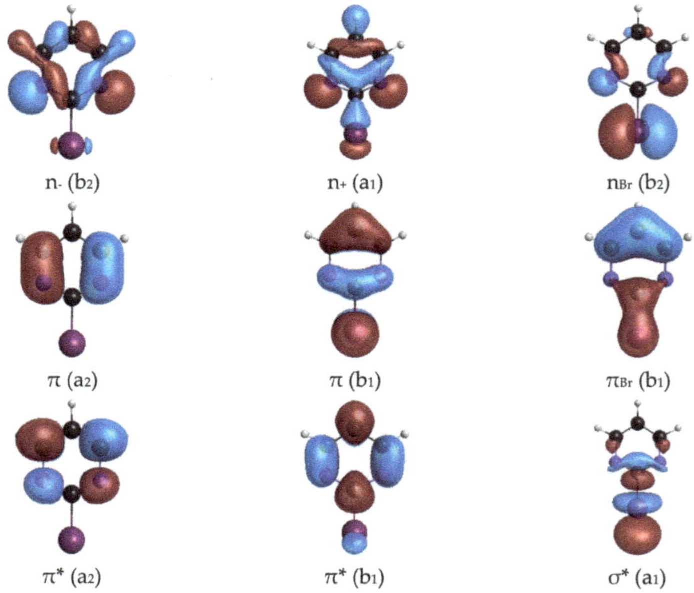

Figure 2. Typical hole and particle natural transition orbitals for the electronic transitions of 2-bromopyrimidine.

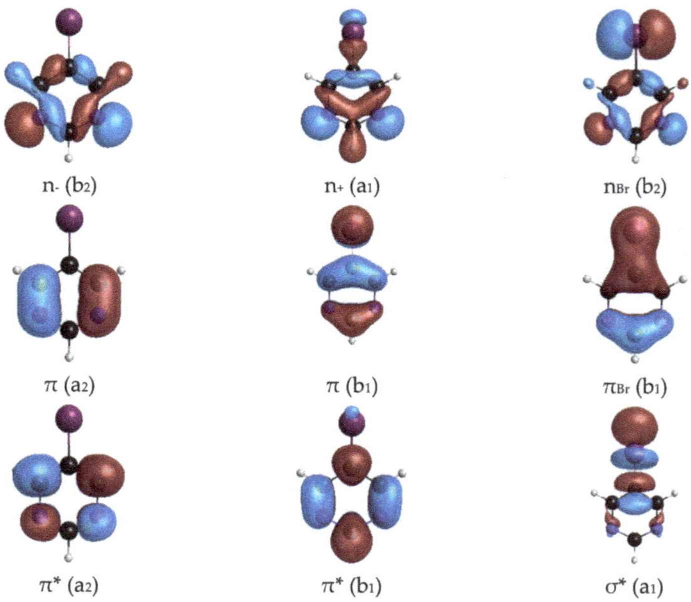

Figure 3. Typical hole and particle natural transition orbitals for the electronic transitions of 5-bromopyrimidine.

Figure 4 shows the measured high-resolution VUV photoabsorption spectra of 2BrPyr (top panel) and of 5BrPyr (bottom panel) in the 3.7–10.8 eV photon energy range. Such high resolution measurements allowed us to resolve several fine features. The computed absorption cross sections are presented in Figures 5 and 6, where they are compared with the experimental results. Agreement between experimental and theoretical cross sections are fairly good, both in magnitude and in shape. In light of this, we have interpreted the measured absorption bands largely based on the quantum chemical calculations. Tables 1 and 2 summarize the computed vertical excitation energies and state characters, for each of the 40 lower-lying singlet excited states plus two higher-lying states presenting large oscillator strengths. The dominant character of the transitions was assigned by visually inspecting the corresponding natural transitions orbitals [46]. In Figure S1, we also compare the cross sections computed with the NEA method and with the simpler vertical approximation, which only takes the ground-state equilibrium geometry into account. Once the appropriate sampling of the ground state geometry is performed via the NEA method, the comparison with respect to the experiment improves significantly, for the whole energy range covered.

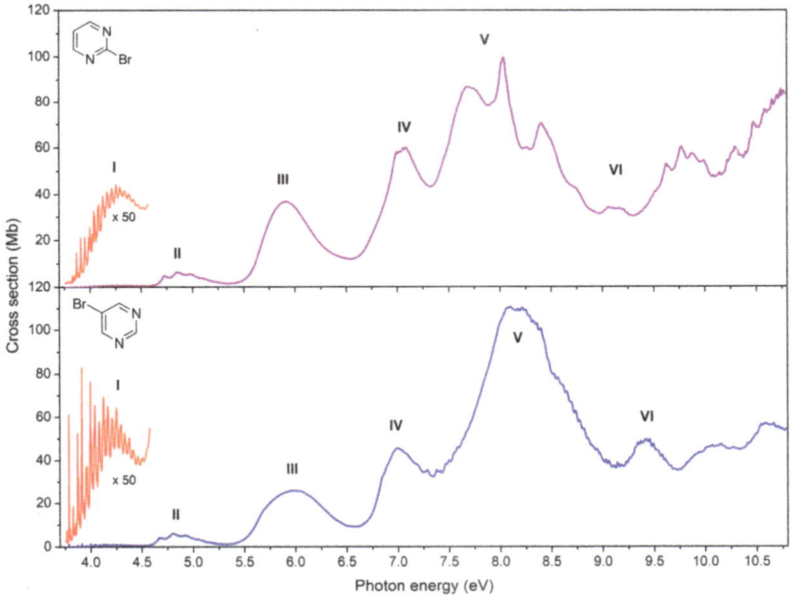

Figure 4. High-resolution VUV photoabsorption spectrum of 2-bromopyrimidine (**top** panel) and 5-bromopyrimidine (**bottom** panel) in the 3.7–10.8 eV photon energy range. Appropriate band labels (I–VI) are also given. See text for details.

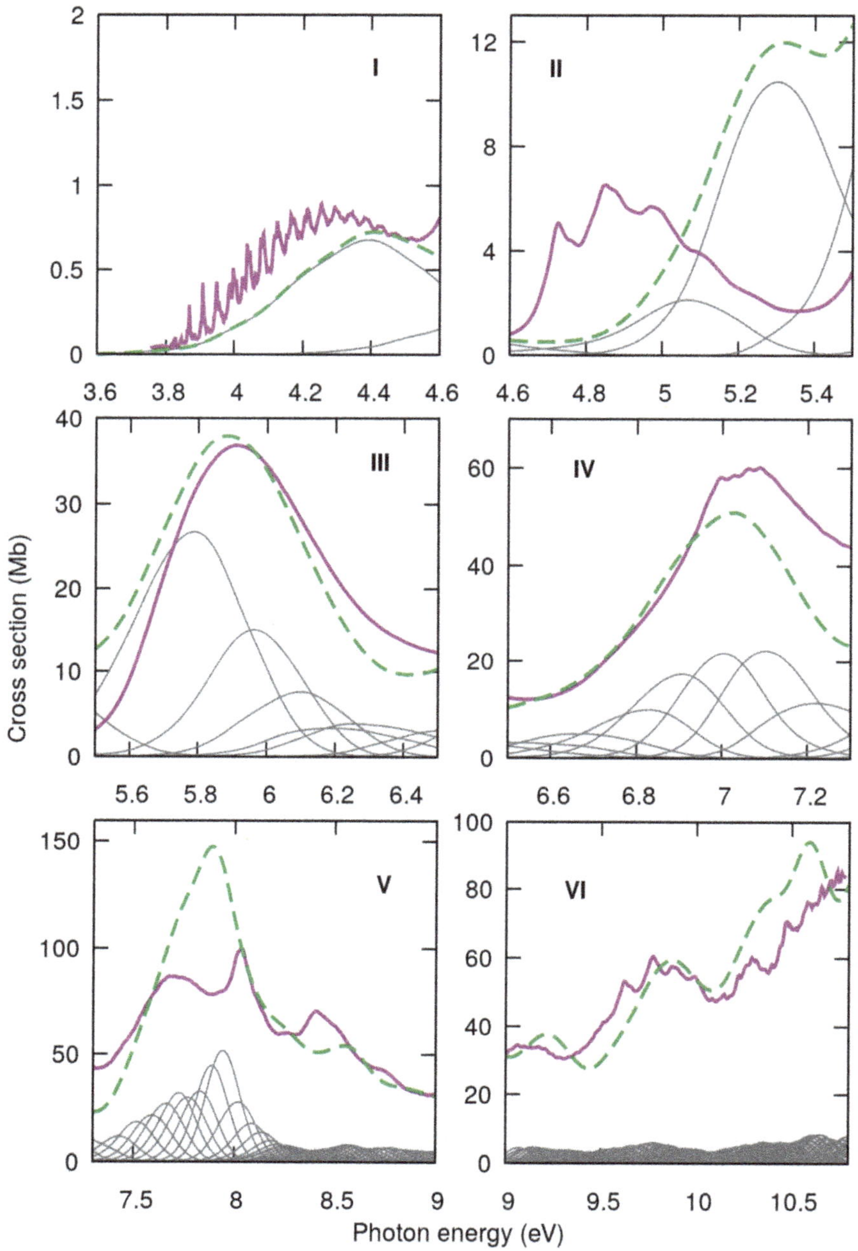

Figure 5. Comparison between measured (purple) and computed (green) absorption cross sections for 2-bromopyrimidine in the 3.7–10.8 eV photon energy range, separated into six absorption bands (**I–VI**). Individual contributions of the adiabatic excited states in grey.

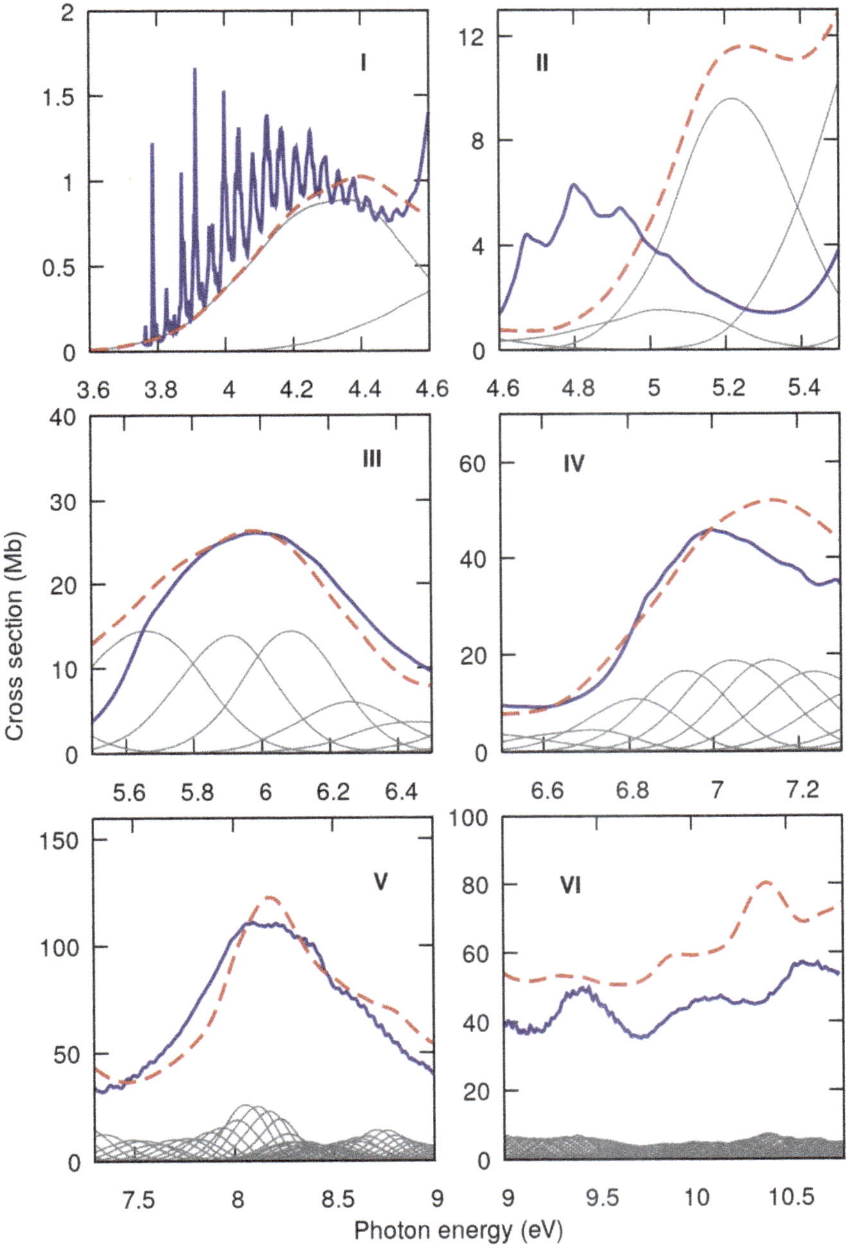

Figure 6. Comparison between measured (blue) and computed (red) absorption cross sections for 5-bromopyrimidine in the 3.7–10.8 eV photon energy range, separated into six absorption bands (**I–VI**). Individual contributions of the adiabatic excited states in grey.

Table 1. Calculated vertical excitation energies and vertical oscillator strengths (f_0) of 2-bromopyrimidine, computed at the CAM-B3LYP/aug-cc-pVDZ+2s2p2d level of theory, compared with the experimental data. In the main character column, we present the dominant hole and particle natural transition orbitals and their corresponding occupation numbers.

2-bromopyrimidine						
Experimental			Theoretical			
Band	Energy (eV)	Cross Section (Mb)	State	Energy (eV)	f_0	Main Character
I	4.25	0.9	$1B_1$	4.55	0.0037	0.99 $n_-(b_2)\ \pi^*(a_2)$
II	4.85	6.5	$1A_2$	4.98	0	0.99 $n_-(b_2)\ \pi^*(b_1)$
			$1B_2$	5.40	0.0343	0.89 $\pi(b_1)\ \pi^*(a_2)$
III	5.91	36.9	$1A_1$	6.03	0.2446	0.95 $\pi(b_1)\ \pi^*(b_1)$
			$2A_2$	6.15	0	0.97 $n_+(a_1)\ \pi^*(a_2)$
			$2B_1$	6.34	0.0014	0.86 $\pi(b_1)\ \sigma^*_{Br}(a_1)$
			$3A_2$	6.34	0	0.99 $n_{Br}(b_2)\ \pi^*(b_1)$
			$3B_1$	6.42	0.0006	0.87 $n_{Br}(b_2)\ \pi^*(a_2)$
			$2B_2$	6.45	0.0014	0.98 $n_{Br}(b_2)\ \sigma^*_{Br}(a_1)$
IV	7.09	60.2	$4B_1$	6.53	0.0050	0.96 $n_+(a_1)\ \pi^*(b_1)$
			$3B_2$	6.97	0.0001	0.87 $n_-(b_2)\ 3s(a_1)$
			$5B_1$	7.00	0.0450	0.97 $\pi(b_1)\ 3s(a_1)$
			$4B_2$	7.02	0.0014	0.71 $\pi_{Br}(b_1)\ \pi^*(a_2)$ + 0.28 $\pi(a_2)\ \pi^*(b_1)$
			$2A_1$	7.24	0.1241	0.60 $\pi(a_2)\ \pi^*(a_2)$ + 0.33 $\pi_{Br}(b_1)\ \pi^*(b_1)$
			$5B_2$	7.33	0.0002	0.88 $n_-(b_2)\ \sigma^*_{Br}(a_1)$
			$6B_1$	7.50	0.0114	0.92 $\pi(b_1)\ 3p_z(a_1)$
V	8.04	99.9	$6B_2$	7.56	0.0000	0.74 $n_{Br}/n_-(b_2)\ 3p_z(a_1)$ + 0.24 $n_{Br}/n_-(b_2)$ $3s/\sigma^*_{Br}(a_1)$
			$3A_1$	7.62	0.0045	0.88 $n_-(b_2)\ 3p_y(b_2)$
			$4A_1$	7.75	0.1902	0.61 $\pi(b_1)\ 3p_x(b_1)$ + 0.23 $\pi_{Br}(b_1)\ \pi^*(b_1)$
			$4A_2$	7.75	0	0.96 $\pi(b_1)\ 3p_y(b_2)$
			$5A_2$	7.80	0	0.95 $n_-(b_2)\ 3p_x(b_1)$
			$5A_1$	7.87	0.2166	0.45 $\pi(b_1)\ 3p_x(b_1)$ + 0.30 $\pi_{Br}(b_1)\ \pi^*(b_1)$
			$7B_2$	7.91	0.0098	0.63 $n_-(b_2)\ 4s(a_1)$ + 0.33 $n_{Br}(b_2)\ 3s(a_1)$
			$8B_2$	8.00	0.4315	0.51 $\pi(a_2)\ \pi^*(b_1)$ + 0.33 $\pi_{Br}(b_1)\ \pi^*(a_2)$
			$7B_1$	8.07	0.0001	0.95 $\pi(b_1)\ 3d_{x2}/3d_{z2}(a_1)$
			$9B_2$	8.18	0.0192	0.85 $n_-(b_2)\ 3d_{x2\text{-}y2}(a_1)$
			$6A_2$	8.24	0	0.95 $\pi(b_1)\ 3d_{yz}(b_2)$
			$6A_1$	8.29	0.0098	0.71 $n_-/n_{Br}(b_2)\ 3d_{yz}(b_2)$ + 0.19 $\pi(b_1)\ 3d_{xz}(b_1)$
			$8B_1$	8.30	0.0163	0.96 $\pi(b_1)\ 3d_{y2}(a_1)$
			$10B_2$	8.31	0.0004	0.68 $n_{Br}/n_-(b_2)\ \sigma^*_{Br}/3d_{z2}(a_1)$
			$7A_1$	8.32	0.0093	0.73 $\pi(b_1)\ 3d_{xz}(b_1)$ + 0.17 $n_-/n_{Br}(b_2)\ 3d_{yz}(b_2)$
			$7A_2$	8.36	0	0.81 $n_{Br}/n_-(b_2)\ 3d_{xz}(b_1)$
			$9B_1$	8.36	0.0003	0.93 $\pi(b_1)\ 3d_{y2}(a_1)$
			$11B_2$	8.40	0.0112	0.93 $n_-/n_{Br}(b_2)\ 4s/\sigma^*_{Br}(a_1)$
			$10B_1$	8.42	0.0002	0.96 $n_-/n_{Br}(b_2)\ 3d_{xy}(a_2)$
			$12B_2$	8.44	0.0021	0.99 $\pi(b_1)\ 3d_{xy}(a_2)$
			$8A_1$	8.45	0.0001	0.49 $n_+(a_1)\ \sigma^*_{Br}/4s(a_1)$ + 0.34 $n_{Br}/n_-(b_2)\ 3p_y(b_2)$
			$11B_1$	8.48	0.0014	0.95 $\pi(b_1)\ 4s(a_1)$
			$9A_1$	8.52	0.0050	0.87 $\pi(b_1)\ 4p_x(b_1)$
			$8A_2$	8.53	0	0.96 $\pi(b_1)\ 4p_y(b_2)$
			$11A_1$	8.69	0.0700	0.38 $n_+(a_1)\ \sigma^*_{Br}(a_1)$ + 0.37 $n_{Br}(b_2)\ 3p_y(b_2)$
			$12A_1$	8.75	0.0818	0.64 $n_+(a_1)\ \sigma^*_{Br}(a_1)$ + 0.30 $n_{Br}(b_2)\ 3p_y(b_2)$
			$20B_1$	9.34	0.0319	0.75 $\pi(a_2)\ \sigma^*_{CH}/3p_y(b_2)$
			$22A_1$	9.92	0.0603	0.65 $n_+(a_1)\ 3d_{yz}(a_1)$ + 0.25 $n_{Br}(b_2)\ 3d_{y2\text{-}z2}(b_2)$

The VUV photoabsorption spectrum of each molecule shows an overall similar behavior for energies below 7.3 eV. The major differences between the two isomers are found at higher energies, most noticeably between 7.3 and 9.0 eV. The proposed vibrational assignments are based on the identification in the spectra of a given progression and its comparison to the computed vibrational frequencies, aided by inspecting the character of the vibrational modes and of the orbitals involved in the dominant excitations. The exception is Band I, where the assignments were performed more rigorously, based on a FCHT analysis. The vibrational modes and frequencies are presented in Tables S1 and S3, and were labelled according to the Mulliken convention.

Table 2. Calculated vertical excitation energies and vertical oscillator strengths (f_0) of 5-bromopyrimidine, computed at the LYP/aug-cc-pVDZ+2s2p2d level of theory, compared with the experimental data. In the main character column, we present the dominant hole and particle natural transition orbitals and their corresponding occupation numbers.

5-bromopyrimidine						
Experimental			Theoretical			
Band	Energy (eV)	Cross Section (Mb)	State	Energy (eV)	f_0	Main Character
I	4.21	1.2	$1B_1$	4.48	0.0059	$0.99\ n_{-}(b_2)\ \pi^*(a_2)$
II	4.80	6.3	$1A_2$	4.89	0	$0.99\ n_{-}(b_2)\ \pi^*(b_1)$
			$1B_2$	5.35	0.0349	$0.89\ \pi_r(b_1)\ \pi^*(a_2)$
III	5.98	26.0	$2B_1$	5.89	0.0002	$0.99\ \pi(b_1)\ \sigma^*_{Br}(a_1)$
			$2A_2$	6.03	0	$0.97\ n_{+}(a_1)\ \pi^*(a_2)$
			$1A_1$	6.07	0.2129	$0.94\ \pi(b_1)\ \pi^*(b_1)$
			$3B_1$	6.39	0.0049	$0.99\ n_{+}(a_1)\ \pi^*(b_1)$
			$2B_2$	6.57	0.0003	$0.95\ n_{Br}(b_2)\ \sigma^*_{Br}(a_1)$
IV	7.00	45.6	$4B_1$	6.86	0.0020	$0.98\ n_{Br}(b_2)\ \pi^*(a_2)$
			$5B_1$	6.89	0.0162	$0.97\ \pi(b_1)\ 3s(a_1)$
			$3B_2$	6.98	0.0032	$0.74\ n_{-}(b_2)\ 3s/\sigma^*_{Br}(a_1) + 0.24\ n_{Br}(b_2)\ \sigma^*_{Br}(a_1)$
			$3A_2$	7.10	0	$0.99\ n_{Br}(b_2)\ \pi^*(b_1)$
			$4B_2$	7.23	0.0474	$0.51\ \pi(a_2)\ \pi^*(b_1) + 0.46\ \pi(b_1)\ \pi^*(a_2)$
			$2A_1$	7.34	0.2217	$0.78\ \pi(a_2)\ \pi^*(a_2) + 0.10\ \pi(b_1)\ \pi^*(b_1)$
V	8.10	110.8	$5B_2$	7.36	0.0002	$0.83\ n_{-}(b_2)\ \sigma^*_{Br}(a_1)$
			$3A_1$	7.63	0.0273	$0.73\ n_{-}(b_2)\ 3p_y(b_2) + 0.14\ n_{+}(a_1)\ \sigma^*_{Br}/3s(a_1)$
			$4A_2$	7.66	0	$0.99\ \pi(b_1)\ 3p_y(b_2)$
			$6B_2$	7.67	0.0686	$0.88\ n_{-}(b_2)\ 3p_z(a_1)$
			$6B_1$	7.69	0.0028	$0.96\ \pi(b_1)\ 3p_z(a_1)$
			$4A_1$	7.69	0.0060	$0.88\ \pi(b_1)\ 3p_x(b_1)$
			$5A_2$	7.73	0	$0.97\ n_{-}(b_2)\ 3p_x(b_1)$
			$7B_2$	7.99	0.0056	$0.54\ n_{Br}/n_{-}(b_2)\ 3s(a_1) + 0.25\ n_{Br}/n_{-}(b_2)\ 3d_{x2\text{-}z2}(a_1)$
			$7B_1$	8.06	0.0000	$0.96\ \pi(b_1)\ 3d_{x2}(a_1)$
			$8B_2$	8.13	0.0107	$0.82\ n_{-}(b_2)\ 3d_{x2}(a_1)$
			$9B_2$	8.18	0.3188	$0.43\ \pi_{Br}(b_1)\ \pi^*(a_2) + 0.25\ \pi(a_2)\ \pi^*(b_1)$
			$8B_1$	8.21	0.0014	$0.81\ \pi_{Br}(b_1)\ \sigma^*_{Br}(a_1)$
			$5A_1$	8.28	0.1517	$0.65\ \pi_{Br}(b_1)\ \pi^*(b_1) + 0.25\ \pi(b_1)\ 3d_{xz}(b_1)$
			$9B_1$	8.29	0.0063	$0.88\ \pi(b_1)\ 3d_{y2\text{-}z2}(a_1)$
			$6A_1$	8.31	0.0218	$0.33\ n_{-}(b_2)\ 3d_{yz}(b_2) + 0.33\ \pi(b_1)\ 3d_{xz}(b_1)$
			$10B_1$	8.36	0.0081	$0.98\ \pi(b_1)\ 3d_{z2}(a_1)$
			$6A_2$	8.36	0	$0.97\ \pi(b_1)\ 3d_{yz}(b_2)$
			$7A_2$	8.38	0	$0.96\ \pi(a_2)\ \sigma^*_{Br}(a_1)$
			$7A_1$	8.38	0.0430	$0.45\ \pi(b_1)\ 3d_{xz}(b_1) + 0.28\ n_{-}(b_2)$
			$10B_2$	8.38	0.0249	$0.93\ \pi(b_1)\ 3d_{xy}(a_2)$
			$11B_1$	8.42	0.0016	$0.95\ n_{-}(b_2)\ 3d_{xy}(a_2)$
			$11B_2$	8.45	0.0261	$0.79\ n_{-}(b_2)\ 3d_{y2\text{-}z2}(a_1) + 0.11\ n_{Br}(b_2)\ 3s(a_1)$
			$12B_2$	8.45	0.0001	$0.99\ n_{-}(b_2)\ 4s(a_1)$
			$8A_2$	8.47	0	$0.96\ n_{-}(b_2)\ 3d_{xz}(b_1)$
			$8A_1$	8.48	0.0460	$0.64\ n_{-}(b_2)\ 3d_{yz}(b_2) + 0.27\ n_{+}(a_1)\ \sigma^*_{Br}/3s(a_1)$
			$9A_1$	8.51	0.0021	$0.92\ \pi(b_1)\ 4p_x(b_1)$
			$11A_1$	8.81	0.2800	$0.62\ n_{+}(a_1)\ \sigma^*_{Br}(a_1) + 0.31\ n_{Br}(b_2)\ 3p_y(b_2)$
			$12A_1$	8.92	0.0683	$0.45\ n_{Br}(b_2)\ 3p_y(b_2) + 0.37\ n_{+}(a_1)\ \sigma^*_{Br}(a_1)$

Each of the following subsections is devoted to the discussion of a given absorption band (where energy ranges are loosely defined), while the last subsection concerns Rydberg states. In particular, the discussion will be focused on the effect of the position of the bromine atom on the VUV photoabsorption spectra, and we also compare with previous results for the closely related bromobenzene molecule [47], which are reproduced in Figure S4. A detailed comparison with pyrimidine [7] will be presented in a future paper.

2.1. The 3.7–4.6 eV Photon Energy Range (Band I)

In both isomers, the first absorption has an onset around 3.7 eV and extends up to 4.6 eV, as shown in Figure 7. Among all the bands discussed, this has the lowest intensities, yet the most fine structure. It is ascribed to the first singlet excited state, a $\pi^*(a_2) \leftarrow n_{-}(b_2)$ transition. It involves excitation from the n_{-} non-bonding orbital of the nitrogen atoms, and thus has no parallel with bromobenzene [47].

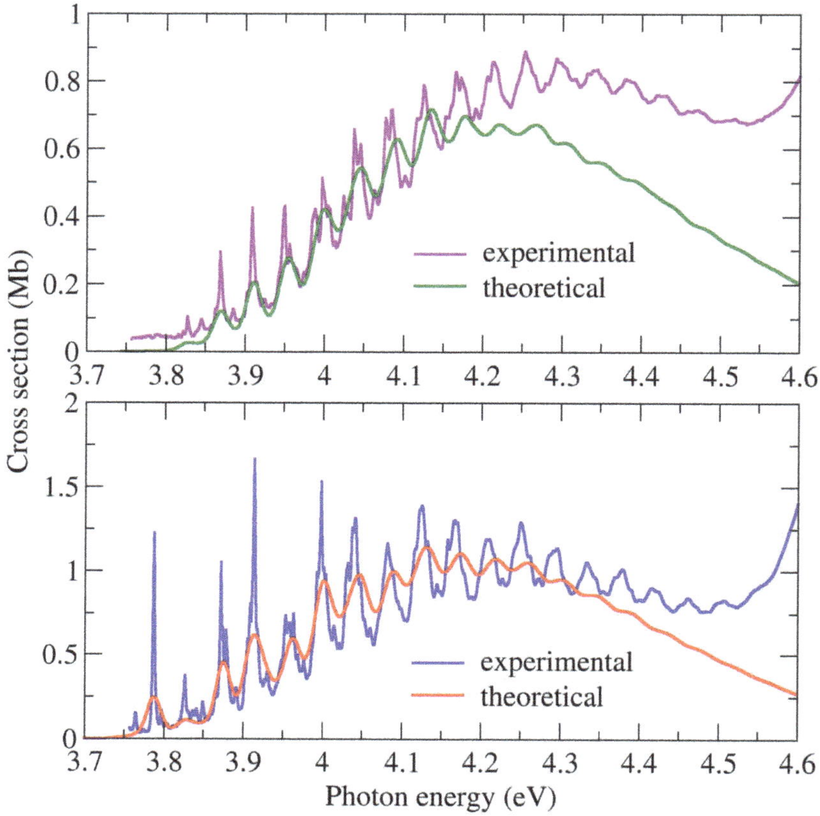

Figure 7. Comparison between measured and computed cross sections for the absorption Band I of 2-bromopyrimidine (**top** panel) and 5-bromopyrimidine (**bottom** panel).

This band is similar in both isomers, however in 5BrPyr it is somewhat more intense, slightly redshifted (~0.1 eV), and with more pronounced vibrational structures. The nuclear ensemble approach calculations reproduce very well the shape and magnitude of the bands, as well as the main observed differences between the two isomers. Importantly, the good agreement stems to a large extent from the ensemble averaging of the excitation energy, which decreases by 0.15 eV with respect to the vertical excitation energy, for both molecules. As pointed out before [48–50] this effect should be taken into account when comparing vertical excitation energies with the energy where absorption has a maximum.

In 2BrPyr, the equilibrium geometry of the first singlet excited state (coordinates in the SI, and normal modes in Table S2) preserves the same symmetry operations as the ground state, and thus belongs to the C_{2v} point group. Therefore, there is a reasonable overlap between ground- and excited-state optimized geometries, and the underlying approximations of the FCHT analysis should hold. In Figure 7 we compare the measured and the computed absorption cross sections for the low energy tail of the band, between 3.75 and 4.15 eV. The theoretical curve was redshifted by 0.108 eV, in order to match with the observed 0–0 transition. In this comparison, we employed an arbitrary linewidth of 2.5 meV (~20.2 cm^{-1}) for the FCHT calculations. Agreement is excellent, given the clear correspondence between most of the structures. This allows us to unequivocally pin down the band origin (the 0–0 transition) to the peak at 3.87 eV, which corresponds to the vibrational ground state of the excited state, thus agreeing with the assignment of Joshi et al. [27]. A number of low intensity hot bands are observed at lower energies, arising mostly from initially excited C-Br rocking (υ_{24}) and stretching (υ_9) modes. Towards higher energies, the peaks obtained with the FCHT analysis appear to be gradually blue shifted with respect to the measurements, which is an indication of further anharmonic effects not considered in the FCHT calculations. In fact, when they are performed without any anharmonic corrections for the excited state, such deviations are magnified. Eventually, the adiabatic Hessian approximation fails to reproduce the energies and shapes of the finer structures, especially above ~4.15 eV. Nevertheless, it still manages to capture the main observed vibrational progressions of the whole band, as shown in Figure 8, where a linewidth of 10 meV (~80.7 cm^{-1}) was employed. A thorough assignment of the fine structures found in the 3.75–4.15 eV range can be found in Table S5. Despite the above described FCHT limitations, we also present in Table S6 the main transitions that should account for the main vibrational progressions up to ~4.1 eV. They comprise successive excitations of modes of a_1 symmetry, mainly ring stretching modes having 700~1200 cm^{-1} frequencies and the C-Br stretching mode (see Table S6 for the full assignments).

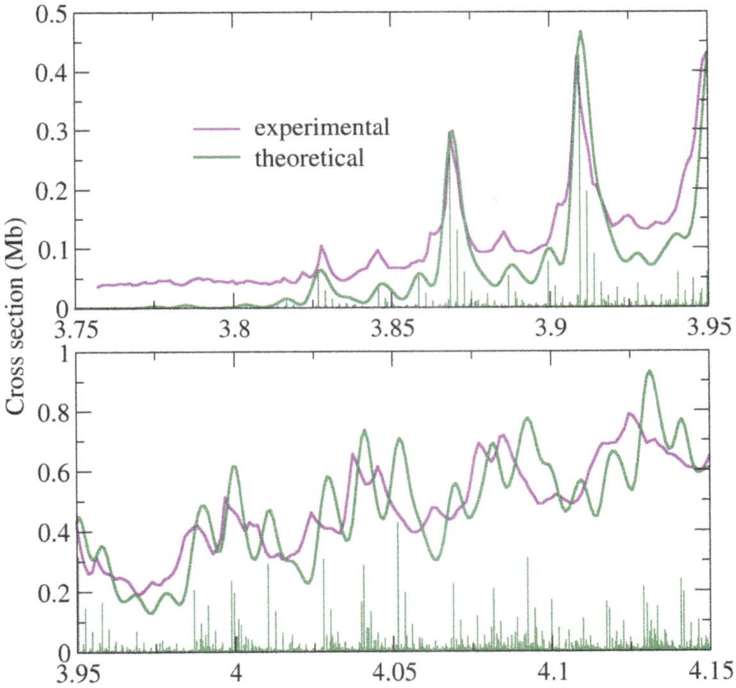

Figure 8. Comparison between measured and computed cross sections for the low energy tail of absorption Band I of 2-bromopyrimidine.

For 5BrPyr, the calculations revealed a pair of degenerate equilibrium geometries for the first excited state (coordinates in the SI, and normal modes in Table S4), which arise from symmetry-breaking of the C_2 axial symmetry. They still preserve a planar structure though, and thus belong to the C_s point group. Considering the existence of two equilibrium geometries and the smaller overlap between the lower-lying ground- and excited-state vibrational wave functions, the FCHT analysis is not expected to reproduce the fine structures observed for 5BrPyr. However, just as in the case of 2BrPyr, the calculations can account for the major vibrational progressions, as also shown in Figure 8 (the most intense computed transitions are presented in Table S7). Here a linewidth of 10 meV was employed, and the computed curve was redshifted by 0.097 eV. The very good comparison between the measured and the computed spectrum allowed us to identify the band origin at 3.79 eV. The main vibrational progressions arise from successive excitations of ring stretching modes presenting frequencies in the 700~1200 cm^{-1} range, and also of the C-Br stretching mode, which overall share the same characters as those identified for 2BrPyr.

2.2. The 4.6–5.4 eV Photon Energy Range (Band II)

Figure 9 shows the second absorption band, found between 4.6 and 5.4 eV. Their magnitude and shape are very similar in both molecules, yet they appear displaced with respect to each other by around 0.05 eV, being lower in 5BrPyr. We assign this band to excitation of the third excited state, a $\pi^*(a_2) \leftarrow \pi(b_1)$ dipole allowed transition, though the second excited state, a dipole forbidden $\pi^*(b_1) \leftarrow n_-(b_2)$ transition, contributes to a lesser extent. The band origin 0–0 is tentatively placed at 4.72 eV for 2BrPyr and at 4.68 eV for 5BrPyr, whereas the observed vibrational progressions of both molecules are assigned to ring stretching (υ_7) and Br rocking (υ_{24}) modes (see Tables 3 and 4). The present calculations overestimate the cross sections by a factor of two, and place the band too high in energy by around 0.5 eV, for both isomers. This disagreement stands out from the much better comparison observed for the other bands. Going from the vertical approximation to the NEA results does not bring any improvement (see Figure S1), and therefore, the problem should be related to the choice of functional (CAM-B3LYP), while basis set effects proved to have a minor role. We notice, though, that the π(b1) and π*(a2) orbitals also take part in other transitions where the comparison with experiment is more favorable. Thus, it is not clear why the theoretical description of this particular band falls short.

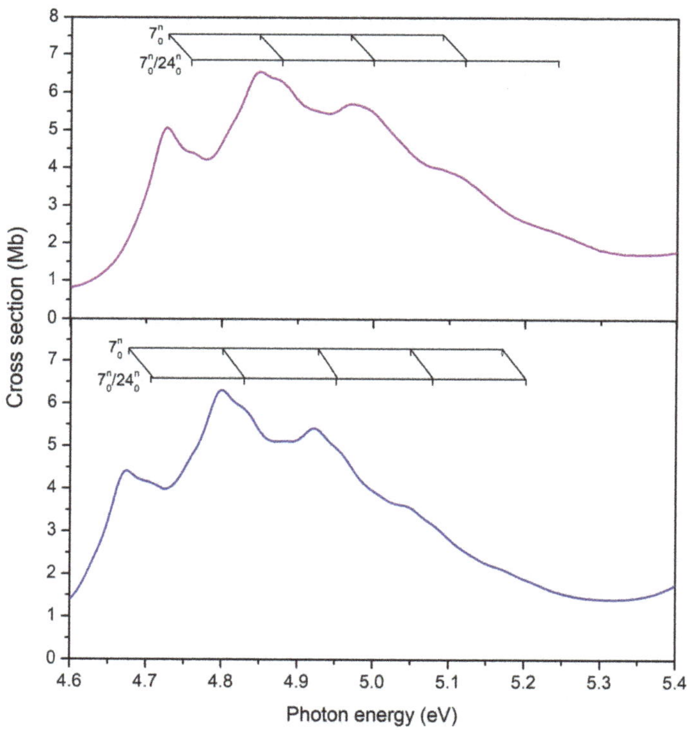

Figure 9. Vibrational progressions in the 4.6–5.4 eV energy range of the absorption Band II of 2-bromopyrimidine (**top** panel) and 5-bromopyrimidine (**bottom** panel).

Table 3. Proposed vibrational assignments in the 4.6–5.4 eV absorption band of 2BrPyr (Band II). ΔE (in eV) represents the energy of one extra quanta in the vibrational progression. The symbol "/" means both modes are involved.

Energy (eV)	Assignment	$\Delta E\left(v_7^{'}\right)$	$\Delta E\left(v_{24}^{'}\right)$
4.73	0_0^0		
4.85	7_0^1	0.12	
4.88 (b)	$24_0^1/7_0^1$	0.12	0.030
5.97 (b)	7_0^2	0.12	
5.00 (s)	$24_0^1/7_0^2$	0.12	0.029
5.09 (b)	7_0^3	0.12	
5.12 (b)	$24_0^1/7_0^3$	0.12	0.029
5.24 (b)	$24_0^1/7_0^4$	0.12	
4.73	0_0^0		
4.85	7_0^1	0.12	
4.88 (b)	$24_0^1/7_0^1$	0.12	0.030
5.97 (b)	7_0^2	0.12	

(b) broad structure; (s) shoulder structure.

Table 4. Proposed vibrational assignments in the 4.6–5.4 eV absorption band of 5BrPyr (Band II). ΔE (in eV) represents the energy of one extra quanta in the vibrational progression. The symbol "/" means both modes are involved.

Energy (eV)	Assignment	$\Delta E\left(v_7^{'}\right)$	$\Delta E\left(v_{24}^{'}\right)$
4.68	0_0^0		
4.70 (b)	24_0^1		0.03
4.80	7_0^1	0.13	
4.83 (b)	$24_0^2/7_0^1$	0.12	0.03
4.92	7_0^2	0.12	
4.95 (b)	$24_0^3/7_0^2$	0.13	0.03
5.05	7_0^3	0.12	
5.07 (b)	$24_0^4/7_0^3$	0.12	0.03
5.17 (b)	7_0^4	0.13	
5.20 (b)	$24_0^5/7_0^4$	0.13	0.03

(b) broad structure.

In bromobenzene, the analogous $\pi^*(a_2) \leftarrow \pi(b_1)$ state was also found in the same energy range, where it accounts for the first absorption band [47]. There are, however, two key differences. First, the cross sections are considerably larger in the bromopyrimidines (having a maximum of 6–7 Mb) than in bromobenzene (around 1 Mb). Second, the vibrational structures are much narrower in the latter, pointing to considerably longer-excited state lifetimes. In this energy range, the photodissociation mechanism of bromobenzene has been suggested to involve intersystem crossings from the bright singlet state to dissociative triplet states [51,52]. Assuming a similar mechanism takes place in the bromopyrimidines, significantly larger intersystem crossing rates from the $\pi^*(a_2) \leftarrow \pi(b_1)$ state would be expected.

2.3. *The 5.4–6.5 eV Photon Energy Range (Band III)*

The next absorption band lies between 5.4 and 6.5 eV, peaking at 5.91 eV for 2BrPyr and at 5.98 eV for 5BrPyr. In neither are there distinct vibrational structures (see Figure 4). The band profile is similar in both molecules, though the magnitudes differ from a maximum of 37 Mb in 2BrPyr to 26 Mb in 5BrPyr. The theoretical calculations support that the bulk of the intensity is due to the $\pi^*(b_1) \leftarrow \pi(b_1)$ transition, which have comparable excitation energies and a somewhat larger oscillator strength in 2BrPyr.

We notice, however, a subtle difference between the two spectra. It is more peaked in 2BrPyr and asymmetric in 5BrPyr, where a shoulder around 5.7 eV can be discerned. Considering that the NEA calculations seem to reproduce this feature, it might have an electronic origin, rather than a vibrational one. In fact, while other excitation energies in this range are rather close in both molecules, the $\sigma^*_{Br}(a_1) \leftarrow \pi(b_1)$ transition appears 0.45 eV lower in 2BrPyr than in 5BrPyr. Such a big shift would arise from the $\sigma^*_{Br}(a_1)$ orbital, as discussed in the context of negative ion states [34–36]. The ensemble averaging further decreases the excitation energies by 0.25 eV (in both molecules), besides promoting intensity borrowing from the bright state. The greater sensitivity of this state on the position of the bromine atom could thus explain the observed difference in the shape of this band.

In the case of bromobenzene, the analogous band emerges in the same energy range, also stemming from a dominant $\pi^*(b_1) \leftarrow \pi(b_1)$ transition [47]. Moreover, the vibrational profile observed for bromobenzene is completely lost in the bromopyrimidines. In the latter, the $\sigma^*_{Br}(a_1) \leftarrow \pi(b_1)$ excited state (dissociative along the C-Br bond) lies within the band, whereas in the former the analogous state arises in the high energy tail of the band. Therefore, more favorable couplings between the optically bright state and the dissociative state are expected in the bromopyrimidines, which could explain the lack of vibrational structures.

2.4. The 6.5–7.3 eV Photon Energy Range (Band IV)

This energy range accommodates the fourth absorption band (see Figure 10), having a maximum at 7.09 eV in 2BrPyr, and at 7.00 eV in 5BrPyr. In both, the $\pi^*(a_2) \leftarrow \pi(a_2)$ transition brings most of the intensity to this band, even though other excitations also contribute to some extent, in particular the $3s(a_1) \leftarrow \pi(b_1)$ transition for 2BrPyr, and the $\pi^*(b_1) \leftarrow \pi(a_2) + \pi^*(a_2) \leftarrow \pi(b_1)$ transition for 5BrPyr. Differences in the characters and oscillator strengths of such secondary excitations could explain the observed differences found in the vibrational progressions of each spectrum. In 5BrPyr the band origin 0–0 is tentatively placed at 6.84 eV, and at 6.99 eV for 2BrPyr. The latter develops into two sets of progressions, one has an average energy spacing of 0.13 eV and is tentatively assigned to a ring stretching (υ_7) mode. The other has an average energy spacing of 0.03 eV, and also seems to be present in 5BrPyr, though with an observed spacing of ~0.05 eV. A close inspection of Tables S1 and S3 reveals that no computed mode matches the observed ones. This may indicate that a significant relaxation of the excited states takes place, resulting in rather different frequencies.

In bromobenzene [47], the most intense band of the VUV spectrum is in the 6.3–7.0 eV range, being redshifted by around 0.3 eV with respect to the Band IV of the bromopyrimidines, besides being considerably more intense. As will be clearer in the next section, the intense band of bromobenzene should not only be compared with Band IV of bromopyrimidines, but also with part of Band V. In bromobenzene, the band has been assigned to two major excitations, the dominant one with A_1 symmetry and $\pi^*(a_2) \leftarrow \pi(a_2) + \pi^*(b_1) \leftarrow \pi(b_1)$ character, followed by the $\pi^*(b_1) \leftarrow \pi(a_2) + \pi^*(a_2) \leftarrow \pi(b_1)$ excitation, of B_2 symmetry. Except for the leading term of the former excitation, the other terms are absent in the Band IV of bromopyrimidines, giving rise to the next band, as discussed below.

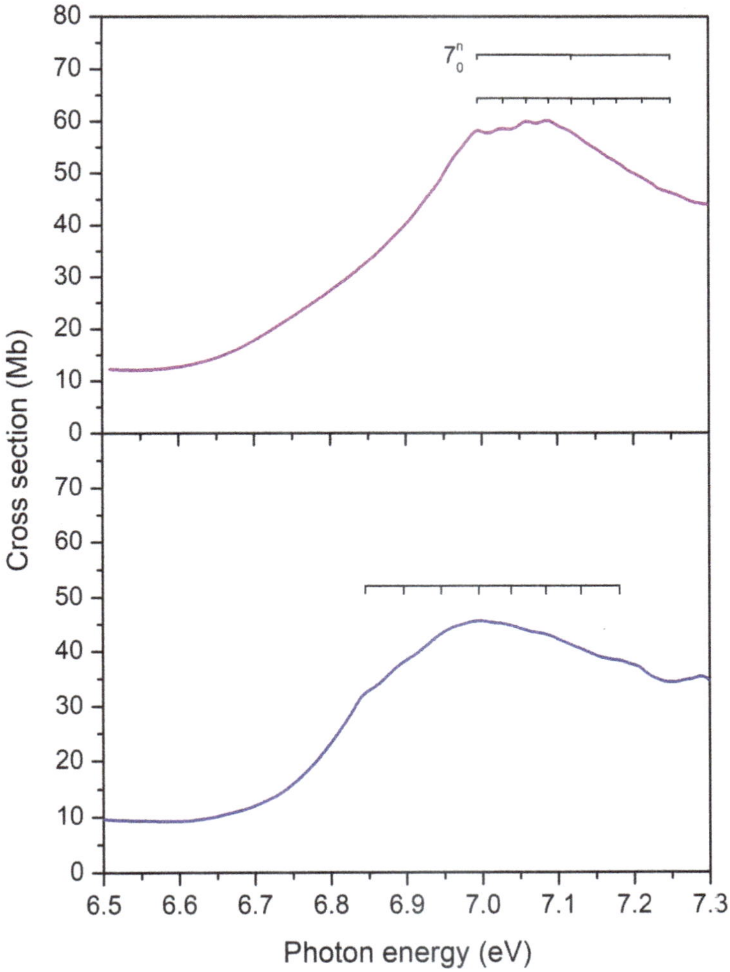

Figure 10. Vibrational progressions in the 6.5–7.3 eV energy range of the absorption Band IV of 2-bromopyrimidine (**top** panel) and 5-bromopyrimidine (**bottom** panel).

2.5. The 7.3–9.0 eV Photon Energy Range (Band V)

The most intense absorption band lies between 7.3 and 9.0 eV, for both bromopyrimidines, where several optically bright excited states are involved (see Figures 5, 6 and 11). In contrast to the four previous bands, here the position of the bromine atom greatly affects the behavior of the cross sections, which present many more discernible features in 2BrPyr, whereas in 5BrPyr the band is considerably broader. The latter also presents overall larger magnitudes and much more pronounced vibrational progressions. Indeed, due to the complexity of the spectra in this energy region, and the fact that some vibrational modes have close frequencies, our vibrational assignments become more tentative for this band.

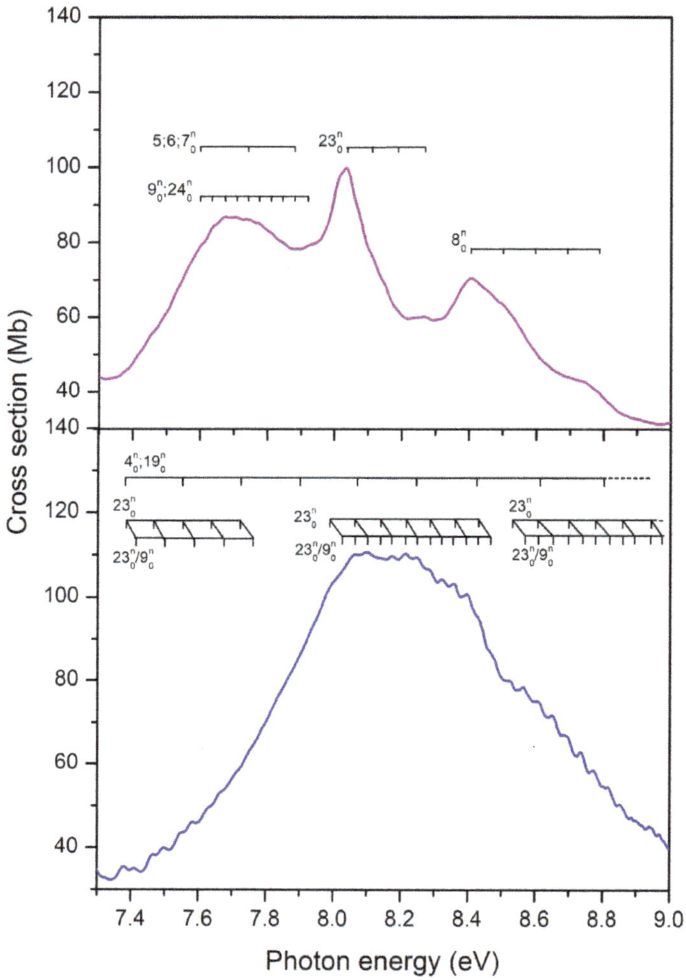

Figure 11. Vibrational progressions in the 7.2–9.0 eV energy range of the absorption Band V of 2-bromopyrimidine (**top** panel) and 5-bromopyrimidine (**bottom** panel).

In 5BrPyr, the band is centered at 8.2 eV, and is assigned to $\pi^*(a_2) \leftarrow \pi_{Br}(b_1) + \pi^*(b_1) \leftarrow \pi(a_2)$ ($f_0 = 0.32$) and $\pi^*(b_1) \leftarrow \pi_{Br}(b_1)$ ($f_0 = 0.15$) excitations, vertically located at 8.18 and 8.28 eV according to the calculations. Two vibrational progressions were identified and tentatively attributed to excitation of the C–Br stretching (υ_9) and ring stretching (υ_{23}) modes (see Table 5). Another progression with an average spacing of 0.180 eV (1452 cm^{-1}) was also identified, and could be related to either υ_4 or υ_{19} ring stretching modes.

Table 5. Proposed vibrational assignments in the 7.3–9.0 eV absorption band of 5BrPyr (Band V). ΔE (in eV) represents the energy of one extra quanta in the vibrational progression. The symbol "/" means both modes are involved; the symbol ";" means two alternative assignments.

Energy (eV)	Assignment	$\Delta E\left(v_4';v_{19}'\right)$	$\Delta E\left(v_9'\right)$	$\Delta E\left(v_{23}'\right)$
7.38	0^0_0			
7.41	9^1_0		0.03	
7.46	23^1_0			0.09
7.50	$9^2_0/23^1_0$		0.03	0.09
7.55	$23^2_0/4^1_0;19^1_0$	0.170		0.09
7.59	$9^3_0/23^2_0$		0.04	0.09
7.63	23^3_0			0.08
7.68	$9^4_0/23^3_0$		0.03	0.09
7.72	$23^4_0/4^2_0;19^2_0$	0.170		0.09
7.76 (s)	$9^5_0/23^4_0$		0.04	0.09
7.90 (s)	$4^3_0;19^3_0$	0.180		
7.99 (s)	0^0_0			
8.03	9^1_0		0.04	
8.06	$9^2_0/23^1_0/4^4_0;19^4_0$	0.160	0.04	0.07
8.10	$9^3_0/23^1_0$		0.04	0.07
8.14 (w)	$9^4_0/23^2_0$		0.04	0.08
8.17	$9^5_0/23^2_0$		0.03	0.08
8.22	$9^6_0/23^3_0$		0.04	0.08
8.25	$9^7_0/23^3_0/4^5_0;19^5_0$	0.190	0.04	0.08
8.29	$9^8_0/23^4_0$		0.05	0.07
8.32	$9^9_0/23^4_0$		0.04	0.07
8.36	$9^{10}_0/23^5_0$		0.04	0.07
8.39	$9^{11}_0/23^5_0$		0.04	0.07
8.43 (s)	$9^{12}_0/23^6_0/4^6_0;19^6_0$	0.170	0.04	0.07
8.47 (s)	$9^{13}_0/23^6_0$		0.04	0.07
8.53	0^0_0			
8.57	9^1_0		0.04	
8.60	$9^2_0/23^1_0/4^7_0;19^7_0$	0.190	0.04	0.08
8.65	$9^3_0/23^1_0$		0.05	0.08
8.70	$9^4_0/23^2_0$		0.05	0.09
8.73	$9^5_0/23^2_0$		0.04	0.09
8.77	$9^6_0/23^3_0$		0.04	0.08
8.82	$9^7_0/23^3_0/4^8_0;19^8_0$	0.210	0.05	0.08
8.85	$9^8_0/23^4_0$		0.04	0.08
8.89	$9^9_0/23^4_0$		0.04	0.07
8.94	$9^{10}_0/23^5_0$		0.05	0.09
8.97	$9^{11}_0/23^5_0$		0.03	0.08

(s) shoulder structure; (w) weak feature.

The analogous band of 2BrPyr presents a broader peak centered at 7.67 eV, followed by a sharp peak at 8.05 eV. A quick inspection of Table 1 indicates three states (vertically at 7.75, 7.87, and 8.00 eV) with large oscillator strengths, having $3p_x(b_1) \leftarrow \pi(b_1)$ (f_0 = 0.19), $\pi^*(b_1) \leftarrow \pi_{Br}(b_1)$ (f_0 = 0.22), and $\pi^*(b_1) \leftarrow \pi(a_2) + \pi^*(a_2) \leftarrow \pi_{Br}(b_1)$ (f_0 = 0.43) characters. While these states should account for the two observed features, we cannot be sure which ones contribute to which peak, in view of their close-lying energies. Two vibrational progressions were tentatively assigned to ring stretching (υ_5, υ_6, υ_7) and Br rocking (υ_{24}) or C-Br stretching (υ_9) modes. At higher energies, a ring deformation (υ_{23}) and a ring stretching (υ_8) were also tentatively assigned (see Table 6).

Table 6. Proposed vibrational assignments in the 7.3–9.0 eV absorption band of 2BrPyr (Band V). ΔE (in eV) represents the energy of one extra quanta in the vibrational progression. The symbol "/" means both modes are involved; the symbol ";" means two alternative assignments.

Energy (eV)	Assignment	$\Delta E\left(v_9';v_{24}'\right)$	$\Delta E\left(v_{23}'\right)$	$\Delta E\left(v_8'\right)$	$\Delta E\left(v_5';v_6';v_7'\right)$
7.6	0_0^0				
7.64 (s)	$9_0^1; 24_0^1$	0.040			
7.67	$9_0^2; 24_0^2$	0.035			
7.71	$9_0^3; 24_0^3$	0.034			
7.74	$9_0^4; 24_0^4/5_0^4; 6_0^4; 7_0^4$	0.034			0.143
7.78 (w)	$9_0^5; 24_0^5$	0.034			
7.81 (w)	$9_0^6; 24_0^6$	0.034			
7.84 (w)	$9_0^7; 24_0^7$	0.034			
7.88 (w)	$9_0^8; 24_0^8/5_0^4; 6_0^4; 7_0^4$	0.036			0.138
7.92 (w)	$9_0^9; 24_0^9$	0.040			
8.04	0_0^0				
8.11 (s)	23_0^1		0.076		
8.19 (s)	23_0^2		0.076		
8.30	23_0^3		0.080		
8.40	0_0^0				
8.50 (s)	8_0^1			0.095	
8.59 (s)	8_0^2			0.095	
8.70 (b)	8_0^3			0.095	
8.79 (b)	8_0^4			0.095	

(b) broad structure; (s) shoulder structure; (w) weak feature.

Clearly, the brightest electronic transitions in both of the bromopyrimidine isomers share the same character, and have similar excitation energies and oscillator strengths. The main contributions concern excitations of $\pi^*(a_2) \leftarrow \pi_{Br}(b_1)$, $\pi^*(b_1) \leftarrow \pi(a_2)$, and $\pi^*(b_1) \leftarrow \pi_{Br}(b_1)$ characters, which account for the large intensity of Band V. The remaining excitation involving these orbitals, $\pi^*(a_2) \leftarrow \pi(a_2)$, is the one responsible for Band IV, as discussed above. Recalling that the same four excitations give rise to the intense absorption peak around 6.3–7.0 eV in bromobenzene [47], we may conclude that the analogous band in both bromopyrimidines is displaced to higher energies, becomes considerably broader, and actually splits into more discernible features. In this sense, the third and most intense band in bromobenzene would correspond to the combination of Band IV and the first half of Band V in the bromopyrimidines, extending over the 6.8–8.2 eV energy range in 2BrPyr and the 6.8–8.5 eV range in 5BrPyr.

Embedded into the broad Band V, the distinct peak at 8.41 eV in 2BrPyr as well as the broad shoulder around 8.7 eV in 5BrPyr arise from $\sigma^*_{Br}(a_1) \leftarrow n_+(a_1) + 3p_y(b_2) \leftarrow n_{Br}(b_2)$ transitions. The more stable $\sigma^*_{Br}(a_1)$ orbital of 2BrPyr explains the observed redshift for this transition, similarly to what we have discussed for the Band III.

2.6. The 9.0–10.8 eV Photon Energy Range (Band VI)

The energy region between 9.0 and 10.8 eV presents no sharp peaks as the cross sections vary less abruptly, displaying instead some broader and several narrow features. Many valence and Rydberg excited states can be populated in this energy regime, which may couple among themselves and develop complicated vibrational progressions, making interpretation of the spectra especially difficult. Despite that, we hope our TDDFT and NEA calculations can provide at least qualitative results at these energies. In fact, the computed cross sections are quite comparable to the measured ones for 2BrPyr, though somewhat overestimated for 5BrPyr. Importantly, the major features of both molecules are correctly reproduced by the calculations.

A close inspection of Figure 12 shows some differences between the two spectra. Three vibrational progressions were identified for 5BrPyr and were tentatively attributed to excitation of the in-plane Br rocking (υ_{24}), a ring stretching (υ_{23}), and the C-Br stretching (υ_9) modes (see Table 7). In 2BrPyr, the vibrational assignments were more challenging, and different vibrational modes were identified (see Figure 12), as summarized in Table 8.

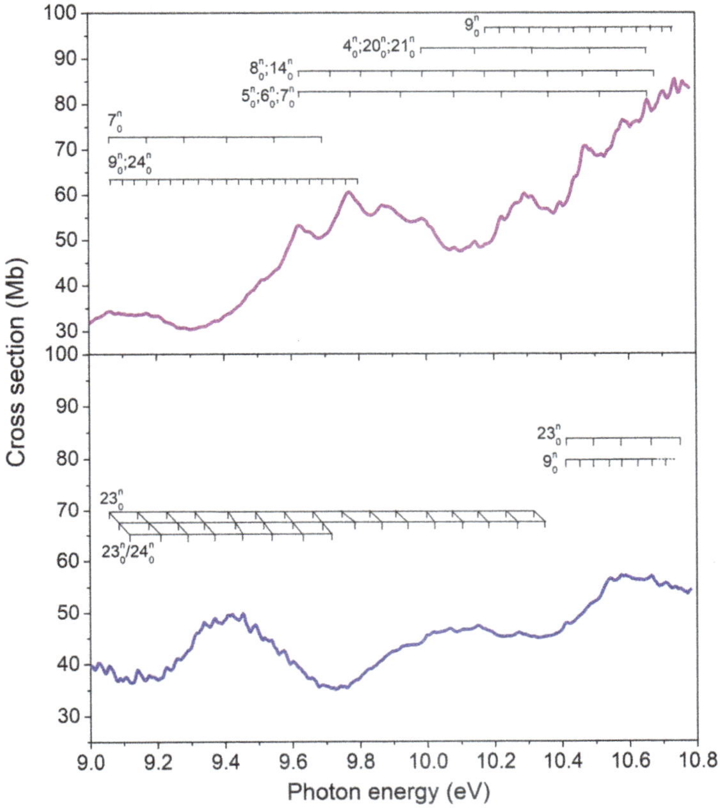

Figure 12. Vibrational progressions in the 9.0–10.8 eV energy range of the absorption Band VI of (**top** panel) and 5-bromopyrimidine (**bottom** panel).

Table 7. Proposed vibrational assignments in the 9.0−10.8 eV absorption band of 5BrPyr (Band VI). ΔE (in eV) represents the energy of one extra quanta in the vibrational progression. The symbol "/" means both modes are involved; the symbol ";" means two alternative assignments.

Energy (eV)	Assignment	$\Delta E\left(v'_9\right)$	$\Delta E\left(v'_{23}\right)$	$\Delta E\left(v'_{24}\right)$
9.06	0^0_0			
9.09	24^1_0			0.03
9.11	24^2_0			0.02
9.14	23^1_0		0.08	
9.17	$23^1_0/24^1_0$		0.08	0.03
9.21	$23^1_0/24^2_0$		0.10	0.04
9.23	23^2_0		0.09	
9.26	$23^2_0/24^1_0$		0.09	0.03
9.29 (s)	$23^2_0/24^2_0$		0.08	0.03
9.32 (s)	23^3_0		0.09	
9.34	$23^3_0/24^1_0$		0.08	0.02
9.37	$23^3_0/24^2_0$		0.08	0.03
9.41	23^4_0		0.09	
9.43	$23^4_0/24^1_0$		0.09	0.02
9.45	$23^4_0/24^2_0$		0.08	0.02
9.49	23^5_0		0.08	
9.52	$23^5_0/24^1_0$		0.09	0.03

Table 7. *Cont.*

Energy (eV)	Assignment	$\Delta E\left(v_9'\right)$	$\Delta E\left(v_{23}'\right)$	$\Delta E\left(v_{24}'\right)$
9.54	$23_0^5/24_0^2$		0.09	0.02
9.57	23_0^6		0.08	
9.60	$23_0^6/24_0^1$		0.08	0.03
9.63 (s)	$23_0^6/24_0^2$		0.09	0.03
9.66	23_0^7		0.09	
9.69	$23_0^7/24_0^1$		0.09	0.03
9.72 (w)	$23_0^7/24_0^2$		0.09	0.03
9.75	23_0^8		0.09	
9.79	$23_0^8/24_0^1$		0.10	0.04
9.83	23_0^9		0.08	
9.86 (s) (b)	$23_0^9/24_0^1$		0.07	0.03
9.90 (b)	23_0^{10}		0.07	
9.95 (b)	$23_0^{10}/24_0^1$		0.07	0.05
10.00 (w)	23_0^{11}		0.10	
10.02	$23_0^{11}/24_0^1$		0.09	0.02
10.07	23_0^{12}		0.07	
10.10	$23_0^{12}/24_0^1$		0.07	0.03
10.15	23_0^{13}		0.08	
10.18	$23_0^{13}/24_0^1$		0.08	0.03
10.24 (w)	23_0^{14}		0.09	
10.27	$23_0^{14}/24_0^1$		0.09	0.03
10.32 (w)	23_0^{15}		0.08	
10.35 (w)	$23_0^{15}/24_0^1$		0.08	0.03
10.41				
10.45 (s)	9_0^1	0.04		
10.49	$23_0^1/9_0^2$	0.04	0.08	
10.54	9_0^3	0.05		
10.57	$23_0^2/9_0^4$	0.05	0.08	
10.61 (w)	9_0^5	0.04		
10.67	$23_0^3/9_0^6$	0.06	0.09	
10.71	9_0^7	0.05		
10.75	$23_0^4/9_0^8$	0.06	0.07	

(b) broad structure; (s) shoulder structure; (w) weak feature.

Table 8. Proposed vibrational assignments in the 9.0–10.8 eV absorption band of 2BrPyr (Band VI). ΔE (in eV) represents the energy of one extra quanta in the vibrational progression. The symbol "/" means both modes are involved; the symbol ";" means two alternative assignments.

Energy (eV)	Assignment	$\Delta E\left(v_9';v_{24}'\right)$	$\Delta E\left(v_7'\right)$	$\Delta E\left(v_5';v_6';v_7'\right)$
9.07	0_0^0			
9.10 (w)	$9_0^1;24_0^1$	0.033		
9.14 (w)	$9_0^2;24_0^2$	0.040		
9.17	$9_0^3;24_0^3/7_0^1$	0.034	0.112	
9.21	$9_0^4;24_0^4$	0.036		
9.24	$9_0^5;24_0^5$	0.040		
9.28	$9_0^6;24_0^6/7_0^2$	0.042	0.110	
9.32	$9_0^7;24_0^7$	0.042		
9.36	$9_0^8;24_0^8$	0.042		
9.41	$9_0^9;24_0^9/7_0^3$	0.037	0.127	
9.45	$9_0^{10};24_0^{10}$	0.033		
9.48 (w)	$9_0^{11};24_0^{11}$	0.035		
9.51	$9_0^{12};24_0^{12}$	0.034		
9.55 (w)	$9_0^{13};24_0^{13}$	0.035		
9.58 (s)	$9_0^{14};24_0^{14}$	0.035		
9.61	$9_0^{15};24_0^{15}$	0.035		
9.65	$9_0^{16};24_0^{16}$	0.035		
9.69 (s)	$9_0^{17};24_0^{17}$	0.035		
9.73 (s)	$9_0^{18};24_0^{18}$	0.035		
9.77	$5_0^1;6_0^1;7_0^1$			0.150
9.93 (s)	$5_0^2;6_0^2;7_0^2$			0.150
10.08	$5_0^3;6_0^3;7_0^3$			0.158
10.23 (s)	$5_0^4;6_0^4;7_0^4$			0.142
10.36	$5_0^5;6_0^5;7_0^5$			0.137
10.52 (b)	$5_0^6;6_0^6;7_0^6$			0.154
10.65	$5_0^7;6_0^7;7_0^7$			0.139

(b) broad structure; (s) shoulder structure; (w) weak feature.

In comparison to bromobenzene [47], the cross sections of both bromopyrimidines are smaller in this energy range, besides having less pronounced features. Both observations could be related to the suppression of Rydberg states, discussed in the next section.

2.7. Rydberg States

Rydberg states are characterized by excitation into diffuse (Rydberg) orbitals, whose energies follow the well-known Rydberg formula, $E_n = \mathrm{IE} - [13.606/(n - \delta)^2]$ [eV], where the excitation energy E_n is estimated from the ionization energy IE, the principal quantum number n, and the quantum defect δ. For each angular momentum (s, p, d, etc.), and for each orbital from which the electron is excited, the Rydberg states form a series converging to the ionization energy associated with this orbital. Such states tend to appear in VUV spectra as sharper peaks than valence excitations, though with varying intensities, and their precise assignment is usually performed with the aid of the Rydberg formula.

Our measured spectra do not display pronounced signatures of Rydberg states, for either 2BrPyr and 5BrPyr, in marked contrast to what was observed for bromobenzene [47]. That being said, there is some evidence that weak structures in the VUV spectra would arise from the $3s(a_1) \leftarrow \pi(b_1)$ Rydberg state. In 2BrPyr, the four small peaks observed in Band IV (starting at 6.99 eV) could actually correspond to a vibrational progression of this Rydberg state. The theoretical calculations further support this assignment, as this state was found at 7.00 eV, with an oscillator strength of 0.045, which, albeit small, is probably sufficiently large to leave a weak mark on the spectrum. The associated quantum defect would be 0.84, considering the experimental ionization energy of 9.93 eV for the $\pi(b_1)$ orbital [36]. In 5BrPyr, the analogous $3s(a_1) \leftarrow \pi(b_1)$ Rydberg state could account for the weak shoulder observed at 6.84 eV (a quantum defect of 0.90, given an ionization energy of 9.93 eV [35]). Again, this is supported by the calculations, which show an excitation energy of 6.89 eV, and an oscillator strength of 0.016. In particular, the lower oscillator strength in 5BrPyr is also consistent with the weaker feature on the spectrum.

According to our calculations, some of the 3p Rydberg states presented strong mixing with valence states, contributing to Band V. In contrast, purer 3p Rydberg states that show up in the same energy range (7.5–8.0 eV) have reduced oscillator strengths, thus leaving no distinct signatures in the spectrum. Finally, the 3d Rydberg states would start emerging in the 8.0–8.5 eV range, most of them with low intensities.

Towards higher energies, a multitude of Rydberg series are expected, though with progressively smaller intensities. In fact, the theoretical calculations (see Tables 1 and 2) indicate that Rydberg transitions become more and more prevailing at higher energies, though with small oscillator strengths, particularly for 2BrPyr. For instance, the $4s(a_1) \leftarrow \pi(b_1)$ state of this isomer would appear at 8.48 eV according to our calculations, with a very low oscillator strength (0.001). The same applies for the other angular momentum and for excitations from other high-lying occupied orbitals. Indeed, at higher energies we have observed no sharp peaks that could be clearly associated with Rydberg states. Rather, the absorption spectra present a large number of closely-lying weak features. Thus, any attempt to locate the Rydberg series based on the Rydberg formula and the ionization energies would be very tentative and not conclusive. To a great extent, that would be a less pertinent task, considering the overall minor contribution of the higher-lying Rydberg states to the photoabsorption cross section of bromopyrimidines.

The scenario is quite distinct for bromobenzene, where much more pronounced structures have been observed [47]. With the support of high-resolution PES data, this allowed several Rydberg states to be clearly assigned. We notice that a similar behavior can be recognized when one compares the VUV spectra of benzene and pyrimidine [7,53]. Such observations suggest that going from benzene to pyrimidine, as well as from bromobenzene to bromopyrimidines, promotes a major suppression of the Rydberg series.

3. Methods

3.1. Experimental Methods

The high-resolution VUV photoabsorption measurements were performed at the AU-UV beamline of the ASTRID2 synchrotron facility at Aarhus University in Denmark. A detailed explanation of the experimental apparatus is given by Palmer et al. [54] and an in-depth description of procedures for measurement and calibration can be found in Duflot et al. [55]. In short, the set-up consists of a gas cell placed on the output of the monochromator where a MgF_2 window was mounted to separate it from the ultra-high vacuum (UHV) of the beam line. The synchrotron radiation passes through a static gas sample and the transmitted light intensity is detected by a photomultiplier tube (PMT). The absolute pressure of the effusive molecular gas is measured with a capacitance manometer (Chell CDG100D). The wavelength is selected using a toroidal dispersion grating with 2000 lines/mm providing a resolution of 0.075 nm, corresponding to 3 meV at the midpoint of the energy range studied. In order to avoid any absorption of O_2 from the air in measurements below 200 nm (energies above 6.20 eV), the small gap between the PMT and the MgF_2 gas cell exit window is evacuated using a scroll pump. For higher wavelength measurements, air is admitted into the gap at atmospheric pressure to allow O_2 to absorb any higher orders of light produced by the monochromator. ASTRID2 operates in "top-up" mode, keeping the stored electron beam current quasi-constant. Absolute photoabsorption cross sections (σ) were obtained at room temperature (~25 °C) by using the Beer−Lambert attenuation law (1),

$$I_t = I_0 \exp(-N\sigma l), \tag{1}$$

where I_t is the light intensity transmitted through the gas sample, I_0 is that transmitted through the evacuated cell, N is the molecular density, and l is the absorption path length (15.5 cm). The range of pressures used for the measurements of 5BrPyr was 0.01 to 0.54 mbar, while for 2BrPyr, which has a lower vapor pressure, the pressure range was 0.02 to 0.11 mbar. The accuracy of the cross section is estimated to be ~10%, which is ensured by recording the VUV spectra in small (5 or 10 nm) sections to allow optimization of pressure according to the local cross-sections, with at least 10 points overlap between the adjoining ranges [55,56].

The 2-bromopyrimidine sample was purchased from FluoroChem with a stated purity of ≥97%, while 5-bromopyrimidine was purchased from Sigma Aldrich with a stated purity of ≥97%. The samples were used with no further purification and degassed before measurement through repeated freeze−pump−thaw cycles.

3.2. Theoretical Methods

The photoabsorption cross sections were computed with the NEA [57,58] in combination with TDDFT calculations. The latter were performed with the Gaussian 16 software [59], while the managing of the NEA calculations were performed as implemented in the Newton-X package [60,61].

We employed the CAM-B3LYP functional, which provided excitation energies comparable to those observed in the measured VUV spectra. Other range separated functionals were explored (LC-ωPBE, ωB97X, and LCBLYP), however overall, they did not perform as well as CAM-B3LYP for the cases of 2BrPyr and 5BrPyr. We used the aug-cc-pVDZ basis set supplemented with a set of 2s2p2d diffuse functions, centered at the carbon atom closest to the bromine atom. For each angular momentum, the exponents were generated by successive divisions of the previous one by four (starting from the most diffuse functions of the aug-cc-pVDZ set). The CAM-B3LYP/aug-cc-pVDZ+2s2p2d level of theory was employed for the geometry optimization, normal mode analysis, and calculation of anharmonic frequencies (see below).

In the NEA, the configurations are sampled according to the Wigner distribution (Gaussians) of a set of independent harmonic potentials, which represent the normal modes at the ground state optimized geometry. The corresponding Wigner distribution at 298 K (experimental condition) was considered for the sampling. Anharmonic corrections for the frequencies were accounted for with the importance sampling algorithm [62], though such effects had a marginal impact on the computed cross sections.

Excitation energies $\Delta E0n(q_i)$ and oscillator strengths $f_{0n}(q_i)$ of the n-th excited state are computed for each sampled configuration q_i. Then, the photoabsorption cross section is obtained as (in atomic units):

$$\sigma(E) = \frac{1}{E}\sum_{n}^{N_s}\frac{1}{N_p(n)}\sum_{i}^{N_p}\Delta E_{0n}(q_i)f_{0n}(q_i)g[E-\Delta E_{0n}(q_i),\delta], \quad (2)$$

where the phenomenological line profile g is represented by a normalized Gaussian function centered at each excitation energy $\Delta E_{0n}(q_i)$. Here we have employed a linewidth of $\delta = 0.1$ eV, which is large enough to damp the artificial undulations of the cross-section curve, yet small enough to keep its actual overall shape. We have used a decreasing number of configurations as we moved higher in energy. For both 2BrPyr and 5BrPyr, the number of sampled configurations was $N_p = 7500$ for the first eight singlet excited states (n = 1–8), $N_p = 5500$ for n = 9–16, $N_p = 3500$ for n = 17–40, $N_p = 2500$ for n = 41–60, $N_p = 1500$ for n = 61–80, and $N_p = 1000$ for n = 81–160. For 2BrPyr, we used an additional set of $N_p = 500$ points for n = 161–170. This represents a total of 701,000 computed excitation energies and oscillator strengths.

While the NEA is able to describe the overall shapes and intensities of the absorption bands, it cannot describe vibrational progressions, which would require treating the nuclei quantum-mechanically. As discussed in Section 2.1, Band I of both bromopyrimidines presents very rich vibrational structures, whose interpretation becomes impractical without support from theory. Given the limitations of the NEA calculations, we have also employed a second theoretical approach, based on FCHT analysis, as performed with the time-independent procedure [63–65]. The excited-state potential energy surface is described with the adiabatic Hessian model, where force constants and vibrational frequencies are computed at the equilibrium geometry of the excited state. Ground state anharmonic frequencies were computed with the generalized second-order vibrational perturbation method [64,66], and from those we also obtained excited state anharmonic frequencies by employing a mode-specific scaling factor built from the Duschinsky matrix [67]. Temperature effects are accounted for by considering excitations from vibrationally excited levels with populations larger than 2%, for a temperature of 298 K, corresponding to the present experimental condition. All the calculations required for the FCHT analysis were performed with the Gaussian 16 software [59].

4. Conclusions

Absolute photoabsorption cross sections obtained with high resolution VUV synchrotron radiation in the 3.7–10.8 eV energy range are reported for 2BrPyr and 5BrPyr. Theoretical cross sections are also presented, as computed within the NEA in combination with TDDFT calculations. The good agreement between measurements and calculations gives strong support for the interpretation and discussion of both obtained VUV spectra.

The present study, together with a previous report on the photoabsorption of fluoranil [50] showed that the NEA can be successfully employed to probe photoabsorption for considerably larger energies, closer to the ionization threshold. While the improved accuracy of the NEA with respect to the simpler vertical approximation is well known for the low energy regime, it remained to be demonstrated in practice for higher energies. However, the overall computational cost increases sharply when the NEA is employed for higher energies, requiring the evaluation of dozens or hundreds of excited states for an extensive set of geometries.

In conclusion, the lowest lying band (3.7–4.6 eV) presents the lowest cross sections, and it was assigned to the first singlet excited state, a $\pi^*(a_2) \leftarrow n_-(b_2)$ transition. With the support of FCHT calculations, we placed the band origin of 2BrPyr at 3.87 eV and of 5BrPyr at 3.79 eV, whereas a number of low intensity hot bands and main vibrational progressions could be assigned to C-Br rocking/wagging modes and to totally symmetric ring stretching modes found with frequencies in the 700~1200 cm^{-1} range, respectively. The Band II (4.6–5.4 eV) was attributed to a $\pi^*(a_2) \leftarrow \pi(b_1)$ transition, which is followed by a third band (5.4–6.5 eV) and assigned to a $\pi^*(b_1) \leftarrow \pi(b_1)$ transition, the latter displaying no vibrational structures. For the three lowest-lying bands, the cross sections of both 2BrPyr and 5BrPyr are rather similar, the first two being slightly red-shifted in 5BrPyr, and the third being somewhat more intense in 2BrPyr. Still in Band III, a weak shoulder was observed for 5BrPyr, and explained by a larger effect of the bromine atom position on the $\sigma^*_{Br}(a_1) \leftarrow \pi_r(b_1)$ transition.

Going up in energy we have the most intense absorption bands, IV (6.5–7.3 eV) and V (7.3–9.0 eV), which should be analyzed together. As a whole, their large intensities stem from the four possible excitations from the $\pi_{Br}(b_1)$ and $\pi(a_2)$ occupied orbitals, to the $\pi^*(b_1)$ and $\pi^*(a_2)$ unoccupied orbitals. The analogous excitations account for the main absorption bands of the related bromobenzene, pyrimidine, and benzene molecules. Compared with bromobenzene, the band of both bromopyrimidines is displaced to higher energies and broadens significantly, such that some individual $\pi^* \leftarrow \pi$ transitions become apparent. Even so, the isomer effect is more substantial in this energy region, affecting the behavior of the individual $\pi^* \leftarrow \pi$ excitations and of other, less intense transitions. Finally, Band VI (9.0–10.8 eV) should involve many more valence and Rydberg excited states, but despite this, broader features could still be interpreted. In this regime, both spectra showed distinct vibrational progressions.

An extensive comparison between the VUV spectra of bromopyrimidines and the available data for bromobenzene revealed a series of similarities and differences. Namely, for the bromopyrimidines we observed a new band (Band I), broader (Band II) or a complete suppression of vibrational structures (Band III), broadened and blue shifted features (Bands IV and V), and a loss of intensity (Bands II and VI). An additional key distinction concerns the Rydberg states, much less pronounced in both bromopyrimidines than in bromobenzene. A similar trend can be noticed when comparing the spectra of pyrimidine and benzene.

This work is the most comprehensive VUV photoabsorption study yet reported for 2BrPyr and 5BrPyr. Considering that bromopyrimidines are known for their potential as radiosensitizers, the detailed picture we have provided on their excited states and VUV photoabsorption spectra could bring important insights on the production of reactive radicals and on radiation damage in general.

Supplementary Materials: The following are available online at https://www.mdpi.com/article/10.3390/ijms22126460/s1.

Author Contributions: Measurements, F.F.d.S. and N.C.J.; funding acquisition, F.F.d.S.; analysis of results, M.M., A.I.L., F.K., J.P.-d.-S., R.R., J.A.; writing—original draft preparation, M.M., A.I.L., F.K.; theoretical calculations, F.K.; writing—review, F.F.d.S., N.C.J., S.V.H. All authors have read and agreed to the published version of the manuscript.

Funding: J.P.-d.-S. acknowledges the Portuguese National Funding Agency FCT-MCTES through PhD grant PD/BD/142768/2018, together with R.R., J.A., M.M. and F.F.d.S. through the researcher grant PTDC/FIS-AQM/31215/2017 and AIL through the researcher grant PTDC/FIS-AQM/31281/2017. This work was also supported by Radiation Biology and Biophysics Doctoral Training Programme (RaBBiT, PD/00193/2012); UIDB/04378/2020 (UCIBIO); and UIDB/00068/2020 (CEFITEC). The research leading to this result has been supported by the project CALIPSOplus under the Grant Agreement 730872 from the EU Framework Programme for Research and Innovation HORIZON 2020.

Data Availability Statement: The data presented in this study are available on request from the corresponding author.

Conflicts of Interest: The authors declare no conflict of interest.

References

1. Alizadeh, E.; Orlando, T.M.; Sanche, L. Biomolecular Damage Induced by Ionizing Radiation: The Direct and Indirect Effects of Low-Energy Electrons on DNA. *Annu. Rev. Phys. Chem.* **2015**, *66*, 379–398. [CrossRef]
2. Boudaïffa, B.; Cloutier, P.; Hunting, D.; Huels, M.A.; Sanche, L. Resonant formation of DNA strand breaks by low-energy (3 to 20 eV) electrons. *Science* **2000**, *287*, 1658–1660. [CrossRef] [PubMed]
3. Huels, M.A.; Boudaïffa, B.; Cloutier, P.; Hunting, D.; Sanche, L. Single, double, and multiple double strand breaks induced in DNA by 3-100 eV electrons. *J. Am. Chem. Soc.* **2003**, *125*, 4467–4477. [CrossRef] [PubMed]
4. Wardman, P. Chemical Radiosensitizers for Use in Radiotherapy. *Clin. Oncol.* **2007**, *19*, 397–417. [CrossRef] [PubMed]
5. Wang, H.; Mu, X.; He, H.; Zhang, X.D. Cancer Radiosensitizers. *Trends Pharmacol. Sci.* **2018**, *39*, 24–48. [CrossRef]
6. Baccarelli, I.; Bald, I.; Gianturco, F.A.; Illenberger, E.; Kopyra, J. Electron-induced damage of DNA and its components: Experiments and theoretical models. *Phys. Rep.* **2011**, *508*, 1–44. [CrossRef]
7. Ferreira da Silva, F.; Almeida, D.; Martins, G.; Milosavljević, A.R.; Marinković, B.P.; Hoffmann, S.V.; Mason, N.J.; Nunes, Y.; Garcia, G.; Limão-Vieira, P. The electronic states of pyrimidine studied by VUV photoabsorption and electron energy-loss spectroscopy. *Phys. Chem. Chem. Phys.* **2010**, *12*, 6717–6731. [CrossRef]
8. Ferreira da Silva, F.; Matias, C.; Almeida, D.; García, G.; Ingólfsson, O.; Flosadóttir, H.D.; Ómarsson, B.; Ptasinska. S.; Puschnigg, B.; Scheier, P.; et al. NCO-, a key fragment upon dissociative electron attachment and electron transfer to pyrimidine bases: Site selectivity for a slow decay process. *J. Am. Soc. Mass Spectrom.* **2013**, *24*, 1787–1797. [CrossRef] [PubMed]
9. Mendes, M.; Pamplona, B.; Kumar, S.; Ferreira da Silva, F.; Aguilar, A.; García, G.; Bacchus-Montabonel, M.-C.; Limão-Vieira, P. Ion-pair formation in neutral potassium-neutral pyrimidine collisions: Electron transfer experiments. *Front. Chem.* **2019**, *7*, 264. [CrossRef]
10. Almeida, D.; Ferreira Da Silva, F.; García, G.; Limão-Vieira, P. Selective bond cleavage in potassium collisions with pyrimidine bases of DNA. *Phys. Rev. Lett.* **2013**, *110*, 1–5. [CrossRef]
11. Itälä, E.; Granroth, S.; Ha, D.T.; Kooser, K.; Levola, H.; Rachlew, E.; Tanzer, K.; Kukk, E. Fragmentation of imidazole, pyrimidine and purine induced by core ionization: Significance of small-scale chemical environment. *J. Photochem. Photobiol. A Chem.* **2018**, *356*, 283–289. [CrossRef]
12. Pandey, R.; Ryszka, M.; da Fonseca Cunha, T.; Lalande, M.; Dampc, M.; Limão-Vieira, P.; Mason, N.J.; Poully, J.C.; Eden, S. Threshold behavior in metastable dissociation of multi-photon ionized thymine and uracil. *Chem. Phys. Lett.* **2017**, *684*, 233–238. [CrossRef]
13. Knight, A.E.W.; Lawburgh, C.M.; Parmenter, C.S. n,π* Fluorescence from selected vibronic levels of pyrimidine vapor: Franck-condon factors excited state anharmonic coupling. *J. Chem. Phys.* **1975**, *63*, 4336. [CrossRef]
14. Luo, Y.; Ågren, H.; Knuts, S.; Jørgensen, P. The two-photon spectrum of pyrimidine. Role of vibronic coupling. *Chem. Phys. Lett.* **1993**, *213*, 357–362. [CrossRef]
15. Plekan, O.; Coreno, M.; Feyer, V.; Moise, A.; Richter, R.; de Simone, M.; Sankari, R.; Prince, K.C. Electronic state resolved PEPICO spectroscopy of pyrimidine. *Phys. Scr.* **2008**, *78*, 058105. [CrossRef]
16. Opara-Kubinska, Z.; Lorkiewicz, Z.; Szybalski, W. Genetic transformation studies. II. Radiation sensitivity of halogen labeled DNA. *Biochem. Biophys. Res. Commun.* **1961**, *4*, 288–291. [CrossRef]
17. Djordjevic, B.; Szybalski, W. Genetics of Human Cell Lines. III. Incorporation of 5-Bromo- and 5-Iododeoxyuridine into the Deoxyribonucleic Acid of Human Cells and Its Effect on Radiation Sensitivity. *J. Exp. Med.* **1960**, *112*, 509–531. [CrossRef] [PubMed]
18. Chomicz, L.; Zdrowowicz, M.; Kasprzykowski, F.; Rak, J.; Buonaugurio, A.; Wang, Y.; Bowen, K.H. How to find out whether a 5-substituted uracil could be a potential DNA radiosensitizer. *J. Phys. Chem. Lett.* **2013**, *4*, 2853–2857. [CrossRef]
19. Ameixa, J.; Arthur-Baidoo, E.; Meißner, R.; Makurat, S.; Kozak, W.; Butowska, K.; Ferreira Da Silva, F.; Rak, J.; Denifl, S. Low-energy electron-induced decomposition of 5-trifluoromethanesulfonyl-uracil: A potential radiosensitizer. *J. Chem. Phys.* **2018**, *149*, 164307. [CrossRef] [PubMed]
20. Dewey, W.C.; Humphrey, R.M. Increase in Radiosensitivity to Ionizing Radiation Related to Replacement of Thymidine in Mammalian Cells with 5-Bromodeoxyuridine. *Radiat. Res.* **1965**, *26*, 538–553. [CrossRef]
21. O'Donoghue, J.A.; Wheldon, T.E. Targeted radiotherapy using Auger electron emitters. *Phys. Med. Biol.* **1996**, *41*, 1973–1992. [CrossRef]
22. Bloomer, W.; Adelstein, S.J. 5-125I-iododeoxyuridine as prototype for radionuclide therapy with Auger emitters *Nature* **1977**, *265*, 620–621. [CrossRef]
23. Wang, S.; Zhao, P.; Zhang, C.; Bu, Y. Mechanisms Responsible for High Energy Radiation Induced Damage to Single-Stranded DNA Modified by Radiosensitizing 5-Halogenated Deoxyuridines. *J. Phys. Chem. B* **2016**, *120*, 2649–2657. [CrossRef]
24. Berezin, V.I.; Berezin, V.V. Characteristic Vibrations of 5-Monosubstituted Pyrimidines. *Chem. Heterocycl. Compd.* **1988**, *24*, 300–303. [CrossRef]
25. Berezin, V.I.; Berezin, V.V. Characteristic vibrations of 2-monosubstituted pyrimidines. *Chem. Heterocycl. Compd.* **1984**, *20*, 102–106. [CrossRef]
26. Ikari, Y.; Sakamoto, H.; Nakama, S.; Nibu, Y.; Shimada, H.; Shimada, R. Normal Vibrations of 5-Chloro-, 5-Bromo-, 5-Methyl-, and 2-Methylpyrimidines. *Chem. Soc. Jpn.* **1990**, *63*, 2891–2898. [CrossRef]
27. Joshi, A.; Tonannavar, J.; Ayachit, N.H.; Shashidhar, M.A.; Rao, K.S. Infra-red and electronic absorption spectra of 2-bromopyrimidine. Spectrochim. *Acta Part A Mol. Spectrosc.* **1987**, *43*, 1017–1021. [CrossRef]
28. Nakama, S.; Shimada, H.; Shimada, R. Polarized Raman and Infrared Spectra of 2-Chloro- and 2-Brimipyrimidines. *Chem. Soc. Jpn.* **1984**, *57*, 2584–2590. [CrossRef]
29. Bolognesi, P.; Mattioli, G.; O'Keeffe, P.; Feyer, V.; Plekan, O.; Ovcharenko, Y.; Prince, K.C.; Coreno, M.; Bonapasta, A.A.; Avaldi, L. Investigation of halogenated pyrimidines by x-ray photoemission spectroscopy and theoretical DFT methods. *J. Phys. Chem. A* **2009**, *113*, 13593–13600. [CrossRef] [PubMed]

30. Bolognesi, P.; O'Keeffe, P.; Ovcharenko, Y.; Coreno, M.; Avaldi, L.; Feyer, V.; Plekan, O.; Prince, K.C.; Zhang, W.; Carravetta, V. Pyrimidine and halogenated pyrimidines near edge x-ray absorption fine structure spectra at C and N K-edges: Experiment and theory. *J. Chem. Phys.* **2010**, *133*, 034302. [CrossRef]
31. Bolognesi, P.; Kettunen, A.; Cartoni, A.; Richter, R.; Tosic, S.; Maclot, S.; Rousseau, P.; Delaunay, R.; Domaracka, A.; Avaldi, L. Selectivity in the photofragmentation of halo-pyrimidines. *J. Phys. Conf. Ser.* **2015**, *635*, 2–3. [CrossRef]
32. Bolognesi, P.; Kettunen, J.A.; Cartoni, A.; Richter, R.; Tosic, S.; Maclot, S.; Rousseau, P.; Delaunay, R.; Avaldi, L. Site- and state-selected photofragmentation of 2Br-pyrimidine. *Phys. Chem. Chem. Phys.* **2015**, *17*, 24063–24069. [CrossRef]
33. Storchi, L.; Tarantelli, F.; Veronesi, S.; Bolognesi, P.; Fainelli, E.; Avaldi, L. The Auger spectroscopy of pyrimidine and halogen-substituted pyrimidines. *J. Chem. Phys.* **2008**, *129*. [CrossRef]
34. Castrovilli, M.C.; Bolognesi, P.; Cartoni, A.; Catone, D.; O'Keeffe, P.; Casavola, A.R.; Turchini, S.; Zema, N.; Avaldi, L. Photofragmentation of halogenated pyrimidine molecules in the VUV range. *J. Am. Soc. Mass Spectrom.* **2014**, *25*, 351–367. [CrossRef]
35. O'Keeffe, P.; Bolognesi, P.; Casavola, A.R.; Catone, D.; Zema, N.; Turchini, S.; Avaldi, L. An experimental and computational study of the valence photoelectron spectra of halogenated pyrimidines. *Mol. Phys.* **2009**, *107*, 2025–2037. [CrossRef]
36. Śmiałek, M.A.; Szymańska, E.; MacDonald, M.; Zuin, L.; Mason, N.J. Photoelectron spectroscopy of brominated derivative of pyrimidine: 2-bromopyrimidine. *Eur. Phys. J. Spec. Top.* **2013**, *222*, 2361–2366. [CrossRef]
37. Modelli, A.; Bolognesi, P.; Avaldi, L. Temporary anion states of pyrimidine and halopyrimidines. *J. Phys. Chem. A* **2011**, *115*, 10775–10782. [CrossRef]
38. Barbosa, A.S.; Bettega, M.H.F. Shape resonances in low-energy-electron collisions with halopyrimidines. *J. Chem. Phys.* **2013**, *139*, 214301. [CrossRef]
39. Cheng, H.Y.; Chen, Y.C.; Lin, C.J.; Liu, W.C.; Hsieh, S.H. Temporary anion states of radiosensitive halopyrimidines: Shape and core-excited resonances. *Comput. Theor. Chem.* **2016**, *1075*, 18–29. [CrossRef]
40. Abouaf, R.; Pommier, J.; Dunet, H. Electronic and vibrational excitation in gas phase thymine and 5-bromouracil by electron impact. *Chem. Phys. Lett.* **2003**, *381*, 486–494. [CrossRef]
41. Kobyłecka, M.; Migani, A.; Asturiol, D.; Rak, J.; Blancafort, L. Benign Decay vs. Photolysis in the Photophysics and Photochemistry of 5-Bromouracil. A Computational Study. *J. Phys. Chem. A* **2009**, *113*, 5489–5495. [CrossRef]
42. Abouaf, R.; Pommier, J.; Dunet, H. Negative ions in thymine and 5-bromouracil produced by low energy electrons. *Int. J. Mass Spectrom.* **2003**, *226*, 397–403. [CrossRef]
43. Abdoul-Carmine, H.; Huels, M.A.; Brüning, F.; Illenberger, E.; Sanche, L. Dissociative electron attachment to gas-phase 5-bromouracil. *J. Chem. Phys.* **2000**, *113*, 2517. [CrossRef]
44. Scheer, A.M.; Aflatooni, K.; Gallup, G.A.; Burrow, P.D. Bond Breaking and Temporary Anion States in Uracil and Halouracils: Implications for the DNA Bases. *Phys. Rev. Lett.* **2004**, *92*, 3–6. [CrossRef] [PubMed]
45. Kossoski, F.; Varella, M.T.D.N. Negative ion states of 5-bromouracil and 5-iodouracil. *Phys. Chem. Chem. Phys.* **2015**, *17*, 17271–17278. [CrossRef] [PubMed]
46. Martin, R.L. Natural transition orbitals. *J. Chem. Phys.* **2003**, *118*, 4775–4777. [CrossRef]
47. Palmer, M.H.; Ridley, T.; Hoffmann, S.V.; Jones, N.C.; Coreno, M.; De Simone, M.; Grazioli, C.; Zhang, T.; Biczysko, M.; Baiardi, A.; et al. Interpretation of the photoelectron, ultraviolet, and vacuum ultraviolet photoabsorption spectra of bromobenzene by ab initio configuration interaction and DFT computations. *J. Chem. Phys.* **2015**, *143*, 164303. [CrossRef]
48. Ferrer, F.J.A.; Cerezo, J.; Stendardo, E.; Improta, R.; Santoro, F. Insights for an accurate comparison of computational data to experimental absorption and emission spectra: Beyond the vertical transition approximation. *J. Chem. Theory Comput.* **2013**, *9*, 2072–2082. [CrossRef]
49. Bai, S.; Mansour, R.; Stojanović, L.; Toldo, J.M.; Barbatti, M. On the origin of the shift between vertical excitation and band maximum in molecular photoabsorption. *J. Mol. Model.* **2020**, *26*, 107. [CrossRef]
50. Pereira-Da-Silva, J.; Mendes, M.; Kossoski, F.; Lozano, A.I.; Rodrigues, R.; Jones, N.C.; Hoffmann, S.V.; Ferreira da Silva, F. Perfluoro effect on the electronic excited states of para-benzoquinone revealed by experiment and theory. *Phys. Chem. Chem. Phys.* **2021**, *23*, 2141–2153. [CrossRef]
51. Tang, B.; Zhu, R.; Tang, Y.; Ji, L.; Zhang, B. Photodissociation of bromobenzene at 267 and 234 nm: Experimental and theoretical investigation of the photodissociation mechanism. *Chem. Phys. Lett.* **2003**, *381*, 617–622. [CrossRef]
52. Chen, S.F.; Liu, F.Y.; Liu, Y.J. An ab initio investigation of the mechanisms of photodissociation in bromobenzene and iodobenzene. *J. Chem. Phys.* **2009**, *131*, 124304. [CrossRef] [PubMed]
53. Dawes, A.; Pascual, N.; Hoffmann, S.V.; Jones, N.C.; Mason, N.J. Vacuum ultraviolet photoabsorption spectroscopy of crystalline and amorphous benzene. *Phys. Chem. Chem. Phys.* **2017**, *19*, 27544–27555. [CrossRef] [PubMed]
54. Palmer, M.H.; Ridley, T.; Hoffmann, S.V.; Jones, N.C.; Coreno, M.; De Simone, M.; Grazioli, C.; Biczysko, M.; Baiardi, A.; Limão-Vieira, P. Interpretation of the vacuum ultraviolet photoabsorption spectrum of iodobenzene by ab initio computations. *J. Chem. Phys.* **2015**, *142*, 134302. [CrossRef] [PubMed]
55. Duflot, D.; Hoffmann, S.V.; Jones, N.C.; Limão-Vieira, P. Synchrotron Radiation UV-VUV Photoabsorption of Gas Phase Molecules. In *Radiation in Bioanalysis: Spectroscopic Techniques and Theoretical Methods, BIOANALYSIS, Volume 8*; Pereira, A.S., Tavares, P., Limão-Vieira, P., Eds.; Springer International Publishing: New York, NY, USA, 2019; pp. 43–81, ISBN 978-3-030-28247-9.
56. Eden, S.; Limão-Vieira, P.; Hoffmann, S.V.; Mason, N.J. VUV photoabsorption in CF3X (X = Cl, Br, I) fluoro-alkanes. *Chem. Phys.* **2006**, *323*, 313–333. [CrossRef]

57. Crespo-Otero, R.; Barbatti, M. Spectrum simulation and decomposition with nuclear ensemble: Formal derivation and application to benzene, furan and 2-phenylfuran. *Theor. Chem. Acc.* **2012**, *131*, 1–14. [CrossRef]
58. Bergsma, J.P.; Berens, P.H.; Wilson, K.R.; Fredkin, D.R.; Heller, E.J. Electronic spectra from molecular dynamics: A simple approach. *J. Phys. Chem.* **1984**, *88*, 612–619. [CrossRef]
59. Frisch, M.J.; Trucks, G.W.; Schlegel, H.B.; Scuseria, G.E.; Robb, M.A.; Cheeseman, J.R.; Scalmani, G.; Barone, V.; Petersson, G.A.; Nakatsuji, H.; et al. Gaussian 16, Revision C.01 2016. Available online: https://gaussian.com/citation/ (accessed on 21 May 2021).
60. Barbatti, M.; Granucci, G.; Ruckenbauer, M.; Plasser, F.; Crespo-Otero, R.; Pittner, J.; Persico, M.; Lischka, H. NEWTON-X: A package for Newtonian dynamics close to the crossing seam. Version 2 2016. Available online: http://www.newtonx.org, (accessed on 1 February 2021).
61. Barbatti, M.; Ruckenbauer, M.; Plasser, F.; Pittner, J.; Granucci, G.; Persico, M.; Lischka, H. Newton-X: A surface-hopping program for nonadiabatic molecular dynamics. Wiley Interdiscip. *Rev. Comput. Mol. Sci.* **2014**, *4*, 26–33. [CrossRef]
62. Kossoski, F.; Barbatti, M. Nuclear Ensemble Approach with Importance Sampling. *J. Chem. Theory Comput.* **2018**, *14*, 3173–3183. [CrossRef]
63. Barone, V. Anharmonic vibrational properties by a fully automated second-order perturbative approach. *J. Chem. Phys.* **2005**, *122*, 014108. [CrossRef]
64. Bloino, J.; Biczysko, M.; Santoro, F.; Barone, V. General approach to compute vibrationally resolved one-photon electronic spectra. *J. Chem. Theory Comput.* **2010**, *6*, 1256–1274. [CrossRef]
65. Santoro, F.; Lami, A.; Improta, R.; Bloino, J.; Barone, V. Effective method for the computation of optical spectra of large molecules at finite temperature including the Duschinsky and Herzberg-Teller effect: The Qx band of porphyrin as a case study. *J. Chem. Phys.* **2008**, *128*, 224311. [CrossRef] [PubMed]
66. Piccardo, M.; Bloino, J.; Barone, V. Generalized vibrational perturbation theory for rotovibrational energies of linear, symmetric and asymmetric tops: Theory, approximations, and automated approaches to deal with medium-to-large molecular systems. *Int. J. Quantum Chem.* **2015**, *115*, 948–982. [CrossRef]
67. Bloino, J.; Biczysko, M.; Crescenzi, O.; Barone, V. Integrated computational approach to vibrationally resolved electronic spectra: Anisole as a test case. *J. Chem. Phys.* **2008**, *128*, 244105. [CrossRef]

International Journal of *Molecular Sciences*

MDPI

Article

Influence of Hypoxia on Radiosensitization of Cancer Cells by 5-Bromo-2′-deoxyuridine

Magdalena Zdrowowicz [1], Paulina Spisz [1], Aleksandra Hać [2], Anna Herman-Antosiewicz [2] and Janusz Rak [1,*]

1 Laboratory of Biological Sensitizers, Faculty of Chemistry, University of Gdańsk, 80-308 Gdańsk, Poland; magdalena.zdrowowicz@ug.edu.pl (M.Z.); paulina.spisz@ug.edu.pl (P.S.)
2 Department of Medical Biology and Genetics, Faculty of Biology, University of Gdansk, 80-308 Gdańsk, Poland; aleksandra.hac@ug.edu.pl (A.H.); anna.herman-antosiewicz@ug.edu.pl (A.H.-A.)
* Correspondence: janusz.rak@ug.edu.pl; Tel.: +48-58-523-51-18

Abstract: Radiotherapy is a crucial cancer treatment, but its outcome is still far from satisfactory. One of the reasons that cancer cells show resistance to ionizing radiation is hypoxia, defined as a low level of oxygenation, which is typical for solid tumors. In the hypoxic environment, cancer cells are 2–3 times more resistant to ionizing radiation than normoxic cells. To overcome this important impediment, radiosensitizers should be introduced to cancer therapy. When modified with an electrophilic substituent, nucleosides may undergo efficient dissociative electron attachment (DEA) that leaves behind nucleoside radicals, which, in secondary reactions, are able to induce DNA damage, leading to cancer cell death. We report the radiosensitizing effect of one of the best-known DEA-type radiosensitizers—5-bromo-2′-deoxyuridine (BrdU)—on breast (MCF-7) and prostate (PC3) cancer cells under both normoxia and hypoxia. MCF-7 and PC3 cells were treated with BrdU to investigate the effect of hypoxia on cell proliferation, incorporation into DNA and radiosensitivity. While the oxygen concentration did not significantly affect the efficiency of BrdU incorporation into DNA or the proliferation of tumor cells, the radiosensitizing effect of BrdU on hypoxic cells was more evident than on normoxic cells. Further mechanistic studies performed with the use of flow cytometry showed that under hypoxia, BrdU increased the level of histone H2A.X phosphorylation after X-ray exposure to a greater extent than under normal oxygenation conditions. These results confirm that the formation of double-strand breaks in hypoxic BrdU-treated cancer cells is more efficient. In addition, by performing stationary radiolysis of BrdU solution in the presence of an •OH radical scavenger, we compared the degree of its electron-induced degradation under aerobic and anaerobic conditions. It was determined that radiodegradation under anaerobic conditions was almost twice as high as that under aerobic conditions.

Keywords: radiosensitizer; radiotherapy; X-ray; modified nucleosides; DNA damage; hypoxia

Citation: Zdrowowicz, M.; Spisz, P.; Hać, A.; Herman-Antosiewicz, A.; Rak, J. Influence of Hypoxia on Radiosensitization of Cancer Cells by 5-Bromo-2′-deoxyuridine. *Int. J. Mol. Sci.* **2022**, *23*, 1429. https://doi.org/10.3390/ijms23031429

Academic Editor: Filipe Ferreira da Silva

Received: 30 November 2021
Accepted: 25 January 2022
Published: 27 January 2022

Publisher's Note: MDPI stays neutral with regard to jurisdictional claims in published maps and institutional affiliations.

1. Introduction

Radiotherapy (RT) continues to be a mainstay in the treatment of a variety of cancers and is used in more than 50% of patients, both for curative and for palliative purposes [1]. Unfortunately, the therapeutic effects of ionizing radiation (IR) are reduced by radioresistance factors [2]. Hypoxia—a common feature of solid tumors that results from an imbalance between oxygen availability and consumption—is defined as one of the most important reasons for radiotherapy failure [3]. Indeed, since the 1950s [4], it has been repeatedly demonstrated that molecular oxygen significantly modifies the effectiveness of radiotherapy [5,6]. The response of the anoxic regions of the tumor to radiotherapy is 2.5–3 times weaker relative to that of well-oxygenated cells. This effect is known as the oxygen enhancement ratio, and it is most commonly explained by the oxygen fixation hypothesis [7,8]. This radiochemical rationale postulates that radical-induced DNA damage can be "fixed" (made permanent and irreparable) by molecular oxygen, rendering DNA damage irreparable. Indeed, if oxygen is available, it can react with DNA damaged by IR ($DNA^{\bullet}$), creating stable organic peroxides ($DNA\text{-}O\text{-}O^{\bullet}$), which are not readily repaired by the cell. In contrast, the damaged biopolymer (DNA-H) is easily repaired in the absence of oxygen, restoring it to the original state [7,9].

One of the ways to overcome radioresistance is by introducing radiosensitizers to the radiation treatment [10]. Radiosensitizers are chemicals or pharmaceuticals that can enhance cancer cell sensitivity to IR by accelerating DNA damage and producing free radicals indirectly [11]. These substances were initially classified into five groups based on the mechanisms of their action: (1) suppressors of endogenous radioprotective substances, (2) cytotoxic substances formed by radiolysis of the radiosensitizer, (3) inhibitors of DNA repair, (4) thymine analogs that can incorporate into DNA and (5) oxygen mimetics that can imitate the action of oxygen [12]. However, the continuous development and innovation of new technologies and strategies of radiosensitization forced the introduction of a new classification into three categories based on sensitizer structures: (1) small molecules, (2) macromolecules and (3) nanomaterials [10].

Modified nucleosides (MN), which are small-molecule radiosensitizers, are geared towards hydrated electrons [13]. In the absence of oxygen, electrons, besides hydroxyl radicals, are the most abundant products of water radiolysis. The former undergo very fast thermalization and become hydrated (e_{hyd}) within a picosecond time scale. Interest in the role of electrons in the context of generating DNA damage and radiosensitization increased when Sanche and coworkers demonstrated the efficient generation of single- and double-strand breaks (SSBs and DSBs) in plasmid DNA under ultrahigh vacuum by electrons with energies below the ionization threshold of DNA [14]. Although native DNA binds solvated electrons in aqueous solutions [15], strand breaks do not occur [16,17], probably due to swift protonation of the resulting anions [18]. In order to counteract this unfavorable (from a radiotherapy viewpoint) situation, one can use substances that "activate" solvated electrons in the context of DNA damage. Such substances can be modified nucleosides, which utilize electrons for their radiosensitizing activity [19]. Electron attachment to electrophilic nucleosides causes efficient dissociative electron attachment (DEA), leading to reactive radicals [20,21]. If such a modified nucleobase is incorporated into the DNA molecule, then the reactive radical, a product of DEA, may enter secondary chemical reactions that lead to a strand break or other type of DNA damage. One of the most thoroughly studied radiosensitizers of this type is 5-bromo-2′-deoxyuridine (BrdU). This is a highly electron-affinic molecule and is susceptible to DEA that leads to the reactive uracil-5-yl radical via the unstable BrdU anion. This derivative is a good substrate for human thymidine kinase and DNA polymerase [22], and thus, BrdU is easily transformed into 5′-triphosphate of 5-bromo-2′-deoxyuridine (BrdUTP) and then effectively incorporated into cellular DNA [23]. It has been demonstrated that the radiolysis of deoxygenated aqueous solutions containing labeled oligonucleotides produces a significant number of single-strand breaks (SSBs). Studies employing native and brominated trinucleotides, TXT (where X = U/BrU, C/BrC, A/BrA or G/BrG), and radical scavengers have offered proof that only the labeled oligonucleotides undergo damage, mainly in the form of SSBs. It was also demonstrated that the presence of BrdU in a DNA strand led to a three-fold increase in the susceptibility of cells to death by high-energy radiation. Despite the fact that BrdU has been the subject of both in vitro [24,25] and in vivo [26,27] experiments for many decades, comprehensive and direct studies showing the effect of hypoxia on the radiosensitizing properties of BrdU are still lacking. In this paper, we show the influence of the oxygenation level on the most important factors and processes determining the sensitizing properties of BrdU. Using both chemical and biological methods, cell proliferation, incorporation of BrdU into DNA and the radiosensitivity of prostate and breast cancer cells were compared under hypoxic and normoxic conditions.

2. Results and Discussion

2.1. Radiolysis

The formation of the uracil-5-yl radical as a result of DEA plays a crucial role in the radiosensitization mechanism of modified uracils [28]. In order to determine the influence of oxygenation on electron-induced degradation of BrdU, stationary radiolysis coupled with chromatographic analysis was performed. The studied solutions of BrdU were irradiated in the presence of *tert*-butanol to scavenge hydroxyl radicals. The chromatograms corresponding to the above-mentioned experiment, which was performed in anaerobic and aerobic conditions, are presented in Figure 1A,B, respectively. As expected, the final stable product of electron attachment to BrdU was 2′-deoxyuridine (dU), whose identity was confirmed by LC–MS analysis (Figure S1 in Supplementary Material). The yields of the electron-induced decomposition of BrdU were compared for both studied systems. These quantities were calculated based on the peak area of the substrate (BrdU) in the irradiated and non-irradiated solutions of the same initial concentration of BrdU. The degradation yield of BrdU under anaerobic conditions is almost twice as large compared to the normal oxygen concentration. This experiment shows unequivocally that under oxygenated conditions, solvated electrons are effectively scavenged by molecular oxygen. Since hydrated electrons have much more hampered access to the BrdU incorporated into cellular DNA than to BrdU dissolved in water, DEA to BrdU-labeled cellular DNA under aerobic conditions is probably negligible.

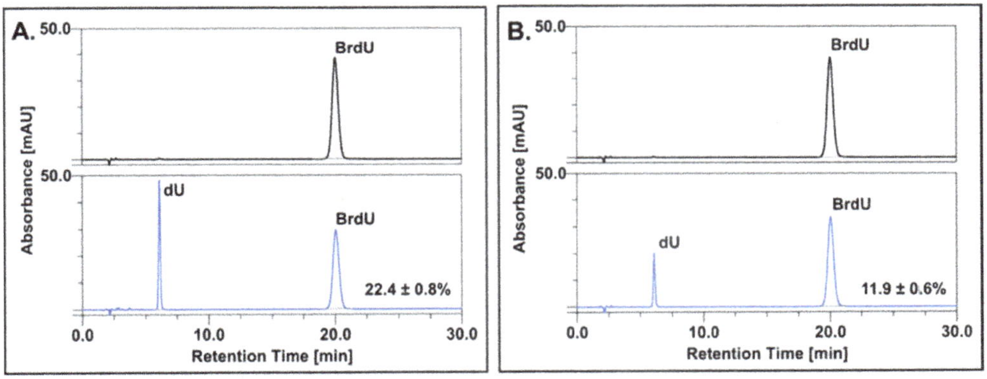

Figure 1. HPLC traces for a solution of BrdU before (black chromatogram) and after irradiation with a dose of 140 Gy (blue chromatogram) under anaerobic (**A**) and aerobic (**B**) conditions.

2.2. Incorporation of BrdU into Genomic DNA

DNA labeled with electron-affinic modified nucleosides that are prone to dissociative electron attachment should be, in contrast to unmodified DNA, sensitive to solvated electrons [21]. Nucleobase radicals formed within the DEA process are able to trigger serious DNA damage (strand breaks, intra- and interstrand cross-links), leading to lethal effects. In order to assess the usefulness of the studied MN, the efficiency of its incorporation into genomic DNA should be determined. Indeed, the level of radiosensitization correlates with the extent to which the analog is incorporated into DNA [29,30]. Therefore, the incorporation of BrdU into the genomic DNA of breast and prostate cancer cells was assessed in hypoxic and normoxic conditions since one could hypothesize that the oxygen status of the cell influences the efficiency of radiosensitizer incorporation into genomic DNA.

MCF-7 and PC3 cells were treated for 24 or 48 h with BrdU at a concentration of 1 µM or 10 µM. Purified DNA was enzymatically digested and analyzed by HPLC (Figure S2 in Supplementary Materials). The results in Table 1 show that the extent of labeling is dependent on the incubation time, analog concentration and cell line, but the oxygenation level does not significantly affect the studied process. The highest level of incorporation was found in the case of 48 h incubation with 10 μM BrdU, which was equal to 24.3%/23.7% (hypoxia/normoxia) for the MCF-7 cell line and was equal to 23.2%/24.2% (hypoxia/normoxia) for PC3 cells.

Table 1. Incorporation of BrdU [%] [a] applied at concentrations of 1 μM or 10 μM into genomic DNA. The experiment was performed in three replicates.

	Hypoxia		Normoxia	
	1 μM BrdU	**10 μM BrdU**	**1 μM BrdU**	**10 μM BrdU**
PC3, 24 h incubation	1.4 ± 0.5	11.2 ± 1.8	1.4 ± 0.3	12.1 ± 1.8
PC3, 48 h incubation	3.6 ± 0.1	23.2 ± 1.6	3.5 ± 0.1	24.2 ± 2.0
MCF-7, 24 h incubation	3.8 ± 0.5	16.6 ± 1.9	4.0 ± 0.5	16.1 ± 1.1
MCF-7, 48 h incubation	4.4 ± 0.4	24.3 ± 0.9	4.8 ± 0.8	23.7 ± 2.0

[a] The percentage values were calculated based on HPLC peak areas corresponding to BrdU and dT (the ratios of BrdU area and the sum of BrdU and dT areas, corrected for the extinction coefficients of dT and BrdU, respectively) in the chromatograms of DNA samples digested to nucleosides.

2.3. *Viability of Cells Labeled with BrdU*

To determine the influence of hypoxic conditions on the cellular cytotoxicity of the studied sensitizer, the MTT test was performed. The assay was conducted for two cancer cell lines—PC3 and MCF-7—under normoxic and hypoxic conditions. The cytotoxicity of BrdU was examined at six concentrations: 0 (for the control) 100, 10, 1, 0.1 and 0.01 μM. The obtained results (Figure S3 in Supplementary Materials) show that the highest reduction in viability is observed in cells incubated for 72 h with 100 μM BrdU. For the PC3 cell line, the absorbance level measured at 570 nm wavelength was equal to 0.85 ± 0.045 for normoxic conditions and 0.86 ± 0.015 for hypoxic conditions compared to the control (1.08 ± 0.025 and 0.90 ± 0.011, respectively). For the MCF-7 cell line, a similar dependence was observed. The decrease in absorbance for the highest concentration after 72 h incubation was equal to 1.25 ± 0.050 for the normoxic conditions and 1.13 ± 0.034 for the hypoxic ones compared to the control (1.69 ± 0.052 and 1.74 ± 0.028, respectively). Thus, the viability, expressed as % of control (Figure 2), was reduced to 78.9 ± 4.5% and 95.5 ± 1.5% for the PC3 line under 21% and 1.5% oxygen, respectively, and 73.7 ± 5.0% and 65.1 ± 3.4% for the MCF-7 line under 21% and 1.5% oxygen, respectively. Surprisingly, these results suggest that BrdU is less cytotoxic to PC3 cells under hypoxic conditions than under normoxic conditions. In contrast, the sensitivity of the MCF-7 cell line to the studied derivative is slightly higher under hypoxic conditions, which might be connected to the fact that MCF-7 cells, contrary to PC3 cells, possess wild-type p53.

Furthermore, the shapes of the viability curves suggest that in the case of the PC3 cells, in the first 24 h, incubation with BrdU resulted in the inhibition of proliferation, which was characteristic for both oxygen variants. In the case of the MCF-7 cell line, the number of cells increased almost exponentially (Figure S3 in Supplementary Materials).

This experiment confirmed the low/acceptable cytotoxicity of BrdU in both cell lines. The results show that in the case of PC3, sensitivity to BrdU under hypoxic conditions, which are characteristic of solid tumors, was lower than under normoxic conditions. Moreover, the obtained results for the 48 h incubation with BrdU allowed for the selection of the radiosensitizer concentrations used in the next experiments, e.g., for the DSBs formation analysis. The selected concentrations were 1 μM and 10 μM. In the PC3 line incubated for 48 h with 1 μM BrdU, the viability was equal to 100.8 ± 2.9% (normoxic conditions) and to 98.6 ± 1.3% (hypoxic conditions). For a 10 μM concentration of BrdU, these values were equal to 99.17 ± 2.6% and 98.15 ± 1.6%, respectively. For the MCF-7 cells treated with the tested compound, the viability was reduced to 88.8 ± 5.6% (normoxic condition) and to 90.1 ± 1.6% (hypoxic condition). For the higher concentration, the viability values were equal to 79.0 ± 2.4% and 81.00 ± 3.2%, respectively.

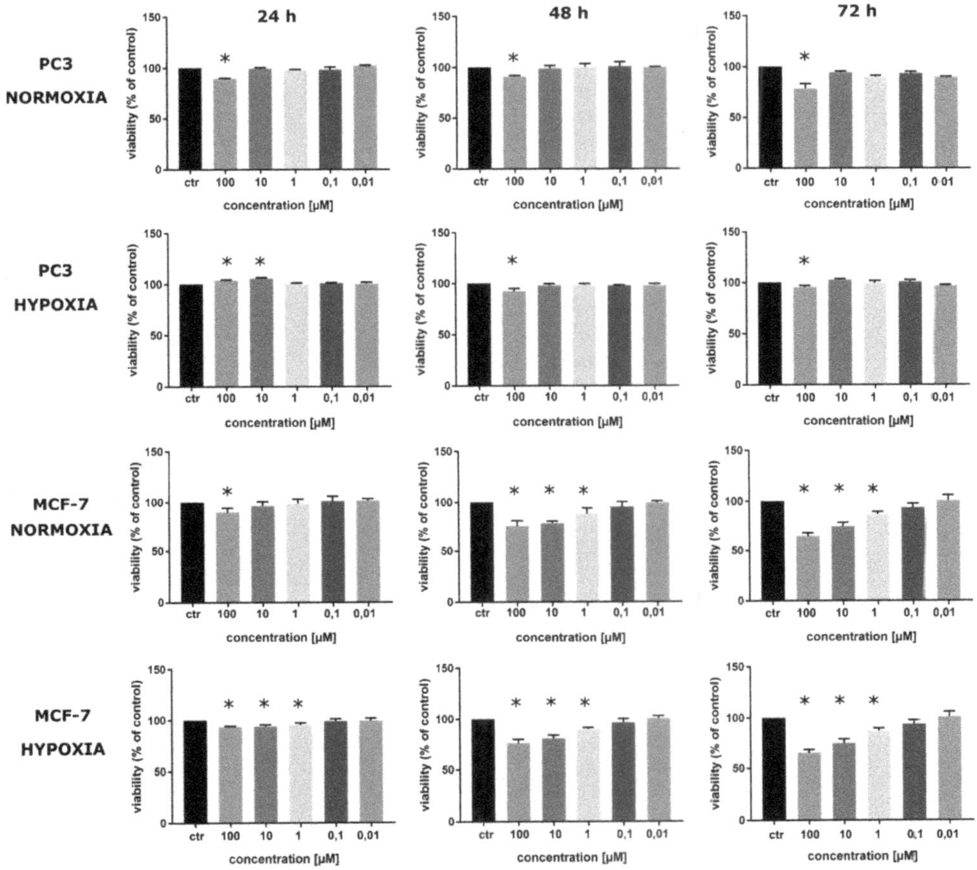

Figure 2. The viability of MCF-7 and PC3 cells after 24, 48 and 72 h treatment with BrdU in a range of concentrations from 0 to 100 µM under hypoxic and normoxic conditions. Results are shown as mean ± standard deviation of three independent experiments performed in triplicate. The statistical significance of the difference between treated culture compared with control (untreated culture) was determined by one-way ANOVA followed by Tukey's multiple comparison test: * $p < 0.05$.

2.4. Clonogenic Survival

To study the influence of the oxygenation level on in vitro radiosensitization by BrdU, PC3 cells were incubated with the tested analog under hypoxic or normoxic conditions and subsequently irradiated. Figure 3 shows the survival fractions of cells irradiated with different doses under normoxia or hypoxia. As expected, in the absence of BrdU, hypoxia reduced the radiosensitivity of PC3 cells in comparison to normoxic conditions. The oxygen enhancement ratio (OER), representing the radiosensitizing effect of oxygen for a survival fraction (SF) equal to 0.5, was 1.92 (0 µM BrdU hypoxia vs. 0 µM BrdU normoxia). In cells pretreated with BrdU, the radiation response of hypoxic cells was enhanced to a much greater extent in comparison to normoxic cells. The sensitizer enhancement ratio (SER), representing the radiosensitizing properties of BrdU, was equal to 3.47 (0 µM BrdU hypoxia vs. 10 µM BrdU hypoxia) under hypoxia and 1.47 under normoxia (0 µM BrdU normoxia vs. 10 µM BrdU normoxia). Considering the values of SF, it can be observed that for pretreatment with the studied analog at a concentration of 10 µM, the survival of PC3 cells irradiated with 0.5 Gy was reduced from 75.8% to 60.7% under normoxia and from 86.7% to 51.6% under hypoxia. With a higher dose of 2 Gy after incubation with BrdU, SF was reduced from 28.6% to 16.5% under normoxia and from 52.3% to 11.1% under hypoxia. In order to compare the survival curves for cells treated with BrdU in normoxic and hypoxic conditions, the doses reducing the SF to 0.5 for both curves were estimated. For the curve corresponding to normoxic BrdU-treated cells, this parameter is equal to 0.742 Gy, and in the case of the curve corresponding to hypoxic cells treated with BrdU, this dose is equal to 0.606 Gy. Furthermore, based on survival curves, parameters for cellular radiosensitivity, such as α (coefficient for linear killing) and β (coefficient for quadratic killing) values, were calculated by fitting using the linear–quadratic model ($SF = \exp(-1 \times (\alpha D + \beta D^2)$, where SF is the survival fraction and D is the dose). Average α and β values were as follows: 0.188 and 0.067 (untreated cells under hypoxia), 1.192 and −0.08 (cells treated with 10 µM BrdU under hypoxia), 0.609 and 0.021 (cells untreated under normoxia) and 0.955 and −0.029 (cells treated with 10 µM BrdU under normoxia). Since from a radiobiological point of view, negative values of the β parameter are not realistic [31], the fitting for curves corresponding to the survival of BrdU-treated cells was performed with β set to zero (see Figure S4). This procedure resulted in a slight change in α values: 1.101 (cells treated with 10 µM BrdU under hypoxia) and 0.919 (cells treated with 10 µM BrdU under normoxia). Regardless of the fitting procedure used, one should note that radiosensitization by BrdU is mainly due to an increase in the α linear parameter, which indicates an increase in the number of lethal events directly caused by BrdU. This observation remains in agreement with previous research [32], which suggested that an important mechanism of radiosensitization in cells substituted with halogenated pyrimidines involves an increase in effective DNA DSBs, and with the results of our studies on DSB formation discussed in Section 2.5. It should also be noted that the α value was enhanced by a factor of 6.3 (the ratio of α for the curve corresponding to cells treated with BrdU and untreated ones) under hypoxia and only 1.5 under normoxia. In general, the comparison of the survival curves (Figure 3) clearly shows that the oxygen level dramatically changes the radiosensitivity of the cancer cells pretreated with BrdU. The analog curves for the MCF-7 cell lines were not measured since, for the concentration of radiosensitizer suitable for PC3 (10 µM), BrdU was determined to be too toxic for reliable performance of the clonogenic test.

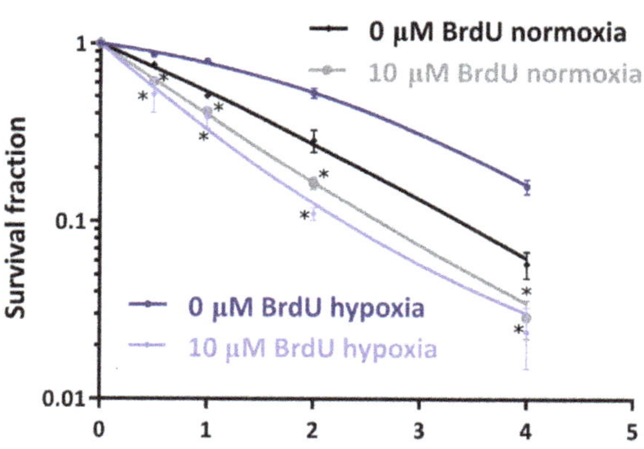

Figure 3. Dose–response curves for PC3 prostate cancer cells not treated or treated with BrdU (10 µM) under normoxia and hypoxia. Experiments were performed at least in duplicate, and the results are expressed as mean ± standard deviation. * Significant difference compared with the untreated variants, $p < 0.05$.

2.5. Double-Strand Break Formation

DSBs play a crucial role in radiation-induced cell killing. DNA damage resulting from BrdU and IR treatments was investigated by measuring the level of γH2A.X by flow cytometry (Figures 4 and S5 in Supplementary Materials). The experiments were conducted on breast and prostate cancer cell lines, and the results were compared both in hypoxic and normoxic conditions. As expected, the number of DSBs (i.e., γH2A.X-positive cells) was significantly lower under hypoxic conditions compared to the normoxic ones for untreated breast and prostate cancer cells. With increasing BrdU concentration, the number of DSBs for irradiated cells increased. Moreover, in hypoxic cells pretreated with BrdU, the number of DSBs was larger than in the pretreated normoxic cells. For example, after incubation of normoxic PC3 cells with 10 μM BrdU and irradiation with a dose of 2 Gy, the level of γH2A.X increased from 40.44% (for the untreated control) to 57.45%. In hypoxic PC3 cells, treatment with 10 μM BrdU caused an increase in γH2A.X-positive cells from 34.49% (for the untreated and irradiated control) to 75.61%. In the case of the MCF-7 cell line, a similar relationship was observed. The results of cytometric analysis of H2A.X phosphorylation confirm that BrdU-mediated radiosensitization in vitro is dependent on the oxygenation level and is significantly greater under hypoxic conditions.

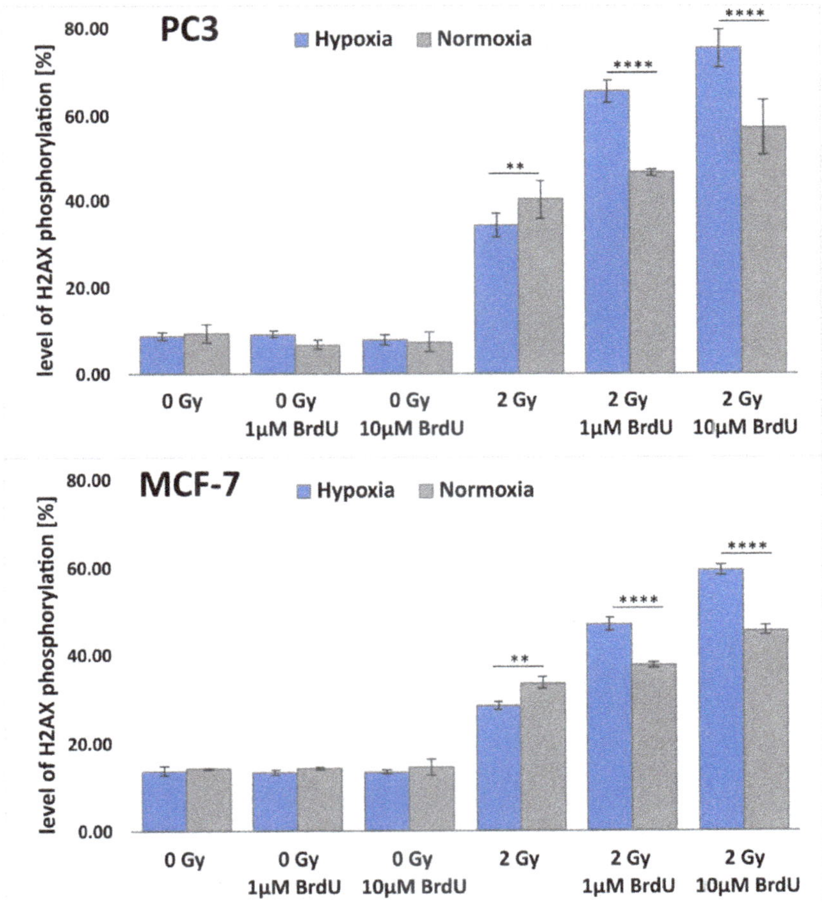

Figure 4. Flow cytometric analysis of H2A.X phosphorylation. γH2A.X was measured 1 h after irradiation. Results are shown as mean ± standard deviation of at least three independent experiments. The statistical significance was determined by one-way ANOVA followed by Tukey's multiple comparison test: **** $p < 0.0001$, ** $p < 0.01$.

2.6. DNA Damage Signaling

Phosphorylation of histone variant H2A.X at Ser-139 (modification termed γH2A.X) is primarily mediated by ataxia-telangiectasia mutated (ATM) kinase in response to DSBs and is crucial for further signaling to proteins engaged in DNA repair [33,34]. ATM is activated by autophosphorylation at Ser-1981 [35]. Among its other important substrates is checkpoint kinase 2 (Chk2), which, upon phosphorylation at Thr-68, becomes activated and inhibits cell cycle progression [36], and p53, which, when phosphorylated at Ser-15, is stabilized and may act as a transcription factor that further enhances the cell cycle block and DNA repair or induces cell death [37,38].

The status of these proteins in cells exposed to BrdU and/or ionizing radiation in normoxic or hypoxic conditions was investigated. HIF-1α was used as a marker of hypoxia, and as shown in Figure 5A, this protein was stabilized in cells cultured under 1.5% oxygen. In both cell lines, radiation (5 Gy) alone enhanced (although not at a statistically significant level) the phosphorylation of ATM and its substrates, Chk2 and H2A.X, to a higher level under normoxia than under hypoxia (Figure 5B,C). For instance, irradiation of PC3 cells elevated p-ATM by c.a. 2-fold under normoxia with no effect under hypoxia, p-Chk2 by 1.5-fold under normoxia and 1.1-fold under hypoxia, and p-H2A.X by 3.7-fold under normoxia compared to 2.6-fold under hypoxia (Figure 5B, lines 1 and 4 of respective blots and graphs). However, if cells were pretreated with BrdU, the phosphorylation of checkpoint proteins in irradiated hypoxic cells was comparable to or even higher than that in their normoxic counterparts (Figure 5B,C). For instance, in PC3 cells exposed to 10 μM BrdU compared to untreated controls, p-ATM was elevated by 2.4-fold and 3.2-fold; p-Chk2 was elevated by 1.8- and 1.7-fold; and p-H2A.X was elevated by 5.2- and 11.6-fold under normoxia and hypoxia, respectively (Figure 5B, line 6 in respective blots and graphs). In MCF-7 cells, which possess wt p53, phosphorylation of p53 at Ser-15 was also higher in irradiated cells cultured in normoxic compared to hypoxic conditions (1.7- vs. 0.9-fold), unless cells were labeled with BrdU. In this case, phosphorylation of p53 was higher in irradiated hypoxic (differences reached statistical significance) than normoxic cells compared to the respective cells not treated with IR (Figure 5C, lines 4–6 of p-p53 blots and graphs). These results indicate that under hypoxia, DNA damage signaling is indeed impaired, which is revealed by the lower activation level of the main checkpoint kinases, ATM and Chk2, in irradiated hypoxic cells compared to irradiated normoxic cells. Importantly, BrdU incorporated into DNA sensitizes hypoxic cells to IR and leads to the efficient activation of ATM, Chk-2 and p53 (in MCF-7 cells), which may result in proliferation inhibition or cell death with comparable efficiency to that observed in normoxic cells (Figure 3, for comparison).

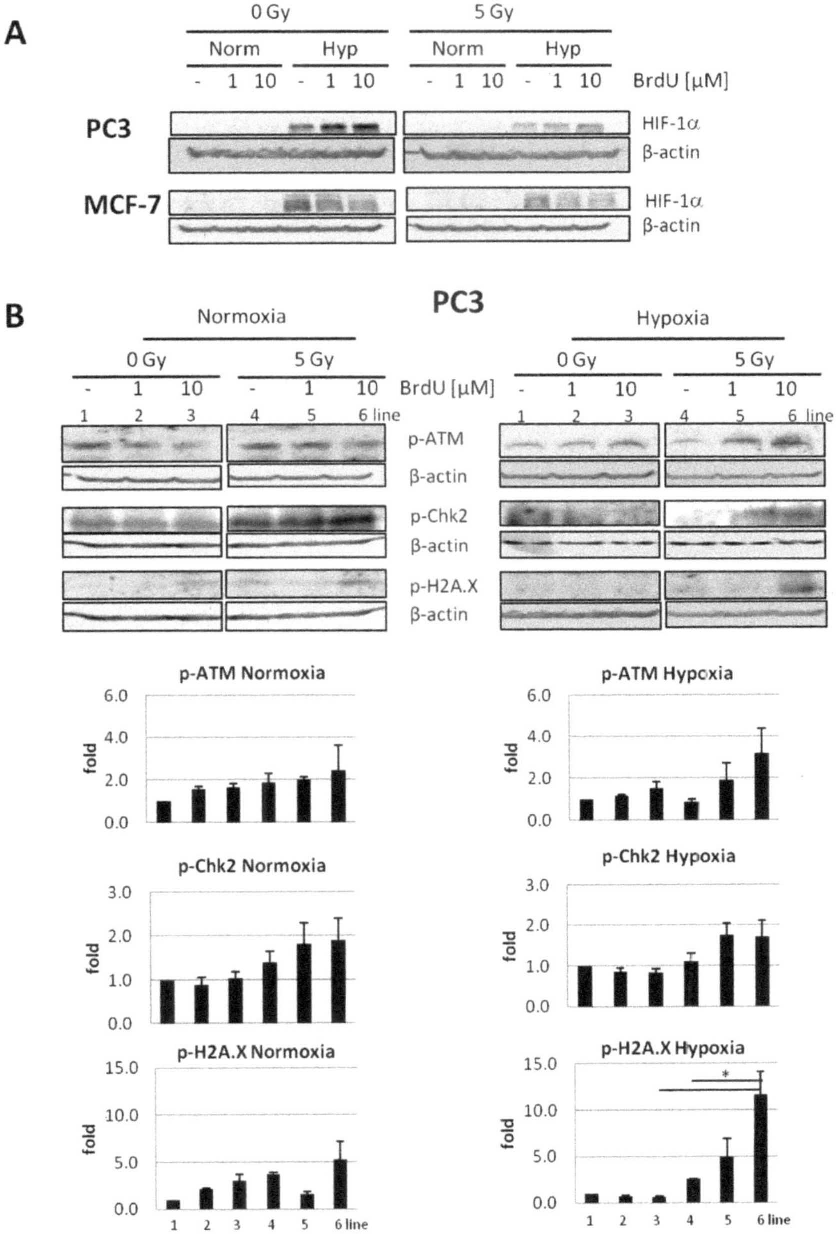

Figure 5. *Cont.*

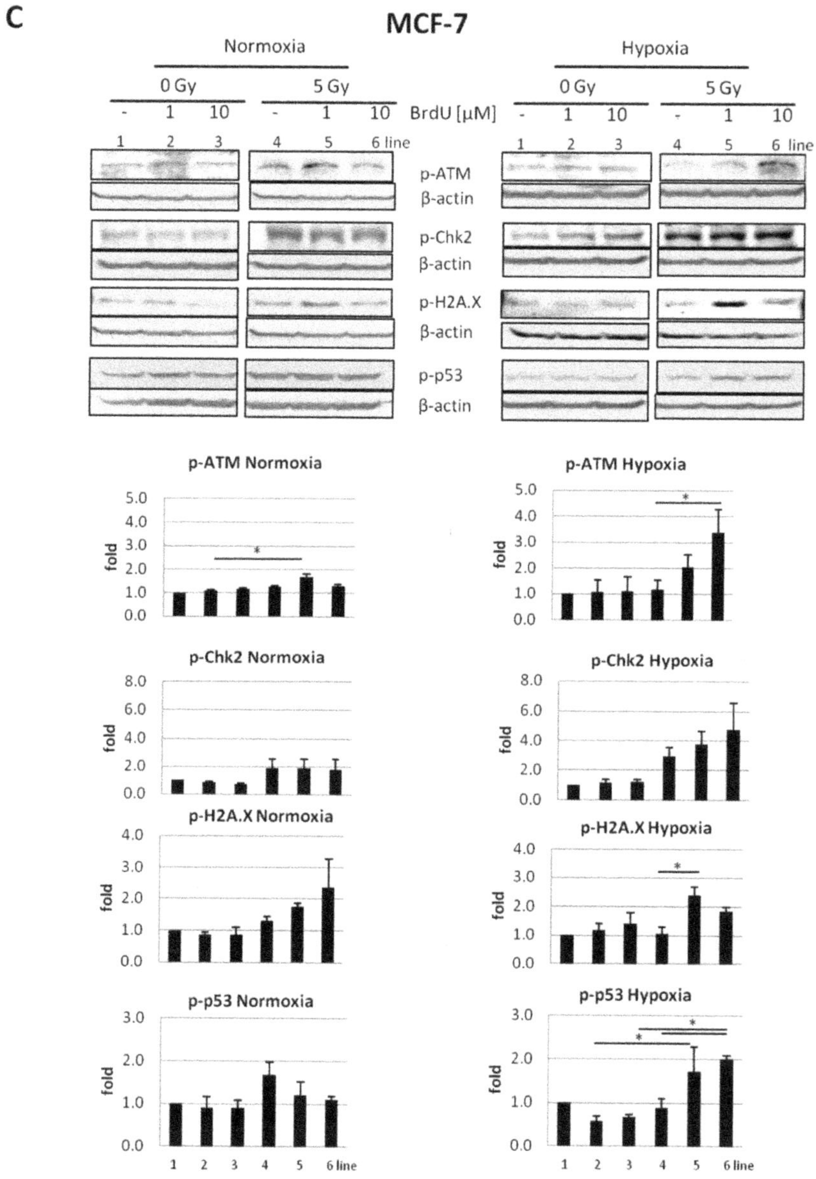

Figure 5. Effect of treatment with BrdU and/or exposure to ionizing radiation of cells in normoxic or hypoxic conditions. (**A**) Investigation of HIF-1α hypoxia marker in MCF-7 and PC3 cells untreated/treated (1 or 10 µM) with BrdU and unexposed/exposed to ionizing radiations (5 Gy) under hypoxia and normoxia conditions. Status of p-ATM, p-Chk2, p-H2A.X and p-p53 in (**B**) PC3 and (**C**) MCF-7 cells untreated/treated (1 or 10 µM) with BrdU and unexposed/exposed to ionizing radiations (5 Gy) under hypoxia and normoxia conditions. The blots were stripped and reprobed with anti-β-actin antibodies to normalize for differences in protein loading. Densitometric data, corrected for loading control, are graphed below (mean ± SE; n = 2–3). The statistical significance of differences was determined by one-way ANOVA followed by Sidak's multiple comparison test, and * $p < 0.05$.

3. Materials and Methods

3.1. Radiolysis

Radiolysis of BrdU solution at a concentration of 10^{-4} M containing 0.03 M *tert*-BuOH as a scavenger of $^{\bullet}OH$ radicals and phosphate buffer (10 mM, pH = 7.0) was carried out in a Cellrad X-ray cabinet (Faxitron X-ray Corporation, Tucson, AZ, USA). The irradiated samples were exposed to 140 Gy (4.14 Gy·min^{-1}, 130.0 kV, 5 0 mA, filter: 0.5 mm aluminum). One part of the sample was deoxygenated by purging with argon for 3 min. A reversed-phase HPLC method was employed for analysis of the irradiated and non-irradiated samples of BrdU with the use of a Dionex UltiMate 3000 system with Diode Array Detector (Dionex Corporation, Sunnyvale, CA, USA). The effluents were monitored at 260 nm. Separation by HPLC was achieved using a C18 column (Wakopak Handy ODS, 4.6 × 150 mm, 5 μm particle size and 100 Å pore size, Fujifilm Wako Chemicals, Osaka, Japan) with an isocratic elution of 0.1% HCOOH and 1% ACN. The studied samples were analyzed in triplicate. The chemicals used were purchased from Merck Millipore (Darmstadt, Germany).

3.2. Cell Culture

MCF-7 cells were grown in RPMI medium supplemented with 10% FBS and antibiotics at a concentration of 100 U·mL^{-1}, and PC3 cells were cultured in F12K medium supplemented with 10% FBS and antibiotics at a concentration of 100 U·mL^{-1}. The cell lines were obtained from ATCC (Manassas, VA, USA) and maintained at 37 °C in a humidified atmosphere with 5% CO_2. All media and supplements were purchased from Gibco (Paisley, UK). The hypoxic group of cells was cultured in 1.5% O_2 in an incubator with oxygen control (CB 60, Binder GmbH, Tuttlingen, Germany). The plates with cells were irradiated using a Cellrad X-ray cabinet. To achieve hypoxic conditions, irradiation and all manipulations were performed in a nitrogen atmosphere.

3.3. Incorporation of BrdU into DNA

MCF-7 and PC3 cell lines were seeded into plates and incubated at 37 °C and 5% CO_2 overnight. After that, the medium was replaced with fresh medium, and the cells were treated with BrdU at a concentration of 0 μM (control), 1 μM or 10 μM. Plates with cells were incubated with the compound in normoxic (21% of O_2) or hypoxic conditions (1.5% of O_2) for 48 h. After this time, the cells were removed from plates, and isolation of DNA was carried out according to the protocol provided by the manufacturer (GeneMATRIX Cell Culture DNA Purification Kit, EURX, Gdańsk, Poland). Then, purified DNA was enzymatically digested by the simultaneous action of nuclease P1, spleen phosphodiesterase, DNase I, snake venom phosphodiesterase (SVP) and bacterial alkaline phosphatase (BAP). Finally, the solution was chloroform-extracted to remove the enzymes, and the aqueous layer was lyophilized, resuspended in water and subjected to HPLC analysis. The isolated and digested DNA samples were analyzed with the use of the reversed-phase HPLC method on a Dionex UltiMate 3000 system with Diode Array Detector. The detection of effluents was carried out at 260 nm. Separations were performed using a Wakopak Handy ODS column (C18, 4.6 × 150 mm, 5 μm; 100 Å), 0.1% HCOOH in water as the mobile phase (isocratic elution) and a flow rate of 1 mL·min^{-1}. The data were obtained from three experiments.

3.4. Viability Assay

To determine the cytotoxicity level of BrdU under hypoxic and normoxic conditions, the MTT assay was carried out. Cells from PC3 and MCF-7 lines were seeded into 96-well plates at a density of 4000 per well and incubated at 37 °C and 5% CO_2 overnight. After that, the medium was replaced with fresh medium, and cells from both lines were treated with BrdU at 7 concentrations: 0 (for the control) 10^{-4}, 10^{-5}, 10^{-6}, 10^{-7}, 10^{-8} and 10^{-9} M. Next, plates with treated cells were incubated under normoxic (21% O_2) or hypoxic (1.5% O_2) conditions for 24, 48 and 72 h. After this time, 25 μL of an aqueous solution of MTT salt (4 mg mL^{-1}) was added to each well and incubated for 4 h. Then, 200 μL of the medium with the MTT salt was removed, and DMSO was added. The absorbance was measured at 570 nm (660 nm was the reference wavelength) with an EnSpire microplate reader (PerkinElmer, Waltham, MA, USA). The vitality of the control was taken as 100%. The obtained results were analyzed with the GraphPad Prism 7 software (San Diego, CA, USA). The statistical significance was calculated using a one-way analysis of variance (ANOVA) followed by Dunnett's multiple comparison test. The data were obtained from three experiments, and each treatment condition was assayed in triplicate. The differences were considered significant at $p < 0.05$.

3.5. Clonogenic Assay

Cells were plated at a density of 10^6 cells per 60 mm dish and treated with BrdU at concentrations of 10 µM (PC3). After 48 h of treatment under normoxic (21% of O_2) or hypoxic (1.5% of O_2) conditions, the cells were irradiated with different doses (0–6 Gy). Delayed plating (after 6 h) was performed; cells were detached with Accutase and plated at a density of 800 cells per 100 mm dish. After 16 days, the resulting colonies were fixed with 6.0% (v/v) glutaraldehyde and 0.5% crystal violet. Stained colonies were counted manually, and colony size was assessed using an inverted fluorescence microscope (Olympus, IX73, Tokyo, Japan). The experiment was carried out in duplicate. Non-irradiated cultures were used as controls. The radiosensitizing effect of oxygen or BrdU is represented by the OER and SER, respectively (ID50 (−oxygen or treatment)/ID50 (+oxygen or treatment), where ID50 means the radiation dose causing 50% growth inhibition, determined by interpolation using the GraphPad Prism software). The obtained results were analyzed with the GraphPad Prism software using the linear–quadratic model. The statistical evaluation was calculated using a one-way analysis of variance (ANOVA) followed by Sidak's multiple comparison test.

3.6. Histone H2A.X Phosphorylation Assay

Cells were treated with BrdU at a concentration of 0 µM (control), 1 µM or 10 µM and incubated (37 °C, 5% CO_2) for 48 h under hypoxic (1.5% O_2) or normoxic (21% O_2) conditions. After this time, plates with cells were irradiated with 2 Gy and incubated for 1 h. Then, the cells were dissociated with Accutase solution, fixed, permeabilized and stained according to the manufacturer's protocol (FlowCellect^{TM} Histone H2A.X Phosphorylation Assay Kit, Luminex, Austin, TX, USA) and then analyzed by flow cytometry (Guava easyCyte^{TM}, Merck, Darmstadt, Germany). Untreated cultures were used as controls. The data were obtained from at least three experiments. The statistical significance of differences between the studied groups was determined by one-way ANOVA followed by Tukey's multiple comparison test.

3.7. Western Blotting

Cells were treated with BrdU at a concentration of 0 µM (control), 1 µM or 10 µM and incubated (37 °C, 5% CO_2) for 48 h under hypoxic (1.5% O_2) or normoxic (21% O_2) conditions. After this time, plates with cells were irradiated with 5 Gy and incubated for 1 h. Cells were collected, washed with ice-cold phosphate-buffered saline (PBS) and lysed in a solution containing 50 mM Tris, 1% Triton X-100, 150 mM NaCl, 0.5 mM EDTA and protease and phosphatase inhibitor cocktails (Roche Diagnostics Poland, Warsaw, Poland). The lysates were cleared by centrifugation at 15,500× g at 4 °C for 20 min. Proteins were resolved on 4–12% Bis-Tris gels (Thermo Fisher Scientific, Waltham MA, USA) by SDS–polyacrylamide gel electrophoresis (SDS-PAGE) with MES Running Buffer (Thermo Fisher Scientific, Waltham, MA, USA) and semi-dry transferred onto PVDF membrane (Thermo Fisher Scientific, Waltham, MA, USA). The membrane was blocked with 5% nonfat dry milk in PBS and incubated with the desired primary antibodies overnight at 4 °C. Antibodies against HIF-1α, p-ATM (Ser-1981), p-Chk2 (Thr-68), p-H2A.X (Ser-139) and p-p53 (Ser-15) were purchased from Cell Signaling Technology (Danvers, MA, USA). Next, the membrane was extensively washed with PBS and incubated with appropriate secondary antibodies for 1 h at room temperature. The immunoreactive bands were detected with enhanced chemiluminescence reagent (Clarity ECL Substrate; Biorad, Hercules, CA, USA) on X-ray films (Fujifilm, Tokyo, Japan). The blots were stripped and reprobed with anti-β-actin antibodies to normalize for differences in protein loading. The experiment was performed in 2 or 3 replicates.

4. Conclusions

Hypoxia is one of the main factors that reduce the effectiveness of radiotherapy. This fact was emphasized by Baumann and colleagues in their recent review regarding the radioresistance of neck and head cancers [39]. The authors suggested that the combination of hypoxia-sensitizing drugs with targeted radiation of hypoxic cells will become a gold standard of radiotherapy. Thus, the key role of a sensitizing drug in such a combined treatment prompted our studies on the effect of hypoxia on the cellular response of cells treated with BrdU, one of the best-known radiosensitizers, to IR. For comparison, all experiments were also performed under normoxic conditions.

One of the main properties of thymidine analogs as radiosensitizers is their effective electron-induced degradation. In this work, we demonstrated that the formation of uridine, as the main stable product of DEA to BrdU, was two-fold more effective under hypoxia, which reflects the fact that solvated electrons readily react with molecular oxygen to form peroxyl anion radicals that are much more inert than hydrated electrons. These data confirm that the pool of solvated electrons is higher at low oxygen levels. To perform their role, radiosensitizers with high electron affinity must be incorporated into DNA. As expected, the oxygen level did not affect the DNA labeling process. The incorporation of 5-bromo-2′-deoxyuridine depended on the time of treatment with radiosensitizer, its concentration and the cell line. On the other hand, hypoxia reduced the effectiveness of radiotherapy, which was confirmed by the clonogenic test. The results show that without BrdU, the cell response to IR (including activation of DNA damage signaling) was reduced under hypoxic conditions compared to normoxic cells. The situation was completely different in cells treated with BrdU, as hypoxia significantly increased the sensitivity of those cells to ionizing radiation. These results correlate with the formation of DSBs. For untreated cells, the DSB level after IR exposure was lower under hypoxic conditions. The opposite was observed for cells labeled with BrdU, where low oxygen concentration led to an increase in DSBs and DNA damage signaling.

Our studies confirm that BrdU is a hypoxic radiosensitizer; i.e., it works much better under hypoxia, which is one of the well-known obstacles for efficient radiotherapy.

Supplementary Materials: The following supporting information can be downloaded at: https://www.mdpi.com/article/10.3390/ijms23031429/s1.

Author Contributions: Conceptualization, M.Z. and J.R.; methodology, M.Z., P.S., A.H. and A.H.-A.; investigation, M.Z., P.S. and A.H.; data curation, M.Z., P.S. and A.H.; writing—original draft preparation, M.Z., P.S., A.H., A.H.-A. and J.R.; writing—review and editing, M.Z., P.S., A.H., A.H.-A. and J.R.; funding acquisition, J.R. All authors have read and agreed to the published version of the manuscript.

Funding: This research was funded by Polish National Science Centre under grant no. 2014/14/A/ST4/00405 (J.R.).

Institutional Review Board Statement: Not applicable.

Informed Consent Statement: Not applicable.

Data Availability Statement: Research data are stored in an institutional repository and will be shared upon request to the corresponding author.

Conflicts of Interest: The authors declare no conflict of interest.

References

1. De Ruysscher, D.; Niedermann, G.; Burnet, N.G.; Siva, S.; Lee, A.W.; Hegi-Johnson, F. Radiotherapy Toxicity. *Nat. Rev. Dis. Primers* **2019**, *5*, 13. [CrossRef]
2. Peters, L.J.; Withers, H.R.; Thames Jr, H.D.; Fletcher, G.H. Keynote Address—The Problem: Tumor Radioresistance in Clinical Radiotherapy. *Int. J. Radiat. Oncol. Biol. Phys.* **1982**, *8*, 101–108. [CrossRef]
3. Rockwell, S.; Dobrucki, I.T.; Kim, E.Y.; Marrison, S.T.; Vu, V.T. Hypoxia and Radiation Therapy: Past History, Ongoing Research, and Future Promise. *Curr. Mol. Med.* **2009**, *9*, 442–458. [CrossRef] [PubMed]
4. Gray, L.H.; Conger, A.; Ebert, M.; Hornsey, S.; Scott, O.C.A. The Concentration of Oxygen Dissolved in Tissues at the Time of Irradiation as a Factor in Radiotherapy. *Br. J. Radiol. Suppl.* **1953**, *26*, 638–648. [CrossRef] [PubMed]
5. Evans, S.M.; Koch, C.J. Prognostic Significance of Tumor Oxygenation in Humans. *Cancer Lett.* **2003**, *195*, 1–16. [CrossRef]
6. Wilson, W.R.; Hay, M.P. Targeting Hypoxia in Cancer Therapy. *Nat. Rev. Cancer* **2011**, *11*, 393–410. [CrossRef]
7. Hall, E.; Giaccia, A. *Radiobiology for the Radiologist*, 6th ed.; Lippincott William and Wilkins: Philadelphia, PA, USA, 2006.
8. Bertout, J.; Patel, S.; Simon, M. The Impact of O_2 Availability on Human cancer. *Nat. Rev. Cancer* **2008**, *1974*, 1967–1974. [CrossRef]
9. Grimes, D.R.; Partridge, M.A. Mechanistic Investigation of the Oxygen Fixation Hypothesis and Oxygen Enhancement Ratio. Biomed. *Phys. Eng. Express.* **2015**, *1*, 045209. [CrossRef]
10. Wang, H.; Mu, X.; He, H.; Zhang, X.D. Cancer Radiosensitizers. *Trends Pharmacol. Sci.* **2018**, *39*, 24–48. [CrossRef]
11. Gong, L.; Zhang, Y.; Liu, C.; Zhang, M.; Han, S. Application of Radiosensitizers in Cancer Radiotherapy. *Int. J. Nanomed.* **2021**, *16*, 1083. [CrossRef]
12. Fowler, J.F.; Adams, G.E.; Denekamp, J. Radiosensitizers of Hypoxic Cells in Solid Tumours. *Cancer Treat. Rev.* **1976**, *3*, 227–256. [CrossRef]
13. Zdrowowicz, M.; Chomicz-Mańka, L.; Butowska, K.; Spisz, P.; Falkiewicz, K.; Czaja, A.; Rak, J. DNA Damage Radiosensitizers Geared Towards Hydrated Electrons. In *Practical Aspects of Computational Chemistry*; Leszczynski, J., Shukla, M.K., Eds.; Springer: Berlin/Heidelberg, Germany, 2021.
14. Boudaïffa, B.; Cloutier, P.; Hunting, D.; Huels, M.A.; Sanche, L. Resonant Formation of DNA Strand Breaks by Low-Energy (3 to 20 eV) Electrons. *Science* **2000**, *287*, 1658–1660. [CrossRef] [PubMed]
15. Scholes, G. *Effects of Ionizing Radiation on DNA. Physical, Chemical and Biological Aspects*; Hüttermann, J., Köhnlein, W., Teoule, R., Eds.; Springer: Berlin/Heidelberg, Germany, 1978; p. 153.
16. Nabben, F.J.; Karman, J.P.; Loman, H. Inactivation of Biologically Active DNA by Hydrated Electrons. *Int. J. Radiat. Biol. Relat. Stud. Phys. Chem. Med.* **1982**, *42*, 23–30. [CrossRef] [PubMed]
17. Wang, W.; Sevilla, M.D. Protonation of Nucleobase Anions in Gamma-Irradiated DNA and Model Systems. Which DNA Base is the Ultimate Sink for the Electron? *Radiat. Res.* **1994**, *138*, 9–17. [CrossRef]
18. Kohanoff, J.; McAllister, M.; Tribello, G.A.; Gu, B. Interactions Between Low Energy Electrons and DNA: A Perspective from First-Principles Simulations. *J. Phys. Condens. Matter* **2017**, *29*, 383001. [CrossRef]
19. Rak, J.; Chomicz, L.; Wiczk, J.; Westphal, K.; Zdrowowicz, M.; Wityk, P.; Żyndul, M.; Makurat, S.; Golon, Ł. Mechanisms of Damage to DNA Labeled with Electrophilic Nucleobases Induced by Ionizing or UV Radiation. *J. Phys. Chem. B* **2015**, *119*, 8227–8238. [CrossRef]
20. Chomicz, L.; Zdrowowicz, M.; Kasprzykowski, F.; Rak, J.; Buonaugurio, A.; Wang, Y.; Bowen, K.H. How to Find out Whether a 5-Substituted Uracil Could be a Potential DNA Radiosensitizer. *J. Phys. Chem. Lett.* **2013**, *4*, 2853–2857. [CrossRef]
21. Chomicz-Mańka, L.; Wityk, P.; Golon, Ł.; Zdrowowicz, M.; Wiczk, J.; Westphal, K.; Żyndul, M.; Makurat, S.; Rak, J. Consequences of Electron Attachment to Modified Nucleosides Incorporated into DNA. In *Handbook of Computational Chemistry*; Leszczynski, J., Kaczmarek-Kedziera, A., Puzyn, T., Papadopoulos, M.G., Shukla, M.K., Eds.; Springer: Berlin/Heidelberg, Germany, 2017.
22. Lee, L.-S.; Cheng, Y. Human Deoxythymidine Kinase. 2. Substrate Specificity and Kinetic Behaviors of the Cytoplasmic and Mitochondria Isozymes Derived from the Blast Cells of the Acute Myelocytic Leukemia. *Biochemistry* **1976**, *15*, 3686–3690. [CrossRef]
23. Goz, B. The Effects of Incorporation of 5-halogenated Deoxyuridines into the DNA of Eukaryotic Cells. *Pharmacol. Rev.* **1977**, *29*, 249–272.
24. Erikson, R.L.; Szybalski, W. Molecular radiobiology of human cell lines. V. Comparative radiosensitizing properties of 5-halodeoxycytidines and 5-halodeoxyuridines. *Radiat. Res.* **1963**, *20*, 252–262. [CrossRef]
25. Kinsella, T.J.; Dobson, P.P.; Mitchell, J.B.; Fornace, A.J., Jr. Enhancement of X Ray Induced DNA Damage by Pre-treatment with Halogenated Pyrimidine Analogs. *Int. J. Radiat. Oncol. Biol. Phys.* **1987**, *13*, 733–739. [CrossRef]
26. Prados, M.D.; Scott, C.; Sandler, H.; Buckner, J.C.; Phillips, T.; Schultz, C.; Urtasun, R.; Davis, R.; Gulin, P.; Cascino, T.; et al. A Phase 3 Randomized Study of Radiotherapy Plus Procarbazine, CCNU, and Vincristine (PCV) with or without BUdR for the Treatment of Anaplastic Astrocytoma: A Preliminary Report of RTOG 9404. *Int. J. Radiat. Oncol. Biol. Phys.* **1999**, *45*, 1109–1115. [CrossRef]
27. Groves, M.D.; Maor, M.H.; Myers, C.; Kyritsi, A.P.; Jaeckle, K.A.; Yung, W.K.A.; Sawaya, R.E.; Hess, K.; Bruner, J.M.; Peterson, P.; et al. A Phase II Trial of High-Dose Bromodeoxyuridine with Accelerated Fractionation Radiotherapy Followed by Procarbazine, Lomustine and Vincristine for Glioblastoma Multiforme. *Int. J. Radiat. Oncol. Biol. Phys.* **1999**, *45*, 127–135. [CrossRef]

28. Spisz, P.; Zdrowowicz, M.; Makurat, S.; Kozak, W.; Skotnicki, K.; Bobrowski, K.; Rak, J. Why Does the Type of Halogen Atom Matter for the Radiosensitizing Properties of 5-Halogen Substituted 4-Thio-2′-Deoxyuridines? *Molecules* **2019**, *24*, 2819. [CrossRef] [PubMed]
29. Iliakis, G.; Kurtzman, S.; Pantelias, G.; Okayasu, R. Mechanism of Radiosensitization by Halogenated Pyrimidines: Effect of BrdU on Radiation Induction of DNA and Chromosome Damage and its Correlation with Cell Killing. *Radiat. Res.* **1989**, *119*, 286–304. [CrossRef]
30. Dillehay, L.E.; Thompson, L.H.; Carrano, A.V. DNA-Strand Breaks Associated with Halogenated Pyrimidine Incorporation. *Mutat. Res.* **1984**, *131*, 129–136. [CrossRef]
31. Van Leeuwen, C.M.; Oei, A.; Crezee, J.; Bel, A.; Franken, N.; Stalpers, L.; Kok, H. The Alfa and Beta of Tumours: A Review of Parameters of the Linear-quadratic Model, Derived from Clinical Radiotherapy Studies. *Radiat. Oncol.* **2018**, *13*, 96. [CrossRef]
32. Franken, N.; Oei, A.; Kok, H.; Rodermond, H.; Sminia, P.; Crezee, J.; Stalpers, L.; Barendsen, G. Cell Survival and Radiosensitisation: Modulation of the Linear and Quadratic Parameters of the LQ Model. *Int. J. Oncol.* **2013**, *42*, 1501–1515. [CrossRef]
33. Petrini, J.H.; Stracker, T.H. The Cellular Response to DNA Double-Strand Breaks: Defining the Sensors and Mediators. *Trends Cell Biol.* **2003**, *13*, 458–462. [CrossRef]
34. Rogakou, E.P.; Pilch, D.R.; Orr, A.H.; Ivanova, V.S.; Bonner, W.M. DNA Double-Stranded Breaks Induce Histone H2AX Phosphorylation on Serine. *J. Biol. Chem.* **1998**, *273*, 5858–5868. [CrossRef]
35. Bakkenist, C.J.; Kastan, M.B. DNA Damage Activates ATM through Intermolecular Autophosphorylation and Dimer Dissociation. *Nature* **2003**, *421*, 499–506. [CrossRef]
36. Bartek, J.; Falck, J.; Lukas, J. CHK2 Kinase—A Busy Messenger. *Nature Rev. Mol. Cell Biol.* **2001**, *2*, 877–886. [CrossRef]
37. Banin, S.; Moyal, L.; Shieh, S.; Taya, Y.; Anderson, C.W.; Chessa, L.; Smorodinsky, N.I.; Prives, C.; Reiss, Y.; Shiloh, Y.; et al. Enhanced Phosphorylation of p53 by ATM in Response to DNA Damage. *Science* **1998**, *281*, 1674–1677. [CrossRef] [PubMed]
38. Siliciano, J.D.; Canman, C.E.; Taya, Y.; Sakaguchi, K.; Appella, E.; Kastan, M.B. DNA Damage Induces Phosphorylation of the Amino Terminus of p53. *Genes Dev.* **1997**, *11*, 3471–3481. [CrossRef] [PubMed]
39. Baumann, R.; Depping, R.; Delaperriere, M.; Dunst, J. Targeting Hypoxia to Overcome Radiation Resistance in Head & Neck Cancers: Real Challenge or Clinical Fairytale? *Expert Rev. Anticancer. Ther.* **2016**, *16*, 751–758. [PubMed]

International Journal of
Molecular Sciences

MDPI

Article

Effect of Humid Air Exposed to IR Radiation on Enzyme Activity

Olga I. Yablonskaya [1,*], Vladimir L. Voeikov [2], Kirill N. Novikov [2], Ekaterina V. Buravleva [2], Valeriy A. Menshov [1] and Aleksei V. Trofimov [1]

1 Emanuel Institute of Biochemical Physics of Russian Academy of Sciences, 119334 Moscow, Russia; vinoman66@mail.ru (V.A.M.); avt_2003@mail.ru (A.V.T.)
2 Faculty of Biology, Lomonosov Moscow State University, 119234 Moscow, Russia; v109028v1@yandex.ru (V.L.V.); kirniknov@yandex.ru (K.N.N.); b_u_k_a@mail.ru (E.V.B.)
* Correspondence: olga.yablonsky@gmail.com

Abstract: Water vapor absorbs well in the infra-red region of the electromagnetic spectrum. Absorption of radiant energy by water or water droplets leads to formation of exclusion zone water that possesses peculiar physico-chemical properties. In the course of this study, normally functioning and damaged alkaline phosphatase, horseradish peroxidase and catalase were treated with humid air irradiated with infrared light with a wavelength in the range of 1270 nm and referred to as coherent humidity (CoHu). One-minute long treatment with CoHu helped to partially protect enzymes from heat inactivation, mixed function oxidation, and loss of activity due to partial unfolding. Authors suggest that a possible mechanism underlying the observed effects involves altering the physico-chemical properties of aqueous media while treatment of the objects with CoHu where CoHu acts as an intermediary.

Keywords: IR radiation; horseradish peroxidase; catalase; alkaline phosphatase; antioxidant; vapor; coherent domain

Citation: Yablonskaya, O.I.; Voeikov, V.L.; Novikov, K.N.; Buravleva, E.V.; Menshov, V.A.; Trofimov, A.V. Effect of Humid Air Exposed to IR Radiation on Enzyme Activity. *Int. J. Mol. Sci.* **2022**, *23*, 601. https://doi.org/10.3390/ijms23020601

Academic Editor: Filipe Ferreira da Silva

Received: 3 December 2021
Accepted: 3 January 2022
Published: 6 January 2022

1. Introduction

In our recent research we the observed changes in physicochemical properties of aqueous systems after non-chemical treatment with humid air exposed to IR radiation that was called coherent humidity (CoHu) [1]. These changes were the redox potential, dielectric constant values alterations along with a change in buffer properties of water and buffers. This indicates of charge distribution alteration and the presence of heterogeneity in the treated aqueous medium, which is typical to exclusion zone (EZ) water [2,3]. Exclusion zone, formed on the interfaces of water and hydrophilic surfaces, excludes particles and solutes—the property to which it owes its name.

Proteins play a key role in the function of cells and tissues. Catalytic, transport, structural, protective, regulatory, and motor functions are the most common and global functions of proteins. Composed of hundreds of amino acids, proteins possess a complex, three-dimensional structure. Their constituent amino acid sequences are encoded in the genome, and the components of protein biosynthesis form one of the most impressive example of machinery in nature.

Protein oxidation is a well-known reason for aging [4]. As proteins are responsible for managing a vast number of cell functions, their oxidation leads to a gradual accumulation of faults and cellular failures. Proteins are oxidized by reactive oxygen species (ROS), such as hydrogen peroxide, singlet oxygen, organic peroxides, nitric oxide, and hydroxyl and superoxide radicals. These compounds are highly reactive and are capable of damaging both small and large organic molecules due to their possession of valent, unpaired electrons allowing for nucleophilic attack of nearby cellular components. An imbalance in ROS generation and its elimination leads to oxidative stress [5]. Oxidation of a protein leads to the formation of chemical radicals within amino acids themselves and increases the chance of structural deformation.

Protein stability is one of the parameters that allow cells to withstand oxidative stress and premature aging. Cysteine residues are especially sensitive to oxidation because the thiol group in cysteine can be oxidized in several ways to form sulfenic acid (SOH), a disulfide bond (S-S), and sulfinic (SO_2H) or sulfonic acids (SO_3H). The latter two are formed irreversibly [6,7]. Thiol groups of cysteine residues are secured against oxidation and other external influences because they play an important role in the protected catalytic and regulatory sites within proteins and enzymes [8]. However, the structural rigidity of proteins lessens as they age, which may lead to the exposure of catalytic sites and thiol groups to oxidation and other damaging factors.

There are chemical and physical ways to protect biological macromolecules from premature damage. Infrared light treatment as a physical influence has shown particularly interesting results. It has been demonstrated that in vivo treatment with near-infrared light at 670 nm reduced the negative effects of traumatic injury to the central nervous system in the form of oxidative stress in the optic nerve, secondary to a decrease in ROS production and metabolism and resulting in restored visual function [9,10]. Other research suggests that brief (2 min, 24 s) 670-nm LED treatments in the course of photobiomodulation may enhance recovery from retinal injury and other ocular diseases in which mitochondrial dysfunction is postulated to play a role [11]. Infrared LED (855 nm) irradiation resulted in a decrease in ROS production following the induction of oxidative stress in human dental pulp cells by lipopolysaccharide without a significant effect on cellular metabolism [12]. The use of a 950 nm LED in a pilot study that investigated the analgesic efficacy of light-emitting diode therapy suggested a general analgesic effect of LED irradiation in favor of the experimental group, although it was not statistically proven [13]. The same authors demonstrated enhanced wound healing in case of treatment with the red-light emitting diode [14]. Lee S.Y.C. et al. concluded that 1072 nm infrared light had a photobiomodulation effect which resulted in an enhanced immune response to cutaneous bacterial infection in mice and significantly increased expression of key immunomodulatory genes [15]. In another study, irradiation with a 904 nm low-level laser reduced inflammatory cell migration in mice with lypopolysaccharide-induced peritonitis in a dose-dependent manner [16].

Danno et al. found that IR radiation suppresses UV-B induced sunburn [17]. Menezes and coworkers have suggested that preliminary irradiation with IR light, which happens during sunrise, is a natural mechanism of cell protection against the solar UV radiation, and have shown that visible to near IR radiation at 400–2000 nm protects human dermal fibroblasts from UV-A and UV-B cytotoxicity [18].

Interestingly, IR also increases the width of exclusion zones (EZ) in water that emerge at natural and artificial hydrophilic surfaces spontaneously [19,20]. It is quite possible that "base-line" environmental IR radiation fuels the formation of EZ [21]. Thus, all cellular membranes and surfaces of macromolecules can be considered as hydrophilic interfaces where EZ originates in living organisms that consist mostly of water.

Since the mid-20th century, it has been postulated that the protein folding process, which is essential for establishing a protein's function, is caused by an opposite shift in entropy between polypeptide and solvent [22,23]. In protein folding conformational entropy loss occurs along with translational entropy gain in the solvent. Protein molecules fold under the influence of thermal pressure, which differs at the protein surface and in the bulk solvent. The proper natural tertiary and quaternary structures are obtained and maintained at maximum translational entropy in the surrounding aqueous media. The loss of conformational entropy consequent of protein folding is compensated for by the increase in hydration entropy, which results from the 'freeing' of water previously held in order by the hydrophobic residues at the aqueous interface, ultimately allowing the protein to fold.

In this study, we have proceeded to investigate the ability of humid air illuminated with IR affect the activity of enzymes in aqueous medium that were exposed to damaging factors (heat, oxidation, and chemical denaturants).

2. Results

2.1. Establishing Enzymatic Activity Controls

Our first task was to find out whether CoHu has any effect upon normally functioning (non-inactivated) enzymes. For that purpose, all enzyme samples were divided in 3 groups as shown in the legend. The data presented in Figure 1A show the effect of CoHu on normally functioning alkaline phosphatase, Figure 1B—horseradish peroxidase and Figure 1C—the effect of CoHu on intact catalase. For catalase (C), residual quantity of H_2O_2 (catalase substrate) in the catalase-assisted degradation reaction is shown, which reflects the activity of the enzyme. The figure shows that CoHu does not have any significant impact on undamaged enzyme activity. Data from this figure will be used as intact enzyme control in other experiments.

2.2. Heating Inactivation Test

The results for HRP are presented in Figure 2. After heating the activity of HRP decreased 4.5 times compared to intact enzyme, the mean of 0.308 a.u. against 1.360 a.u (Figure 1B). Samples treated with CoHu before and after inactivation had higher activity than the rest of the samples. Moreover, treatment of the enzyme before inactivation had a stronger effect than after, which indicates a possible protective effect. The mean increase in peroxidase activity in the end of the reaction in samples treated with CoHu before and after inactivation was 24.9% and 14.9%, respectively, compared to inactivated control. In Figure 2 the p value between the end points marked with * and *** is 0.0287, between * and ** p is 0.0485, and between ** and *** p is 0.0366.

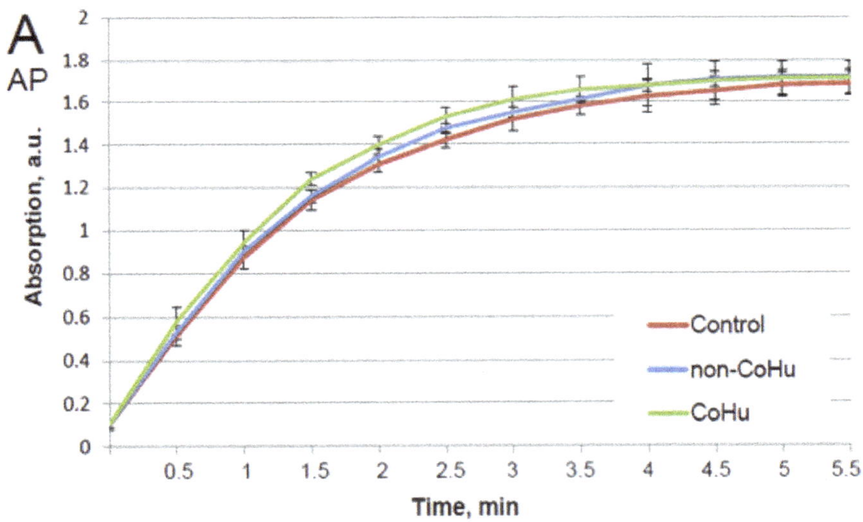

Figure 1. *Cont.*

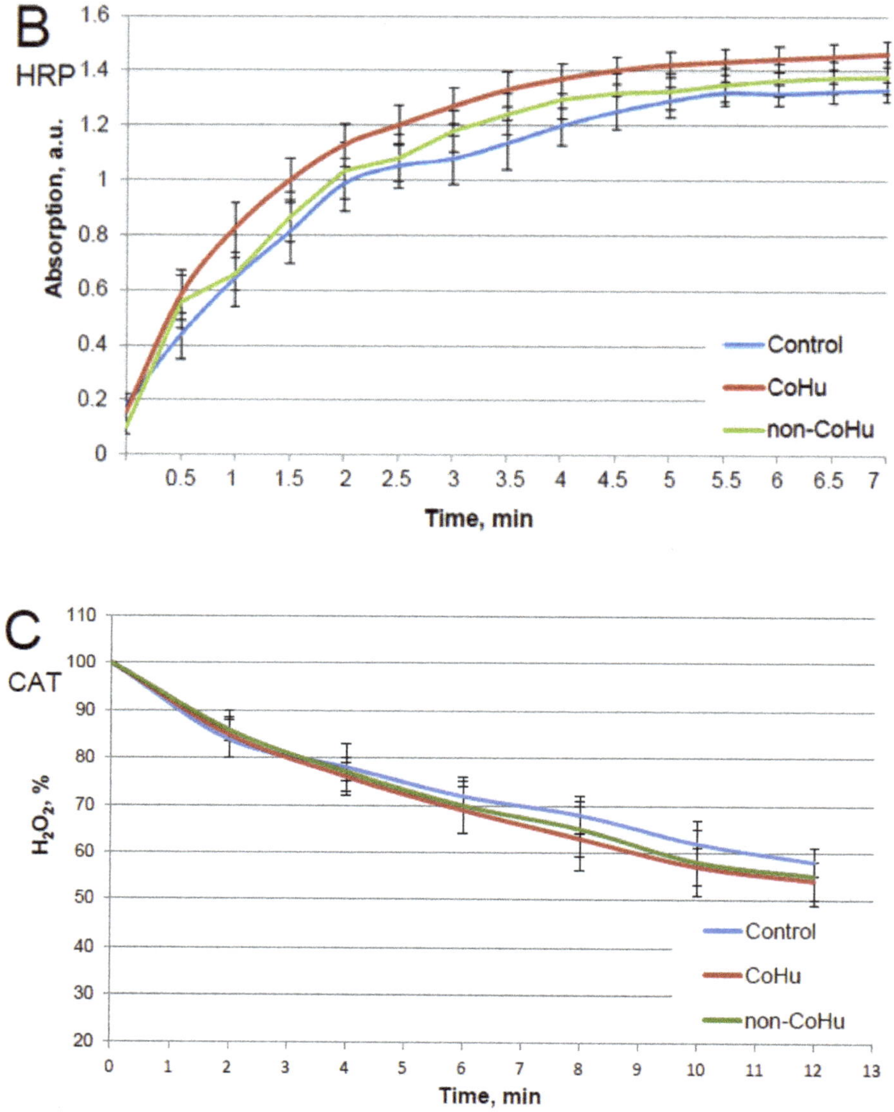

Figure 1. Effect of CoHu on alkaline phosphatase ((**A**), AP), HRP (**B**) and catalase ((**C**), CAT) activity in vitro. Data presented as mean arbitrary units of 7–8 measurements ± SD. Control—the activity of enzymes samples that were not treated with humid airstram. CoHu—the activity of enzymes that were treated with CoHu, and non-CoHu—the activity in control samples.

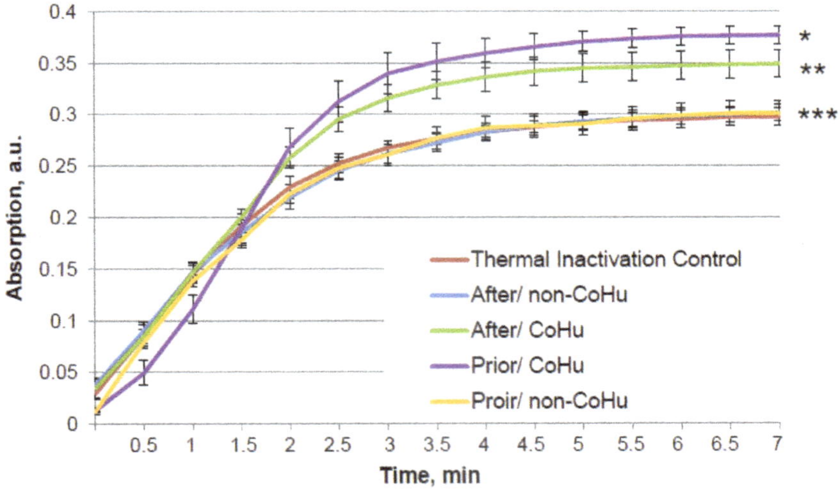

Figure 2. The effect of CoHu on thermally inactivated HRP. Data presented as mean arbitrary units of 8 measurements ± SD. Prior/non-CoHu means that samples were treated with non-CoHu before thermal inactivation. After/CoHu means that samples were treated with CoHu mode after thermal inactivation. Prior/CoHu means that samples were treated with CoHu before thermal inactivation. *, ** and ***—Lines with different superscripts differ significantly ($p < 0.05$).

The same type of test was performed with alkaline phosphatase. Heating of alkaline phosphatase resulted in a 7-fold loss of its activity. However, Figure 3 shows that treatment of inactivated enzyme samples with CoHu after inactivation had the most remarkable restoring effect. The activity of the enzyme in all other samples that were treated with CoHu was also higher than in heat-inactivated controls that had only 14% of the activity of intact enzyme (ref. Figure 1A). The mean increase in alkaline phosphatase activity in the end of the reaction in samples treated with CoHu after inactivation was 130% in the last point of the measurement compared to inactivated control, which was 32% of the activity of intact enzyme. Optical density of the product of the reaction in samples that were treated with CoHu before heating differs from inactivated control ($p = 0.032$). However, there is no significant difference between samples treated with non-CoHu.

Figure 4 shows the dynamics of changes in the content of hydrogen peroxide during an enzymatic reaction. It was shown that in thermally inactivated controls the substrate is not much consumed (83.5% of H_2O_2 remains versus 57% of substrate in the reaction with intact enzyme), but in the samples treated with CoHu, the substrate consumption is increased, therefore, the enzyme activity is increased. Activity comparison was based on the data of the last points of the plots. Treatment with CoHu before or after thermal inactivation allowed to partially restore the catalytic functions of the enzyme (+17.8% and +13.4%, respectively). There is no statistically significant difference between data of these two groups of samples (t-test, $p > 0.05$), however there are reliable differences between these two groups together, control and samples that were treated with non-CoHu.

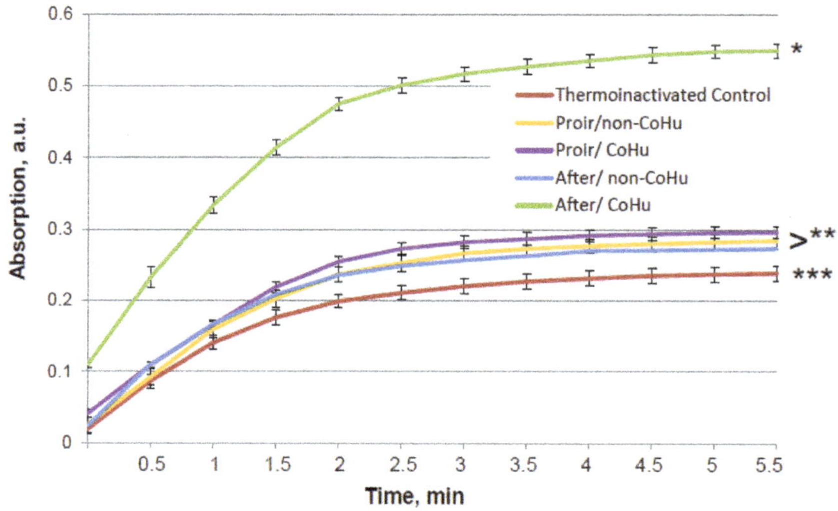

Figure 3. Effect of CoHu on thermally inactivated alkaline phosphatase activity. Data presented as mean arbitrary units ± SD of 10 measurements. Prior/non-CoHu means that samples were treated with non-CoHu before thermal inactivation. After/CoHu means that samples were treated with CoHu after thermal inactivation. Prior/CoHu means that samples were treated with CoHu before thermal inactivation. *, ** and ***—Lines with different superscripts differ significantly ($p < 0.05$).

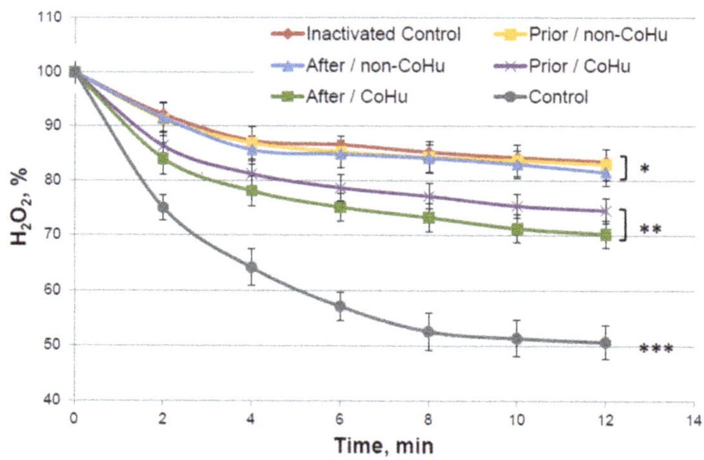

Figure 4. Residual quantity of H_2O_2 (catalase substrate) in the catalase-assisted degradation reaction. Data reflect the activity of the enzyme. *, ** and ***—Lines with different superscripts differ significantly.

2.3. Oxidative Inactivation Test

Figure 5 shows that mixed function oxidation (MFO) decreased HRP activity 9.8 times. Treatment with non-CoHu did not cause changes in activity compared to oxidized control. The activity of samples treated with CoHu before oxidation made it possible to save the mean of 45.3% of HRP activity (450% of oxidized control), while samples treated after inactivation restored 42.4% (421% of oxidized control). In Figure 5 the p value between end points marked with * and **** is 0.0087, between end points marked with ** and **** p is 0.0479, and between ** and *** p is 0.0106. Thus, the enzymatic activity of HRP was either partially protected or restored by treatment of the samples with CoHu both prior to and after oxidation. The pH values were controlled in the samples before measurements.

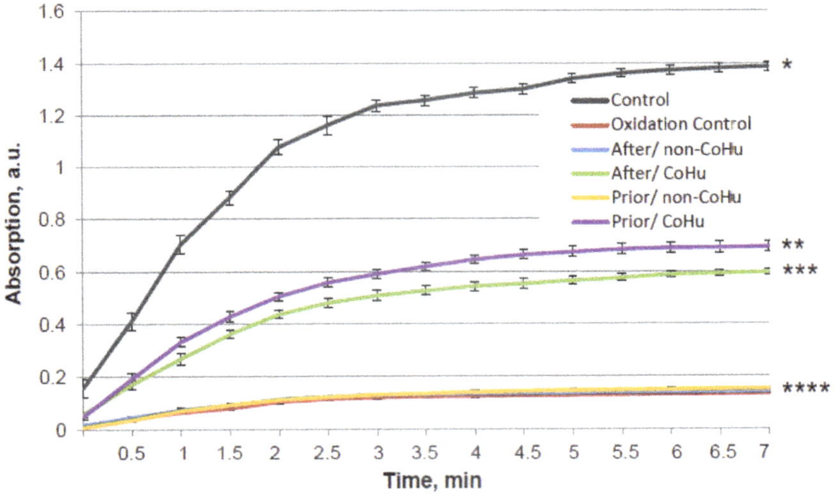

Figure 5. The effect of CoHu on oxidized HRP. Data presented as mean arbitrary units of 10 measurements ± SD. Prior/non-CoHu means that samples were treated with non-CoHu before inactivation. After/CoHu means that samples were treated with CoHu after inactivation. Prior/CoHu means that samples were treated with CoHu before inactivation. *, **, *** and ****—Lines with different superscripts differ significantly ($p < 0.05$).

In Figure 6, it is shown that treatment with CoHu before and after oxidation had a notable protective effect on the activity of alkaline phosphatase. Only 6.4% of alkaline phosphatase activity remained after oxidation in oxidized control samples (red line). Treatment with non-CoHu led to a slight increase of activity. Treatment with CoHu after inactivation helped to restore 19.8% of the intact enzyme activity (309% of oxidized control), while treatment before oxidation partially protected the enzyme and saved a mean of 18.1% of intact enzyme activity (282% of oxidized control). In Figure 6 the p value between lines marked with * and ** is 0.0481, between lines marked with ** and **** p is 0.0187, and between lines marked with *** and **** p is 0.0494. In Figure 6 the activity of the intact enzyme control is not shown to pay more attention to the details in the activity of inactivated samples. The control activity values can be found in Figure 1A.

Figure 7 presents data of the MFO test with catalase. In this figure substrate consumption of intact enzyme is presented (grey line). Decrease of hydrogen peroxide concentration is dependent on the activity of the enzyme. The higher the hydrogen peroxide concentration in the end of the reaction—the lower the activity of the enzyme.

By the 12-th minute of the measurement a mean of 59% of H_2O_2 remains in the medium of the undamaged enzyme, while oxidized catalase leaves 87.12% of the substrate unconsumed. Samples that were treated with non-CoHu did not differ from oxidized controls. A reliably increased consumption of H_2O_2 was observed in samples that were treated with CoHu after oxidation (81.55%, +5.57% to oxidized control). Samples treated with CoHu after inactivation are reliably different from the samples treated with non-CoHu ($p = 0.0455$). Treatment before oxidation tends to preserve the enzyme but not significantly ($p = 0.068$). Samples treated with CoHu before inactivation (purple line) are not reliably different either from the samples treated with non-CoHu or samples treated with CoHu after inactivation (green line).

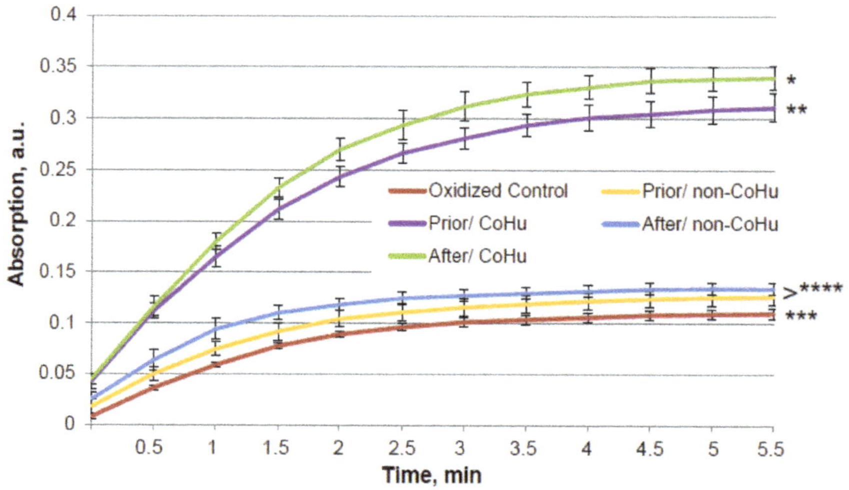

Figure 6. Effect of CoHu on oxidized alkaline phosphatase activity. Data presented as mean arbitrary units ± SD of 10 measurements. Prior/non-CoHu means that samples were treated with non-CoHu before inactivation. After/CoHu means that samples were treated with CoHu after inactivation. Prior/CoHu means that samples were treated with CoHu before inactivation. *, **, *** and ****—Lines with different superscripts differ significantly ($p < 0.05$).

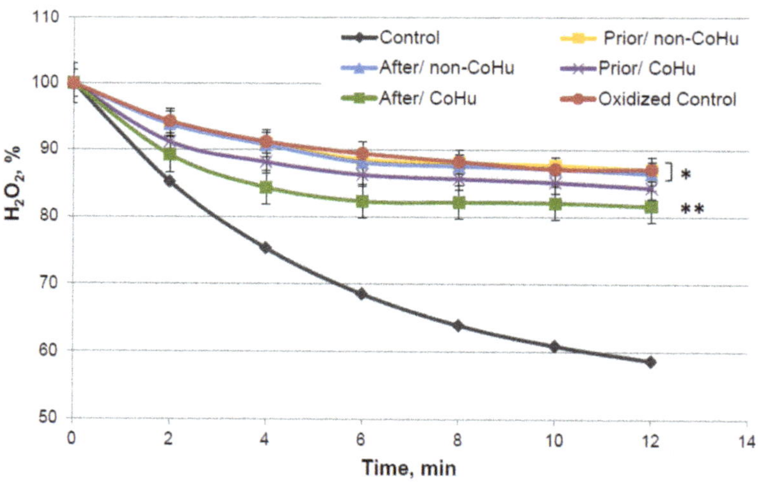

Figure 7. Residual quantity of H_2O_2 (catalase substrate) in the catalase-assisted degradation reaction. Data reflect the activity of the enzyme. * and **—Lines with different superscripts differ significantly ($p < 0.05$).

2.4. Chemical Denaturant Test

Testing whether treatment with humidified air irradiated by IR could partially restore the activity of denatured enzymes was the next objective of this study.

The results for HRP are presented in Figure 8. The grey line represents normal HRP activity as a control. After the unfolding, 32.2% of HRP activity remained (red line). Treatment with non-CoHu did not differ from denatured controls. A restoring effect was observed in response to the use of CoHu both before (53.4% of intact enzyme activity and 166% of denatured control) and after unfolding (60.9% of intact enzyme activity and 189% of denatured control). Therefore, treatment after denaturation had a greater impact. In

Figure 9 the p value between lines marked with ** and *** is 0.0297, between lines marked with *** and **** p is 0.0213, and between lines marked with * and ** p is 0.0076.

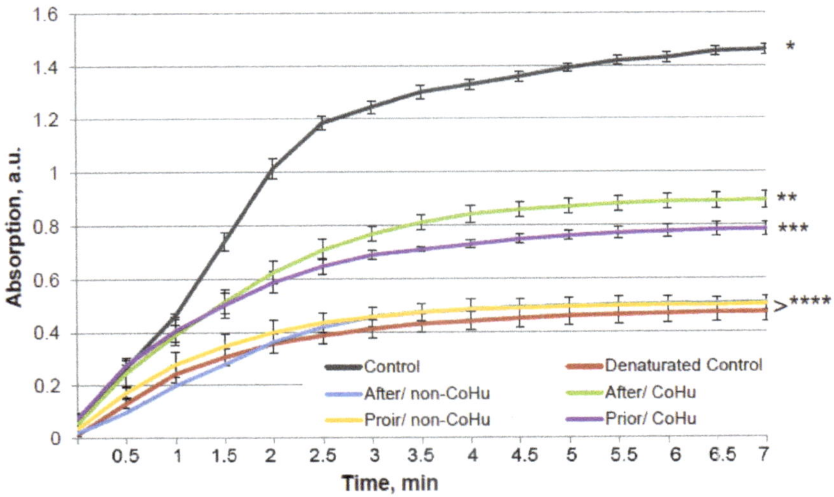

Figure 8. The effect of CoHu on denatured HRP. Data presented as mean arbitrary units ± SD of 8 measurements. Prior/non-CoHu means that samples were treated with non-CoHu before inactivation. After/CoHu means that samples were treated with CoHu after inactivation. Prior/CoHu means that samples were treated with CoHu before inactivation. *, **, *** and ****—Lines with different superscripts differ significantly ($p < 0.05$).

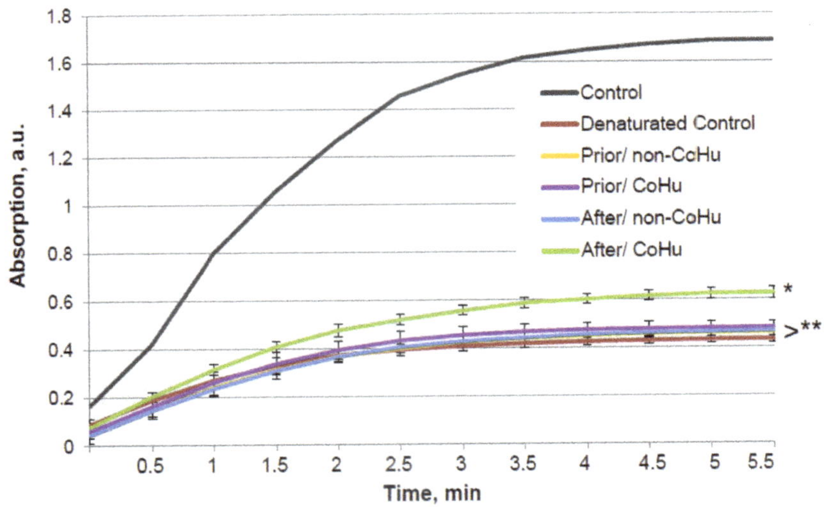

Figure 9. Effect of CoHu on denatured alkaline phosphatase activity. Data presented as mean arbitrary units ± SD. Control—activity of an intact enzyme not exposed to unfolding buffer. Prior/non-CoHu—activity in samples that were treated with non-CoHu before inactivation. After/CoHu—activity in samples that were treated with CoHu after inactivation. Prior/CoHu—activity in samples that were treated with CoHu before inactivation. * and **—Lines with different superscripts differ significantly ($p < 0.05$).

A similar test was performed with alkaline phosphatase. The results are presented in Figure 9. The six groups of samples are similar to those in the previous experiment. The grey line shows the activity of a portion of normally functioning enzyme. Exposure to the unfolding solution decreased enzymatic activity to 23.9% in the unfolding control (red line). In the case of alkaline phosphatase, treatment with CoHu after unfolding allowed to partially restore the enzyme activity (35.9% of the intact enzyme activity and 150% to the unfolded control activity), while all other test set-ups showed no significant effect.

As it is shown on Figure 10, there is a reliably significant effect of treatment with CoHu after denaturation compared to denatured control and samples that were treated with non-CoHu at 30 min, 60 min and 90 min. In time points marked as *, ** and *** there is reliable difference between the following three groups of samples: samples treated with CoHu after exposure to 2-ME, inactivated with 2-ME controls, and samples treated with non-CoHu ($p < 0.05$). At the last point at 120 min there is no significant difference between all the sample groups.

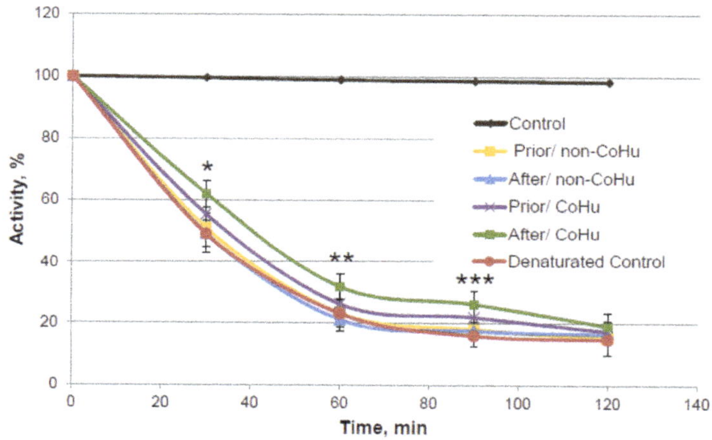

Figure 10. Catalase activity based on the residual quantity of H_2O_2 (catalase substrate) in the catalase-assisted degradation reaction. Control—activity of intact enzyme. Prior/non-CoHu—activity in samples that were treated with non-CoHu before inactivation. After/CoHu—activity in samples that were treated with CoHu after inactivation. Prior/CoHu—activity in samples that were treated with CoHu before inactivation. In time points marked as *, ** and *** there is reliable difference between the groups ($p < 0.05$).

3. Discussion

This pioneering study touches on several little-studied areas at once. First, the exposure of biological objects to light with a wavelength of around 1270 nm has been a rare study object, although the search for biological effects of irradiation with light in the infrared region has long been successful. Second, the effect of such irradiated humidity (CoHu) on the redox abilities of aquatic environments and enzymes, which occupy the basal, molecular level of life organization, has not been previously studied. In other studies, such as those mentioned in the introduction, the objects of observation were cells and tissues, leaving it unclear what molecular substances were responsible for the observed effects. An important detail of this study is that the objects observed herein were not irradiated directly but rather treated with irradiated humidified air (CoHu).

The test results show IR-treated humid airstream does not affect normally functioning enzymes (Figure 1). Treatment of the samples with the non-CoHu airstream did not affect the activity of thermally inactivated peroxidase and catalase. However, treatment of the thermally inactivated samples with CoHu both before and after inactivation significantly increased HRP and catalase activity (Figures 2 and 4). As for alkaline phosphatase, exposure to non-CoHu slightly increased enzyme activity due to the additional air supply. The activity in the samples that were treated with CoHu before inactivation is comparable to the results with peroxidase. It is noteworthy that treatment with CoHu after inactivation markedly increased the activity of alkaline phosphatase (Figure 3).

In tests using mixed function oxidation (MFO), HRP and alkaline phosphatase activity was partially restored when the samples were treated with CoHu, both before and after inactivation (Figures 5 and 6). The enzymes responded slightly differently to treatment with the irradiated humid airstream, but the general trend in enzyme activity is the same for both. Oxidized HRP activity was higher in the case of treatment with CoHu before oxidation, while oxidized alkaline phosphatase activity was higher in case of treatment after oxidation. Oxidized catalase activity was partially restored by treatment with CoHu after inactivation (Figure 7). Treatment with non-CoHu did not reveal any protective or restoring effect in tests with HRP and catalase (Figures 5 and 7).

The chosen enzymes represent different classes with varying action mechanisms and substrates. Therefore, the damaging factors as heat, oxidation and denaturation affect particular molecule parts in every case. In some circumstances damaging effects can combine. Thus, at the base of MFO systems action a 3-step modified metal-catalyzed Haber-Weiss cycle is proposed [24,25]:

$$\text{ascorbate} + 2M^{n+} \rightarrow 2M^{(n-1)+} + \text{dehydroascorbate} + 2H^{+} \quad (1)$$

$$\text{ascorbate} + O_2 \rightarrow \text{dehydroascorbate} + H_2O_2 \quad (2)$$

$$M^{(n-1)+} + H_2O_2 \rightarrow M^{n+} + OH^{\bullet} + OH^{-} \quad (3)$$

In this context, reduced metals and hydrogen peroxide interact to form a highly reactive $OH^{\bullet}$ by a Fenton mechanism. Meanwhile, Cu^{2+} is capable to binding to HRP, leading to a conformational change, causing partial denaturation of HRP to an extent of 54% [26]. Thus, the HRP activity decrease was partially due to oxidation and denaturation combined. Inactivation of HRP by excess amounts of H_2O_2 by mechanism-based pathway has been well described [27,28]. Reversible and irreversible mechanisms of HRP inactivation have been proposed. At H_2O_2 concentrations below 1.0 mM, a reversible mechanism prevails, while at higher H_2O_2 concentrations, irreversible inactivation takes place. This is a plausible explanation for the observed HRP inactivation as hydrogen peroxide is formed in MFO. The regenerative and protective effect of CoHu on HRP under these conditions, therefore, should be associated with its antioxidant properties.

Catalase activity was partially restored when samples were treated with CoHu after inactivation, while treatment before inactivation had a tendency to protect the enzyme (Figure 10). Copper affects the secondary structure when it binds near a heme group in catalase, as well as the binding of Cu^{2+} to bovine liver catalase destroyed H-bonds [29]. We specifically used the least damaging effect and the lowest effective concentration of the MFO so that irreversible damage to the enzymes did not occur, after which there would be no hope of any restoration of function. Treatment of oxidized samples with air humidity irradiated with infrared light has shown good protective and reducing properties with respect to oxidation. The question remains whether this effect can be translated to an entire organism in vivo.

The employed unfolding buffer affected the thiol bonds in proteins. When HRP, alkaline phosphatase and catalase structures were disturbed in the unfolding buffer, the activities of the enzymes were increased by treatment of the samples with CoHu after inactivation (Figures 8–10). HRP was partially preserved even when it was treated with air humidity before exposure to unfolding (Figure 8).

The wavelength range around 1270 nm was not chosen randomly. The idea was to study the effects of a wavelength close to the one of singlet oxygen phosphorescence [24]. To date, many claims have been made about the positive role of reactive oxygen species (ROS) in cell signaling, gene expression [25], cell differentiation, apoptosis, and immune and nervous system functioning [30].

Singlet oxygen, with the oxygen atom in the singlet $^1\Delta_g$ state, occupies a special place among ROS [31]. This short-living ROS can be detected by its emission at 1268 nm during $O_2(^1\Delta_g) \rightarrow O_2\ (^3\Sigma^-{}_g)$ transition in enzymatic reactions and in cells [32]. Singlet oxygen is characterized by a short maximum lifetime of 0.6 μs [33] and, therefore, can only travel short distances, roughly 100 nm, from its place of origin [34]. It has a relatively short list of direct reactions, yet its contribution to cell signaling and oxidative stress correction is crucial. Singlet oxygen is a major agent generated by many different cell types, especially by neutrophils in the course of their immune response to infection. Interestingly, water absorbs intensely in the IR region of 1270 nm [35].

Krasnovskiy and his group have shown that irradiation of water and organic solvents saturated with air at a wavelength of 1268 nm with the power of 30–150 mW led to formation of singlet oxygen [36]. However, the concentration of free oxygen is significantly lower in tissues, which reduces the rate of $O_2(^1\Delta_g)$ formation.

Later Anquez et al. showed that the 1270 nm direct photosensitizer-free laser irradiation initiated singlet oxygen formation in cell culture and induced cancer cell death [37,38]. The lethal effect depended on the radiation wavelength, intensity and oxygen content in the media. However, this publication raised doubts that the observed cell death was due to photoexcitation of dissolved oxygen as the primary absorber of 1270 nm light in the biological environment is water due to the low extinction coefficient for molecular oxygen at 1270 nm [39]. Another argument to support this suggestion is that biological effects appear not only at irradiation at 1270 nm exactly, but at the range of about 1200–1300 nm, where water absorbs intensely.

Water plays a crucial role in protein folding, action, and dynamics [40]. From the thermodynamic point of view, the balance between intramolecular forces and hydration energy of water keeps protein structure stable and facilitates conformational changes [41,42]. Data from crystallography, neutron scattering, and simulations allowed proposing that water layer at protein molecule surface is 10% more dense than bulk water [43,44] and it has a reduced freezing point [45]. The content of such non-freezing water was 9–15% in coacervating systems (lactoferrin/ß-lactoglobulin (heteroprotein) coacervate and others). There is evidence of water's role as a moderator and mediator in protein–substrate interactions, in protein folding and in protein-protein interactions mainly by means of H-bonding formation inside the protein molecule or between molecule and water [46].

Over the course of several decades, Dr. Gerald Pollack proved the presence of the "fourth phase of water," which is formed in bulk water after absorption of radiant energy [19]. It is called Exclusion Zone (EZ) water, which is formed at hydrophilic surfaces, such as biological membranes, most macromolecules, and at phase interfaces. Charge separation takes place in it, since negatively charged hydroxide ions are concentrated along the surface, and positively charged H_3O^+ particles are pushed out of the EZ along with particles and microbubbles. Studies have shown that the absorption of UV and the visible light by water leads to the growth of exclusion zones and, in some cases, to its appearance, yet light in the infrared region of the spectrum has been found to be especially effective in impacting EZ water [20]. Because of its property of charge separation, EZ water is suggested to be an energy source for biological processes [47]. Numerous studies in laboratories worldwide show that the interfacial zone differs physically from the bulk zone: it is a more ordered phase of water, appearing to behave like a liquid crystal that can co-exist with the contiguous bulk phase. Negatively charged exclusion zones can be a source of electrons for oxidized molecules, and the low entropy of structured water can equalize the increased entropy of denatured and damaged molecules. Thus, EZs act as universal antioxidants as well. According to G. Pollack, fully built EZs around each protein seem necessary for optimal cellular functioning [48].

Pollack's model is based on electrostatic laws. Though it allows to explain many experimentally established properties of EZ-water not all its features readily follow from it. On the other hand, long before EZ-water was discovered Emilio Del Giudice and Giuliano Preparata have suggested a model of water where it is represented as a two-phase system in which one fraction of water belongs to the so-called Coherent Domains (CDs), where water molecules oscillate in phase, and the remaining fraction of water constitutes a non-coherent phase. Exclusion zone water has many measured properties of theoretical CDs [49,50]. One of the major features of CDs is that they represent reservoirs of quazi-free electrons that are easily excitable and may be donated to appropriate acceptors. The reason for this lies in the fact that the energy of the excited state of the CD is 12.06 eV, which is very close to the ionization energy of water molecules of 12.6 eV [51]. Ability of EZ-water to be the source of electrons was demonstrated experimentally [48]. The coherent state of water may be significantly stabilized in cases where water is attracted to nearby surfaces, thus shielding its coherent state from the disruptive effects of collisions [52]. Therefore EZ-water has many measured properties of theoretical CDs [49]. They both have notable redox properties, electron exchange, and dynamic stability, which provide for the enhanced ability of the water to stabilize macromolecules present in aqueous systems containing CDs against degradation caused by external factors. As regards humid air irradiated with IR in this study it should be mentioned that in a study on aerosols formed at waterfalls, Madl et al. [53] have shown the presence of negatively charged nanometer-sized and up to 100 nm large water clusters that contain millions of water molecules. Both negative surface charge and the size are in accordance with the properties of CD. Thus, these aerosol water clusters can act as a surface, and exposure to IR is expected to corroborate their coherent properties (that is why we denoted IR-treated water vapor "Coherent Humidity, CoHu".

In our study we did not treat protein solutions with IR directly, instead we used irradiated humid air as an intermediate for several reasons. Firstly, irradiation of water with IR leads to its heating while microdroplets in the air lose heat quicker. Secondly, water microdroplets have an exclusion zone water layer on their surface [48], and their total surface is much larger than that of liquid water. And finally, water vapor is reported to absorb IR energy more intensely than liquid water [54] making it an inducer of EZ in bulk liquid water. When a coherent IR LED radiation signal is transferred to air with suspended microdroplets, they begin to oscillate coherently and create coherent domains (CDs), which is suggested to be more effective than irradiating liquid water.

We hypothesize that irradiation of the aquatic environment with light from the infrared region of the spectrum at wavelengths of singlet oxygen phosphorescence leads to formation of exclusion zones and coherent domains in the aqueous media. Further experiments will be conducted to confirm or refute this hypothesis. Thus, singlet oxygen may be a natural source of radiant energy on molecular and cellular levels for building and enhancing EZs.

4. Materials and Methods

4.1. Illumination

CoHu was produced with a NanoVi® device (Eng3 Corporation, Seattle, WA, USA) that humidifies air with water microdroplets 1–10 μm in size using pure distilled water (Figure 11). Humidifier excitation units and a control unit were used. The main feature of the technology is irradiation of the humid airstream with infra-red electromagnetic energy to modify its physico-chemical properties. We have called the modified air humidity coherent humidity (CoHu). This irradiated humid air can be applied for treatment of bulk water or other objects. Unpurified ambient air with initial humidity of 22–26% is used as the carrier gas. Ambient air is pumped (Pump in) and pressed through the diffuser to water in the humidifier glass container. With another pump (out) the humidified air is pressed through the quartz glass tube of the excitation unit equipped with 24 LEDs, a heat sink element and a thermo sensor. Each pump creates an air output of 4 L/min. The humid air (70–85% humidity) goes through a 40 mm long quartz glass tube with an inner diameter of 3 mm. Within glass tube the water vapor absorbs energy that is emitted by the LEDs placed around the tube at a distance of 2.3 mm. The diameter of each LED is 4.7 mm, and they emit IR energy in the range of 1270 nm with a beam cone of 30° and the output power of 700 pW. The control unit allows to switch the LEDs off and to obtain the non-irradiated humid air for control measurements (non-CoHu). Through the 70 cm long flexible plastic outlet with the inner diameter of 3.2 mm CoHu (or non-CoHu) exits the device.

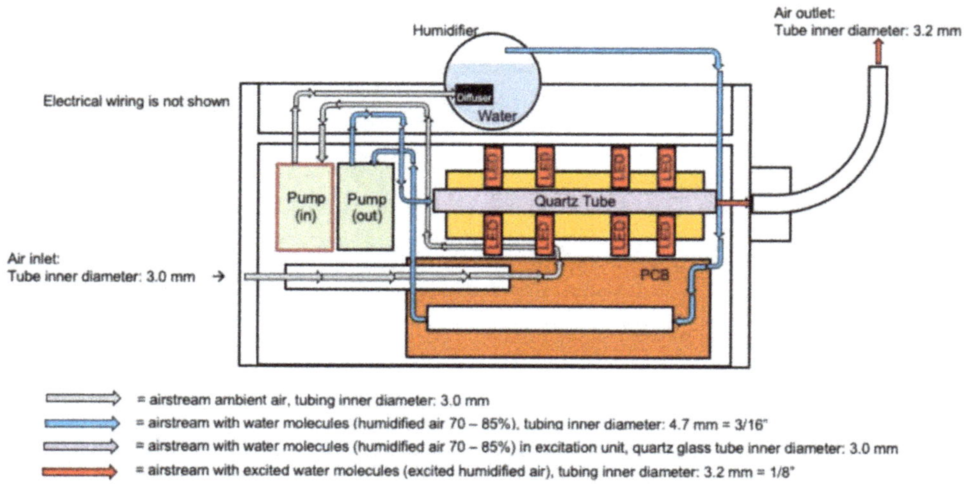

Figure 11. The scheme of the apparatus for CoHu preparation.

Enzyme samples were placed in sterile glass flasks of varied volume, depending on the experimental needs, and humid air supply tube was placed 4 cm above the sample surface at ambient light. Several parallel samples were treated with CoHu and non-CoHu for 1 to 5 min depending on experiment design. After treatment samples were used in the experiments within 3 min. Untreated samples were left as controls.

4.2. Horseradish Peroxidase Activity Measurement In Vitro

Horseradish peroxidase (HRP) type II (RZ > 2.5, 60 U/mg), KH_2PO_4, and hydrogen peroxide 30% (*v*/*v*) were obtained from Merck. Potassium-phosphate buffer solution at 0.05 M (pH 6.8) was used as the solvent system for all other reagents and was prepared in deionized water with resistivity of 18.3 MOhm × cm (Millipore, Merck, Darmstadt, Germany).

Specific activity of peroxidase was determined spectrophotometrically (Specord® Plus Analytic Jena spectrophotometer, Germany) in an enzymatic reaction with 2,2-azino(bis(3-ethylbenzothiazoline-6-sulfonicacid)) (ABTS) [27,55,56]. The chromomorphic biocatalytic reaction can easily be monitored by following formation of the $ABTS^{+\bullet}$ radical at the maximum absorption wavelength of 414 nm. One unit (U) is the amount of HRP that catalyzes the oxidation of 1 μmol of ABTS per minute.

The 0.05 M potassium-phosphate buffer was prepared by the following routine. Solution 1: 75 mL of 0.2 M KH_2PO_4 was prepared by adding 2.04 g of salt to 75 mL of water; Solution 2: 50 mL of 0.2 M KOH was prepared by adding 0.56 g of salt to 50 mL of water; 33.6 mL of Solution 2 was added to 75 mL of Solution 1, and water was added to obtain 300 mL of the final buffer solution.

ABTS 0.025 M stock solution was prepared by dissolving 0.0515 g of ABTS in 4 mL of the buffer. Then, 4 μL of the stock solution was added to 2 mL of reaction medium.

HRP stock solution was prepared by dissolving 0.0125 g of the enzyme in 2 mL (A_{403} = 2.61) of the buffer. Then, the stock solution was diluted 10-fold and 10 μL of this solution was added to 2 mL of the reaction medium.

The reaction medium consisted of 1.9 mL of buffer, 4 μL 30% H_20_2, 4 μL 0.025 M ABTS, and 10 μL HRP. The total reaction lasted 300 s.

In the course of this test set-up, samples containing intact enzyme in buffer were used as controls; samples containing enzyme treated with non-CoHu were also used for comparison against samples of enzyme treated with CoHu. Also, the impact of treatment of samples with coherent and non-coherent humidity was studied both before and after damaging factors were added.

For 1-min long treatment with CoHu or non-CoHu, 10 µL HRP in 1.9 mL of the buffer were put in 5 mL quartz cuvettes, the air outlet of NanoVi® was placed 4 cm above the solution surface. Control samples were not exposed to the treated airstream.

4.3. Alkaline Phosphatase Activity Measurement In Vitro

Materials were purchased from the following sources: Alkaline phosphatase (*E. coli*, M = 190,000), Tris from AppliChem, L-(+)-ascorbic acid, $MgCl_2$, $CuSO_4 \times 5$ H_20 from Merck, $ZnCl_2$, 2-mercaptoethanol, guanidine HCl from Reachem. Ascorbate stock solution was prepared daily in 50 mM Tris buffer, pH 7.4. All solutions were prepared in 50 mM Tris, pH 9.8, dissolved in water and distilled through a Milli-Q water purification system (Merck, Darmstadt, Germany).

Stock solutions of *E. coli* alkaline phosphatase were prepared by dissolving 200 units and 4 mg of the enzyme in 400 µL of Tris buffer, pH 8.4, containing 50 mM Tris, 1 mM $MgCl_2$, 0.1 mM $ZnCl_2$. Protein concentration was calculated from the absorbance at 278 nm by using A1% 1 cm = 7.9.

10 mm quartz spectrophotometer cuvettes with 3 mL of 50 mM Tris buffer were thermostated at 30 °C for 5 min, and a 10 µL aliquot of the enzyme stock solution was added afterward. The reaction was initiated by adding a solution of 4-nitrophenyl phosphate immediately before measuring the optical density at a wavelength of 410 nm (the molar absorption coefficient of 4-nitrophenolate ion is 18,500 $M^{-1} \times cm^{-1}$). Measurements were made against thermostated buffer. The enzyme concentration in all experiments was 3×10^{-12} M, 4-nitrophenyl phosphate—3 mM.

As with peroxidase, we used controls untreated with treated airstream, as well as controls that were treated with non-CoHu. Five mL quartz cuvettes with alkaline phosphatase in buffer were used for 1-min-long treatment with CoHu or non-CoHu. The air outlet was placed 4 cm above the liquid surface.

4.4. Catalase Activity Measurement In Vitro

The catalase assay was developed by Aebi et al., in 1984 [57].

Bovine liver catalase, potassium hydroxide, hydrogen peroxide 30%, KH_2PO_4 and $Na_2HPO_4{\cdot}2H_20$ were purchased from Sigma-Aldrich (St. Louis, MO, USA).

To prepare 50 mM potassium-phosphate buffer with pH 7.0, (a) 6.81 g KH_2PO_4 and (b) 8.90 g $Na_2HPO_4{\cdot}2H_20$ were dissolved in distilled water and made up to 1000 mL each. Solutions (a) and (b) were mixed in the proportion 1:1.5 (v/v). A portion of 0.34 mL 30% hydrogen peroxide was diluted with phosphate buffer to 100 mL.

Bovine liver catalase activity was measured using a UV-Vis spectrophotometer (Specord Analytic Jena, Jena, Germany) using 10 mM H_2O_2 as substrate in 50 mM potassium phosphate buffer (pH 7.0). The procession of hydrogen peroxide was followed at 190 nm. 2.9 mL of 10 mM H_2O_2 was added to a quartz cuvette, then 0.10 mL of catalase solution containing 300 U/mL was pipetted into the cuvette. The obtained solution was immediately mixed by inversion and the decrease in absorbance was accessed by taking one reading per 1 min for 12 min. Residual quantity of H_2O_2 (catalase substrate) in the catalase-assisted degradation reaction was analyzed, which reflects the activity of the enzyme. The higher the hydrogen peroxide concentration in the end of the reaction—the lower the activity of the enzyme.

Treatment with CoHu or non-CoHu was carried out in 2 mL Eppendorf tubes for 1 min with the airstream outlet 4 cm above the surface of enzyme.

4.5. Heating of Enzymes

Five experimental set-ups were used:

1—samples containing heat-inactivated enzyme in a water bath—inactivated control;

2—samples containing the inactivated enzyme treated with non-CoHu prior to inactivation (Prior/non-CoHu);

3—samples containing the inactivated enzyme treated with CoHu prior to inactivation (Prior/CoHu);

4—samples containing the inactivated enzyme treated with non-CoHu after inactivation (After/non-CoHu);

5—samples containing the inactivated enzyme treated with CoHu after inactivation (Prior/CoHu).

Heating was carried out in quartz cuvettes. After heating, enzyme samples were left at room temperature for 20 min for cooling.

Samples containing 1.9 mL of buffer with 10 µL of horseradish peroxidase were heated at 85 °C for 10 min in a water bath. Samples were then cooled at room temperature for 30 min and divided into test groups and heated control. After cooling, the enzymatic reaction was initiated in every tube by adding H_2O_2 and ABTS.

Alkaline phosphatase samples containing 10 µL of enzyme in 3 mL of Tris buffer (pH 8.4) were placed on water bath at 65 °C for 30 min. Then the samples were left at room temperature for 30 min.

Quartz cuvettes with 0.1 mL of catalase were incubated in 2.9 mL potassium-phosphate buffer at 55 °C for 15 min on a water bath.

4.6. Oxidation with Mixed Function Oxidative System

For tests with oxidative inactivation of enzymes enzyme samples were divided into 5 groups:

1—samples containing oxidized enzyme (Oxidized Controls);

2—samples containing oxidized enzyme treated with non-CoHu prior to oxidation (Prior/non- CoHu);

3—samples containing oxidized enzyme treated with CoHu prior to oxidation (Prior/CoHu);

4—samples containing oxidized enzyme treated with non-CoHu after inactivation (After/non-CoHu);

5—samples containing oxidized enzyme treated with CoHu after inactivation (After/CoHu).

In this experiment enzyme stock solutions were exposed to oxidation and treated with Cohu and non-CoHu before or after oxidation for 3 min. Thus, HRP stock solution was separated in two parts. One was the control, and another was oxidized by the addition of 10 mM ascorbic acid (PanReac AppliChem, Darmstadt, Germany) and 0.1 mM $CuCl_2$ (Merck, Darmstadt, Germany). The method of inactivation of enzymes using mixed function solution was performed as previously described by other authors [58,59].

Alkaline phosphatase was incubated for 60 min at 35 °C in 50 mM Tris buffer with 10 mM ascorbate in buffer a 10 mM ascorbate with 0.1 mM $CuCl_2$. Twenty minutes after samples were removed from the thermostat, aliquots were withdrawn from the mixture and assayed for enzyme activity as previously described [60,61]. Each experiment was repeated 10 times. Each compound was added to the enzyme solution immediately after rapid mixing. The reaction was stopped by the addition of EDTA to 0.5 mM.

Buffered catalase samples were incubated with 10 mM ascorbate in buffer with 0.1 mM $CuCl_2$ for 20 min at 25 °C, removed from the thermostat and used in enzymatic activity measurement 20 min afterwards.

4.7. Denaturation of Enzymes

Stock solutions of enzymes were divided in 6 parts. One of them was used for preparation of control samples that were not denatured or treated with vapor. The second was used for preparation of denatured controls that were not treated with vapor. The third was taken to prepare enzyme samples that were denatured and then treated with non-CoHu for 10 min after unfolding. The fourth was used for preparation of enzyme samples that were treated with non-CoHu for 10 min prior to unfolding. The fifth group of samples was treated with CoHu before unfolding, while the sixth was treated with CoHu after unfolding. The stock solution was 5-times more concentrated than required and was diluted 5-fold either in unfolding or basic buffer depending on the test group it belonged to.

A HRP stock solution was prepared by adding 0.025 g of HRP to 2 mL of potassium-phosphate buffer. Denaturation was performed by diluting this stock solution of the peroxidase (12.5 mg/mL) 5-fold into unfolding buffer (25 mM 2-mercaptoethanol (PanReac, AppliChem, Germany), 3.6 M guanidine HCl, and potassium-phosphate buffer at pH 6.8) and incubating for 12 h at 37 °C in 5 mL plastic tubes. A similar method was described by other authors [62,63].

A highly concentrated alkaline phosphatase stock solution was created by dissolving 4 mg of alkaline phosphatase in 80 μL of Tris buffer, which was then diluted 5-fold in denaturation buffer, containing 25 mM 2-mercaptoethanol, 3.6 M guanidine HCl and Tris buffer, as with peroxidase. The enzyme stock solution was incubated for 10 h with denaturation buffer in 2 mL plastic tubes. Then, 10 μL aliquots of the enzyme were withdrawn for estimation of enzyme activity.

A concentrated catalase solution in buffer with 1500 U/mL was diluted 5-fold in denaturation buffer, containing 25 mM 2-mercaptoethanol, 3.6 M guanidine HCl and potassium-phosphate buffer. The enzyme stock solution was incubated for 10 h with denaturation buffer in 2 mL plastic tubes. Aliquots containing 300 units/mL of the enzyme were withdrawn for estimation of enzyme activity at various time points (30 min, 60 min, 90 min and 120 min) after exposure to unfolding buffer.

For controls, the enzymes were dissolved and diluted 5-fold in basic buffers, and enzyme activity was accessed after incubation without any other manipulation of the enzymes.

4.8. Statistical Analysis

Statistical data processing was performed using a PC with Windows 7 software, Statistica Base and Excel 2013. Results are presented as means with their standard error (Student's *t*-test).

Author Contributions: Conceptualization, O.I.Y. and V.L.V.; methodology, O.I.Y.; software, A.V.T.; validation, A.V.T. and V.L.V.; formal analysis, E.V.B. and V.A.M.; investigation, O.I.Y., V.A.M., E.V.B. and K.N.N.; resources, A.V.T. and V.L.V.; data curation, A.V.T.; writing—original draft preparation, O.I.Y.; writing—review and editing, V.L.V.; visualization, E.V.B. and O.I.Y.; supervision, V.A.M. and K.N.N.; project administration, O.I.Y. All authors have read and agreed to the published version of the manuscript.

Funding: This research received no external funding.

Data Availability Statement: Not applicable.

Conflicts of Interest: The authors declare no competing interest.

References

1. Yablonskaya, O.I.; Voeikov, V.L.; Buravleva, E.V.; Trofimov, A.V.; Novikov, K.N. Physicochemical Effects of Humid Air Treated with Infrared Radiation on Aqueous Solutions. *Water* **2021**, *13*, 1370. [CrossRef]
2. Zheng, J.M.; Pollack, G.H. Long-range forces extending from polymer-gel surfaces. *Phys. Rev. E* **2003**, *68*. 031408. [CrossRef] [PubMed]
3. Zheng, J.M.; Chin, W.C.; Khijniak, E.; Khijniak, E., Jr.; Pollack, G.H. Surfaces and interfacial water: Evidence that hydrophilic surfaces have long-range impact. *Adv. Colloid Interface Sci.* **2006**, *127*, 19–27. [CrossRef]
4. Stadtman, E.R. Protein oxidation and aging. *Science* **1992**, *257*, 1220–1224. [CrossRef] [PubMed]
5. Schieber, M.; Chandel, N.S. ROS function in redox signaling and oxidative stress. *Curr. Biol.* **2014**, *24*, 453–462. [CrossRef]
6. Eaton, P. Protein thiol oxidation in health and disease: Techniques for measuring disulfides and related modifications in complex protein mixtures. *Free Radic. Biol. Med.* **2006**, *40*, 1889–1899. [CrossRef] [PubMed]
7. Thomas, J.A.; Mallis, R.J. Aging and oxidation of reactive protein sulfhydryls. *Exp. Gerontol.* **2001**, *36*, 1519–1526. [CrossRef]

8. Pérez, V.I.; Buffenstein, R.; Masamsetti, V.; Leonard, S.; Salmon, A.B.; Mele, J.; Andziak, B.; Yang, T.; Edrey, Y.; Friguet, B.; et al. Protein stability and resistance to oxidative stress are determinants of longevity in the longest-living rodent, the naked mole-rat. *Proc. Natl. Acad. Sci. USA* **2009**, *106*, 3059–3064. [CrossRef]
9. Fitzgerald, M.; Bartlett, C.A.; Payne, S.C.; Hart, N.S.; Rodger, J.; Harvey, A.R.; Dunlop, S.A. Near infrared light reduces oxidative stress and preserves function in CNS tissue vulnerable to secondary degeneration following partial transection of the optic nerve. *J. Neurotrauma* **2010**, *27*, 2107–2119. [CrossRef] [PubMed]
10. Naeser, M.A.; Hamblin, M.R. Traumatic brain injury: A major medical problem that could be treated using transcranial, red/near-infrared LED photobiomodulation. *Laser Surg.* **2015**, *33*, 443–446. [CrossRef]
11. Eells, J.T.; Henry, M.M.; Summerfelt, P.; Wong-Riley, M.T.T.; Buchmann, E.V.; Kane, M.; Whelan, N.T.; Whelan, H.T. Therapeutic photobiomodulation for methanol-induced retinal toxicity. *Proc. Natl. Acad. Sci. USA* **2003**, *100*, 3439–3444. [CrossRef] [PubMed]
12. Montoro, L.A.; Turrioni, A.P.S.; Basso, F.G.; de Souza Costa, C.A.; Hebling, J. Infrared LED irradiation photobiomodulation of oxidative stress in human dental pulp cells. *Int. Endodont. J.* **2014**, *47*, 747–755. [CrossRef]
13. Vinck, E.; Cagnie, B.; Coorevits, P.; Vanderstraeten, G.; Cambier, D. Pain reduction by infrared light-emitting diode irradiation: A pilot study on experimentally induced delayed-onset muscle soreness in humans. *Lasers Med. Sci.* **2006**, *21*, 11–18. [CrossRef]
14. Vinck, E.; Cagnie, B.; Cambier, D.; Cornelissen, M. Do infrared light-emitting diodes have a stimulatory effect on wound healing? From an in-vitro trial to patient treatment. In Proceedings of the Laser Florence 2001: A Window on the Laser Medicine World, Florence, Italy, 15 October 2002; Volume 4903, pp. 156–165.
15. Lee, S.Y.C.; Seong, I.W.; Kim, J.S.; Cheon, K.A.; Gu, S.H.; Kim, H.H.; Park, K.H. Enhancement of cutaneous immune response to bacterial infection after low-level light therapy with 1072 nm infrared light: A preliminary study. *J. Photochem. Photobiol. B Biol.* **2011**, *105*, 175–182.
16. Correa, F.; Martins, R.A.B.L.; Correa, J.C.; Iversen, V.V.; Joenson, J.; Bjordal, J.M. Low-level laser therapy (GaAs λ = 904 nm) reduces inflammatory cell migration in mice with lipopolysaccharide-induced peritonitis. *Photomed. Laser Surg.* **2007**, *25*, 245–249. [CrossRef]
17. Danno, K.; Horio, T.; Imamura, S. Infrared radiation suppresses ultraviolet B-induced sunburn-cell formation. *Arch. Dermatol. Res.* **1992**, *284*, 92–94. [CrossRef]
18. Menezes, S.; Coulomb, B.; Lebreton, C.; Dubertret, L. Non-coherent near infrared radiation protects normal human dermal fibroblasts from solar ultraviolet toxicity. *J. Investig. Dermatol.* **1998**, *111*, 629–633. [CrossRef] [PubMed]
19. Chai, B.; Yoo, H.; Pollack, G.H. Effect of radiant energy on near-surface water. *J. Phys. Chem. B* **2009**, *113*, 13953–13958. [CrossRef] [PubMed]
20. Wang, A.; Pollack, G.H. Effect of infrared radiation on interfacial water at hydrophilic surfaces. *Colloid Interface Sci. Commun.* **2021**, *42*, 100397. [CrossRef]
21. Yu, A.; Carlson, P.; Pollack, G.H. Unexpected axial flow through hydrophilic tubes: Implications for energetics of water. *Eur. Phys. J. Spec. Top.* **2014**, *223*, 947–958. [CrossRef]
22. Brady, G.P.; Sharp, K.A. Entropy in protein folding and in protein–protein interactions. *Curr. Opin. Struct. Biol.* **1997**, *7*, 215–221. [CrossRef]
23. Harano, Y.; Kinoshita, M. Large gain in translational entropy of water is a major driving force in protein folding. *Chem. Phys. Let.* **2004**, *399*, 342–348. [CrossRef]
24. Niedre, M.; Patterson, M.S.; Wilson, B.C. Direct Near-infrared Luminescence Detection of Singlet Oxygen Generated by Photodynamic Therapy in Cells In Vitro and Tissues In Vivo. *Photochem. Photobiol.* **2002**, *75*, 382–391. [CrossRef]
25. Mittler, R.; Vanderauwera, S.; Suzuki, N.; Miller, G.A.D.; Tognetti, V.B.; Vandepoele, K.; Gollery, M.; Shulaev, V.; Van Breusegem, F. ROS signaling: The new wave? *Trends Plant Sci.* **2011**, *16*, 300–309. [CrossRef]
26. Mahmoudi, A.; Nazari, K.; Mohammadian, N.; Moosavi-Movahedi, A.A. Effect of Mn^{2+}, Co^{2+}, Ni^{2+}, and Cu^{2+} on horseradish peroxidase. *Appl. Biochem. Biotechnol.* **2003**, *104*, 81–94. [CrossRef]
27. Arnao, M.B.; Acosta, M.; Del Rio, J.A.; Garcia-Canovas, F. Inactivation of peroxidase by hydrogen peroxide and its protection by a reductant agent. *Biochim. Biophys. Acta Protein Struct. Mol. Enzymol.* **1990**, *1038*, 85–89. [CrossRef]
28. Baynton, K.J.; Bewtra, J.K.; Biswas, N.; Taylor, K.E. Inactivation of horseradish peroxidase by phenol and hydrogen peroxide: A kinetic investigation. *Biochim. Biophys. Acta Protein Struct. Mol. Enzym.* **1994**, *1206*, 272–278. [CrossRef]
29. Hao, F.; Jing, M.; Zhao, X.; Liu, R. Spectroscopy, calorimetry and molecular simulation studies on the interaction of catalase with copper ion. *J. Photochem. Photobiol. B* **2015**, *143*, 100–106. [CrossRef]
30. Maher, P.; Schubert, D. Signaling by reactive oxygen species in the nervous system. *Cell. Mol. Life Sci.* **2000**, *57*, 1287–1305. [CrossRef]
31. Kearns, D.R. Physical and chemical properties of singlet molecular oxygen. *Chem. Rev.* **1971**, *71*, 395–427. [CrossRef]
32. Khan, A.U.; Kasha, M. Singlet molecular oxygen in the Haber-Weiss reaction. *Proc. Natl. Acad. Sci. USA* **1994**, *91*, 12365–12367. [CrossRef]
33. Firey, P.A.; Jones, T.W.; Jori, G.; Rodgers, M.A.J. Photoexcitation of zinc phthalocyanine in mouse myeloma cells: The observation of triplet states but not of singlet oxygen. *Photochem. Photobiol.* **1988**, *48*, 357–360. [CrossRef]
34. Moan, J. On the diffusion length of singlet oxygen in cells and tissues. *J. Photochem. Photobiol.* **1990**, *6*, 343–344. [CrossRef]
35. Curcio, J.A.; Petty, C.C. The near infrared absorption spectrum of liquid water. *J. Opt. Soc. Am.* **1951**, *41*, 302–304. [CrossRef]

36. Krasnovsky, A.A.; Drozdova, N.N.; Ivanov, A.V.; Ambartsumian, R.V. Activation of molecular oxygen by infrared laser radiation in pigment-free aerobic systems. *Biochemistry* **2003**, *68*, 963–966. [PubMed]
37. Anquez, F.; Yazidi-Belkoura, I.E.; Randoux, S.; Suret, P.; Courtade, E. Cancerous cell death from sensitizer free photoactivation of singlet oxygen. *Photochem. Photobiol.* **2012**, *88*, 167–174. [CrossRef] [PubMed]
38. Anquez, F.; Belkoura, I.E.Y.; Suret, P.; Randoux, S.; Courtade, E. Cell death induced by direct laser activation of singlet oxygen at 1270 nm. *Laser Phys.* **2012**, *23*, 025601. [CrossRef]
39. Detty, M.R. Direct 1270 nm irradiation as an alternative to photosensitized generation of singlet oxygen to induce cell death. *Photochem. Photobiol.* **2012**, *88*, 2–4. [CrossRef]
40. Chaplin, M. Do we underestimate the importance of water in cell biology? *Nat. Rev. Mol. Cell Biol.* **2006**, *7*, 861–866. [CrossRef] [PubMed]
41. Maruyama, Y.; Harano, Y. Does water drive protein folding? *Chem. Phys. Lett.* **2013**, *581*, 85–90. [CrossRef]
42. Sessions, R.B.; Thomas, G.L.; Parker, M.J. Water as a conformational editor in protein folding. *J. Mol. Biol.* **2004**, *343*, 1125–1133. [CrossRef]
43. Svergun, D.I.; Richard, S.; Koch, M.H.J.; Sayers, Z.; Kuprin, S.; Zaccai, G. Protein hydration in solution: Experimental observation by x-ray and neutron scattering. *Proc. Natl. Acad. Sci. USA* **1998**, *95*, 2267–2272. [CrossRef] [PubMed]
44. Merzel, F.; Smith, J.C. Is the first hydration shell of lysozyme of higher density than bulk water? *Proc. Natl. Acad. Sci. USA* **2002**, *99*, 5378–5383. [CrossRef]
45. Du, X.; Seeman, D.; Dubin, P.L.; Hoagland, D.A. Nonfreezing water structuration in heteroprotein coacervates. *Langmuir* **2015**, *31*, 8661–8666. [CrossRef]
46. Ball, P. Water as an active constituent in cell biology. *Chem. Rev.* **2008**, *108*, 74–108. [CrossRef]
47. Pollack, G.H. Water, energy and life: Fresh views from the water's edge. *Int. J. Des. Nat. Ecodyn.* **2010**, *5*, 27. [CrossRef] [PubMed]
48. Pollack, G.H. *The Fourth Phase of Water*, 1st ed.; Ebner & Sons Publishers: Seattle, DC, USA, 2013.
49. Del Giudice, E.; Tedeschi, A.; Vitiello, G.; Voeikov, V. Coherent structures in liquid water close to hydrophilic surfaces. *J. Phys. Conf. Ser.* **2013**, *442*, 012028.
50. Del Giudice, E.; Voeikov, V.; Tedeschi, A.; Vitiello, G. The origin and the special role of coherent water in living systems. In *Fields of the Cell*; Daniel, F., Michal, C., Eds.; Research Signpost: Kerala, India, 2015; Chapter 5, pp. 95–111.
51. Arani, R.; Bono, I.; Del Giudice, E.; Preparata, G. QED coherence and the thermodynamics of water. *Int. J. Mod. Phys. B* **1995**, *9*, 1813–1841. [CrossRef]
52. Del Giudice, E.; Spinetti, P.R.; Tedeschi, A. Water Dynamics at the root of metamorphosis in living organisms. *Water* **2009**, *2*, 566–586. [CrossRef]
53. Madl, P.; Del Giudice, E.; Voeikov, V.L.; Tedeschi, A.; Kolarž, P.; Gaisberger, M.; Hartl, A. Evidence of coherent dynamics in water droplets of waterfalls. *Water* **2013**, *5*, 57–68.
54. Carlon, H.R. Aerosol spectrometry in the infrared. *Appl. Opt.* **1980**, *19*, 2210–2218. [CrossRef]
55. Childs, R.E.; Bardsley, W.G. The steady-state kinetics of peroxidase with 2, 2′-azino-di-(3-ethyl-benzthiazoline-6-sulphonic acid) as chromogen. *Biochem. J.* **1975**, *145*, 93–103. [CrossRef]
56. Voeikov, V.L.; Yablonskaya, O.I. Stabilizing effects of hydrated fullerenes C60 in a wide range of concentrations on luciferase, alkaline phosphatase, and peroxidase in vitro. *Electromagn. Biol. Med.* **2015**, *34*, 160–166. [CrossRef]
57. Aebi, H. Catalase In Vitro. *Methods Enzymol.* **1984**, *105*, 121–126.
58. Stadtman, E.R. Ascorbic acid and oxidative inactivation of proteins. *Am. J. Clin. Nutr.* **1991**, *54*, 1125–1128. [CrossRef] [PubMed]
59. Stadtman, E.R.; Wittenberger, M.E. Inactivation of Escherichia coli glutamine synthetase by xanthine oxidase, nicotinate hydroxylase, horseradish peroxidase, or glucose oxidase: Effects of ferredoxin, putidaredoxin, and menadione. *Arch. Biochem. Biophys.* **1985**, *239*, 379–387. [CrossRef]
60. Mordente, A.; Miggiano, G.A.; Martorana, G.E.; Meucci, E.; Santini, S.A.; Castelli, A. Alkaline phosphatase inactivation by mixed function oxidation systems. *Arch. Biochem. Biophys.* **1987**, *258*, 176–185. [CrossRef]
61. Miggiano, G.A.; Mordente, A.; Martorana, G.E.; Meucci, E.; Castelli, A. Characterization of alkaline phosphatase inactivation by ascorbic acid. *Biochim. Biophys. Acta Protein Struct. Mol. Enzymol.* **1984**, *789*, 343–346. [CrossRef]
62. Chang, J.Y. A two-stage mechanism for the reductive unfolding of disulfide-containing proteins. *J. Biol. Chem.* **1997**, *272*, 69–75. [CrossRef]
63. Chardot, T.; Mitraki, A.; Amigues, Y.; Desmadril, M.; Betton, J.M.; Yon, J.M. The effect of phosphate on the unfolding-refolding of phosphoglycerate kinase induced by guanidine hydrochloride. *FEBS Lett.* **1988**, *228*, 65–68. [CrossRef]

International Journal of
Molecular Sciences

MDPI

Article

Pheomelanin Effect on UVB Radiation-Induced Oxidation/Nitration of L-Tyrosine

Alessia Mariano [1], Irene Bigioni [1], Anna Scotto d'Abusco [1], Alessia Baseggio Conrado [2], Simonetta Maina [1], Antonio Francioso [1], Luciana Mosca [1] and Mario Fontana [1,*]

[1] Department of Biochemical Sciences, Sapienza University of Rome, Piazzale Aldo Moro 5, 00185 Rome, Italy; alessia.mariano@uniroma1.it (A.M.); bigioni.1639690@studenti.uniroma1.it (I.B.); anna.scottodabusco@uniroma1.it (A.S.d.); simonetta.maina@yahoo.it (S.M.); antonio.francioso@uniroma1.it (A.F.); luciana.mosca@uniroma1.it (L.M.)

[2] National Heart & Lung Institute, Imperial College London, Norfolk Place, London W2 1PG, UK; a.baseggio-conrado@imperial.ac.uk

* Correspondence: mario.fontana@uniroma1.it; Tel.: +39-06-4991-0948

Abstract: Pheomelanin is a natural yellow-reddish sulfur-containing pigment derived from tyrosinase-catalyzed oxidation of tyrosine in presence of cysteine. Generally, the formation of melanin pigments is a protective response against the damaging effects of UV radiation in skin. However, pheomelanin, like other photosensitizing substances, can trigger, following exposure to UV radiation, photochemical reactions capable of modifying and damaging cellular components. The photoproperties of this natural pigment have been studied by analyzing pheomelanin effect on oxidation/nitration of tyrosine induced by UVB radiation at different pH values and in presence of iron ions. Photoproperties of pheomelanin can be modulated by various experimental conditions, ranging from the photoprotection to the triggering of potentially damaging photochemical reactions. The study of the photomodification of L-Tyrosine in the presence of the natural pigment pheomelanin has a special relevance, since this tyrosine oxidation/nitration pathway can potentially occur in vivo in tissues exposed to sunlight and play a role in the mechanisms of tissue damage induced by UV radiation.

Keywords: pheomelanin; nitrotyrosine; dityrosine; photooxidation; photosensitizer

Citation: Mariano, A.; Bigioni, I.; Scotto d'Abusco, A.; Baseggio Conrado, A.; Maina, S.; Francioso, A.; Mosca, L.; Fontana, M. Pheomelanin Effect on UVB Radiation-Induced Oxidation/Nitration of L-Tyrosine. *Int. J. Mol. Sci.* **2022**, *23*, 267. https://doi.org/10.3390/ijms23010267

Academic Editor: Filipe Ferreira da Silva

Received: 29 November 2021
Accepted: 23 December 2021
Published: 27 December 2021

1. Introduction

Pheomelanin is one of the existing forms of the natural pigment melanin. Melanin is present in the skin in two forms: eumelanin and pheomelanin. Eumelanin is a heterogeneous polymer composed mainly of dihydroxyindole units derived from tyrosinase-catalyzed oxidation of tyrosine or 3,4-dihydroxyphenylalanine (DOPA) to dopaquinone. Compared to eumelanin, pheomelanin structure differs due to non-enzymatic addition of cysteine to dopaquinone during the pathway of pigment biosynthesis. DOPA-derivatives with cysteine, such as 5′-S-cysteinyldopa and in minor amount 2′-S-cystenyldopa, are incorporated into the pigment in the form of 1,4-benzothiazine units (Figure 1) [1]. Before being incorporated into pheomelanin, a minor part of 1,4-benzothiazine units may undergo further structural modifications with formation of benzothiazole moiety which copolymerizes with benzothiazine units [2–4]. Interestingly, slight variations in the monomer composition of pigment polymer skeleton have been shown to determine significant differences in light absorption, antioxidant activity, redox behavior, and metal chelation [5].

It is commonly believed that melanin plays an important role in the modulation of the photochemical reactions that occur in the skin. Numerous experimental and clinical evidences have shown a protective role of eumelanin on the damage triggered by UV irradiation on the skin [6]. A lower incidence of UV-induced skin diseases is observed in individuals with darker skin pigmentation, where eumelanin is present. Conversely, a higher incidence of UV-induced skin diseases was found in red-haired individuals with pale skin and freckles. Traditionally, this UV susceptibility trait has been associated with a high tendency to sunburn and an increased risk of skin tumors and melanoma [7–9]. The damage caused by UV rays would be determined either by the absence of pigmentation or by the photosensitizing properties of the pheomelanin present in the skin of these individuals. Notably, pheomelanin has the capacity to act as photosensitizer by inducing the generation of reactive oxygen species (ROS) upon irradiation with UV light [8,10–14]. Pheomelanin has been observed to increase lipid peroxidation following exposure of liposomes to UV irradiation, suggesting that pheomelanin may act as a pro-oxidant [15]. Analogously to other photosensitizing substances, pheomelanin is able to trigger, following exposure to UV radiation, photochemical reactions capable of modifying and damaging cellular components [16,17]. Furthermore, pheomelanin itself undergoes a photolytic process mediated by the same reactive intermediates [18].

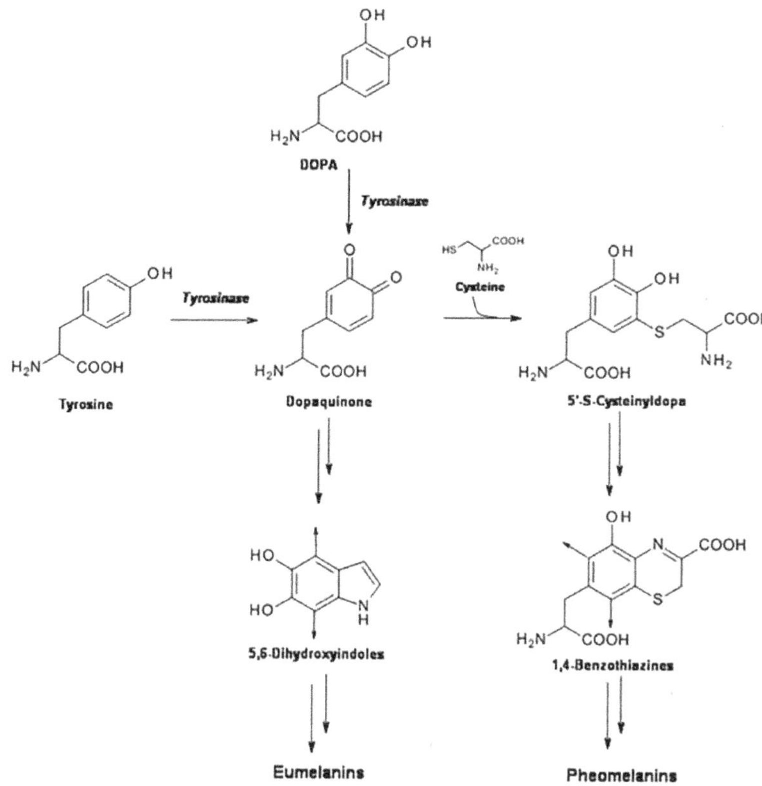

Figure 1. Schematic pathway of eumelanin and pheomelanin biosynthesis.

In particular, by exposure to UV radiation, aromatic rings present in the pheomelanin (Pheo) are excited to the singlet state (^{1}Pheo*) and rapidly converted to the excited triplet state (^{3}Pheo*) [19,20].

The triplet state of pheomelanin can act as photosensitizer triggering photooxidative events by radical-mediated (type I) and singlet oxygen-mediated (type II) mechanisms. The type I mechanism involves free radical formation through the hydrogen atom or electron ransfer by interaction of the triplet excited state of the sensitizer with target molecules (S) or molecular oxygen.

$$^3\text{Pheo}^* + \text{S} \rightarrow \text{Pheo}^{\bullet -} + \text{S}^{\bullet +}$$

Superoxide anion ($O_2^{\bullet-}$) is generated when the pigment in the excited triplet state transfers electrons to molecular oxygen by type I mechanism [10,21].

$$^3\text{Pheo}^* + O_2 \rightarrow \text{Pheo}^{\bullet+} + O_2^{\bullet-}$$

The type II process involves the generation of singlet oxygen ($^1O_2$) by energy transfer from the excited triplet sensitizer to a ground state oxygen molecule [10,12].

$$^3\text{Pheo}^* + O_2 \rightarrow \text{Pheo} + {}^1O_2$$

Recently, pheomelanin has also been implicated in UV-independent pathways of oxidative stress [22].

In this study, the photoproperties of this natural pigment were studied by analyzing the effect of pheomelanin on the oxidation/nitration of tyrosine induced by UVB radiation under different pH values and in presence of iron ions. In particular, pheomelanin effect on UVB-induced oxidation/nitration of tyrosine has been studied at physiological pH and at a weakly acid pH. Under pathophysiological situations, such as inflammation, tissue pH close to 5.5–6 can be found. Moreover, recent studies have shown that acid melanosomal pH suppress melanogenesis, especially eumelanin formation, in melanocytes [23]. Notably, it has been observed that, at pH 5.8, eumelanin biosynthesis is suppressed, while pheomelanin production is enhanced [24].

Following UVB radiation of L-Tyrosine, tyrosyl radical that is generated dimerizes with the formation of 3,3′-dityrosine; in presence of nitrite the photochemical reaction produces tyrosyl radical and reactive nitrogen species which combine to form 3-nitrotyrosine as a further product [25–27] (Figure 2).

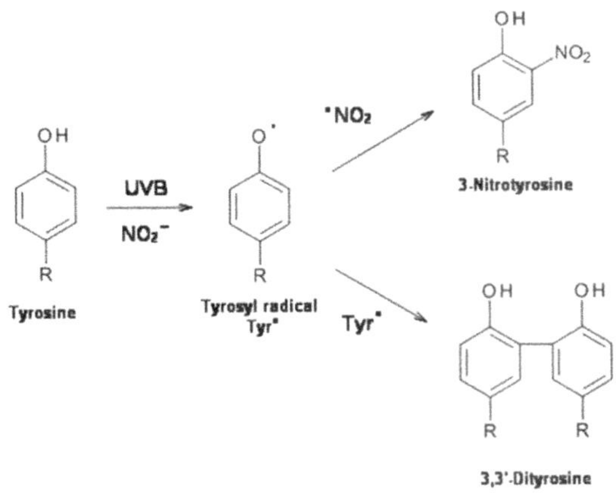

Figure 2. UVB-induced 3-nitrotyrosine and 3,3′-dityrosine formation.

Both 3,3′-dityrosine and 3-nitrotyrosine are considered diagnostic markers of the in vivo production of reactive oxygen and nitrogen species [27–29]. Although reactive nitrogen species, such as peroxynitrite, has been the most widely studied nitrating species, tyrosine nitration occurs also through several alternative routes. At this regard, free tyrosine and tyrosine protein residue nitration can be achieved through mechanisms involving peroxidase/H_2O_2-dependent oxidation of nitrite to nitrogen dioxide radical ($^\bullet NO_2$) [30–32]. In inflammation, myeloperoxidase from activated leukocytes catalyzes tyrosine nitration at high levels [33–36]. Photonitration of tyrosine to 3-nitrotyorosine has been already shown by methylene blue dye and riboflavin as sensitizers [37,38]. Methylene blue-sensitized photomodification of tyrosine in the presence of nitrite occurs mainly through a process which involves singlet oxygen (type II mechanism). Conversely, singlet oxygen plays a minor role in the tyrosine photooxidation/photonitration mediated by riboflavin as sensitizer [38–40].

Interestingly, the oxidation and nitration of tyrosine residues in proteins are considered important post-translational modifications with consequences on the function of proteins and therefore on cellular homeostasis [41–44].

The study of the photomodification of tyrosine in presence of the natural pigment pheomelanin has a special relevance, since this tyrosine oxidation/nitration pathway can potentially occur in vivo in tissues exposed to sunlight and play a role in the mechanisms of tissue damage induced by UV radiation.

2. Results

2.1. UVB Radiation-Induced Photooxidation/Photonitration of L-Tyrosine

The exposure to ultraviolet light (UVB), at room temperature, of a solution containing 1 mM tyrosine leads to the formation of 0.25 ± 0.07 μM at pH 5.5 and 0.13 ± 0.02 μM at pH 7.4 of 3.3′-dityrosine after 30 min of exposure. Tyrosine dimerization was not observed in controls kept in the dark. The exposure of 1 mM tyrosine solution to UVB radiation in the presence of 10 mM nitrite, under the same experimental conditions reported above, leads to 3-nitrotyrosine as a further product in addition to 3,3′-dityrosine (Figure 3). When nitrite is present, 3,3′-dityrosine is 0.08 ± 0.02 μM and 3.60 ± 0.46 μM, at pH 5.5 and pH 7.4, respectively. The amount of 3-nitrotyrosine formed is 2.37 ± 0.4 μM and 1.89 ± 0.15 μM, at pH 5.5 and pH 7.4 respectively, after 30 min of exposure. At low pH values nitrite generates nitrating species which, in the presence of tyrosine, lead to the formation of 3-nitrotyrosine [45]. Control experiments, in which tyrosine and nitrite are incubated in the dark, indicate that, under our experimental conditions, this reaction pathway can contribute minimally to the production of 3-nitrotyrosine only at pH below 3.3.

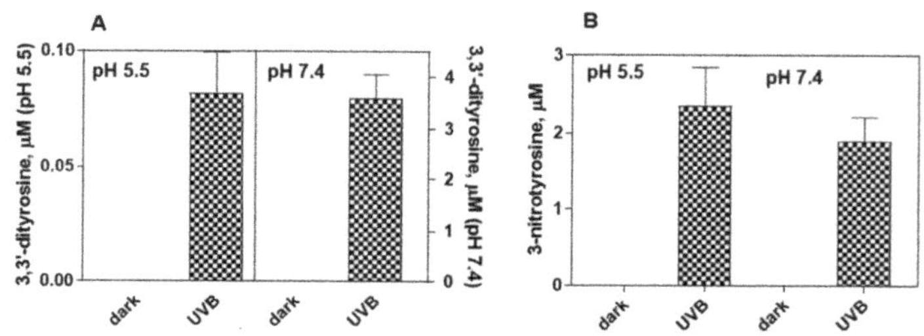

Figure 3. UVB-induced photooxidation/photonitration of tyrosine. A reaction mix containing 1 mM tyrosine in 0.2 M K-phosphate buffer at pH 5.5 or pH 7.4, 0.1 mM DTPA, 10 mM K-nitrite is exposed to UVB radiation. After 30 min of exposure, the reaction is stopped by placing the mixture in the dark and the solution is analyzed by HPLC, to determine the formation of 3,3′-dityrosine (**A**) and 3-nitrotyrosine (**B**), as reported in Materials and Methods. Controls in the dark correspond to the unexposed solution.

2.2. Effect of Pheomelanin on UVB Radiation-Induced Photooxidation/Photonitration of L-Tyrosine

In order to evaluate the photoproperties of pheomelanin on the oxidative/nitrative modifications of tyrosine induced by UVB rays, 1 mM tyrosine and 10 mM nitrite were exposed to UVB radiation in the presence of 4.2 μg/mL synthetic pheomelanin at physiological pH 7.4 and pH 5.5. Pheomelanin was enzymatically prepared from L-Dopa and cysteine as reported in the experimental section. After an exposure of 30 min both the formation of 3,3′-dityrosine and the conversion of tyrosine to 3-nitrotyrosine was assayed. Overall, pheomelanin exerts a photoprotective effect (antioxidant) on the oxidation/nitration of tyrosine induced by UVB radiation (Figure 4). However, at pH 5.5 pheomelanin acts as photosensitizer (prooxidant) in the nitrative modification of tyrosine. As shown in the Figure 5B, pheomelanin does not inhibit the nitration of tyrosine but there is a 60% increase in the formation of 3-nitrotyrosine compared to the control exposed to UVB radiation in the absence of pheomelanin. In control experiments in which pheomelanin alone and nitrite were exposed to UVB radiation, neither nitrotyrosine nor dityrosine were detectable.

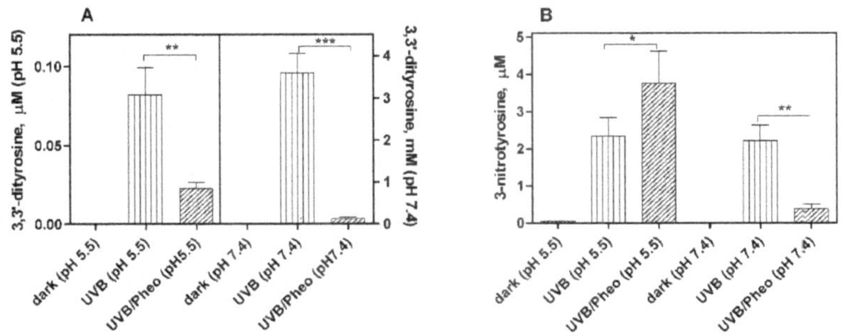

Figure 4. Photooxidation/photonitration of tyrosine by the nitrite/pheomelanin/UVB system. Pheomelanin 4.2 μg/mL is added to reaction mixture containing 1 mM tyrosine, 10 mM K-nitrite, 0.1 mM DTPA in 0.2 M K-phosphate buffer at pH 5.5 or pH 7.4. The solution is exposed to UVB rays for 30 min. The reaction is stopped by placing the mixture in the dark and the supernatant, obtained after centrifugation, is analyzed by HPLC to measure 3,3′-dityrosine (**A**) and 3-nitrotyrosine (**B**), as reported in Materials and Methods. Controls in the dark correspond to unexposed reaction mixtures (pheomelanin/nitrite/tyrosine system). *** $p < 0.001$, ** $p < 0.01$, * $p < 0.05$.

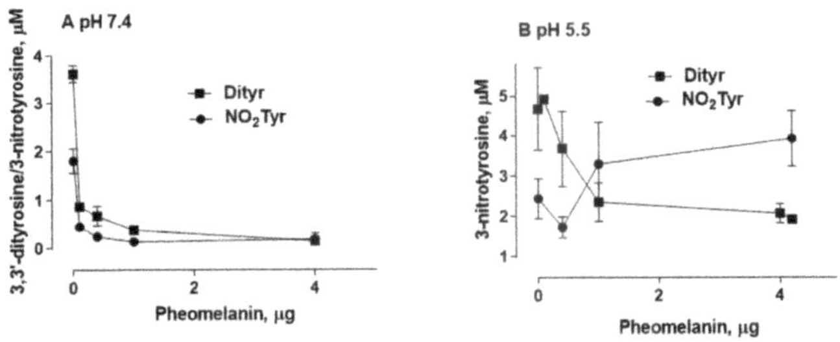

Figure 5. Photooxidation of tyrosine by the nitrite/pheomelanin/UVB system at various synthetic pheomelanin concentrations. Pheomelanin (0.1–4 μg/mL) is added to a reaction mixture containing 1 mM tyrosine, 10 mM K-nitrite, 0.1 mM DTPA in 0.2 M K-phosphate buffer at pH 7.4 (**A**) or pH 5.5 (**B**). The solution is exposed to UVB rays for 30 min. The reaction is stopped by placing the mixture in the dark and the supernatant, obtained after centrifugation, is analyzed by HPLC to determine the formation of 3-nitrotyrosine (•) and 3,3′-dityrosine (■) as reported in Materials and Methods.

Figure 5 shows the formation of 3,3′-dityrosine and 3-nitrotyrosine at various concentrations of pheomelanin (0.1–4 μg/mL) at pH 7.4 and pH 5.5. At all concentrations used, pheomelanin has a dose-dependent photoprotective action on the formation of 3,3′-dityrosine at both pH 5.5 and pH 7.4. The photosensitizing action on the formation of 3-nitrotyrosine at pH 5.5 is observed in the range 0.4–4 μg/mL.

2.3. Photoproperties of Pheomelanin on UVB-Induced Oxidative/Nitrative Modifications of L-Tyrosine: Effect of Fe(III)

It is known that melanins have the ability to bind various metals with the result of modifying their photoproperties [46–49]. In order to evaluate how the presence of metals can influence the oxidative/nitrative modifications of tyrosine, exposure to UVB rays was performed with the addition of Fe(III) to the reaction mixture. Experiments performed in the absence of metal chelator DTPA showed analogous results (data not shown). At pH 5.5, it is observed that the presence of metals influences the photoproperties of pheomelanin by reducing its antioxidant activity against dityrosine formation (Figure 6). Regarding the formation of 3-nitrotyrosine, the photosensitizer effect of pheomelanin is not affected either by the absence of the chelator or by the addition of Fe(III).

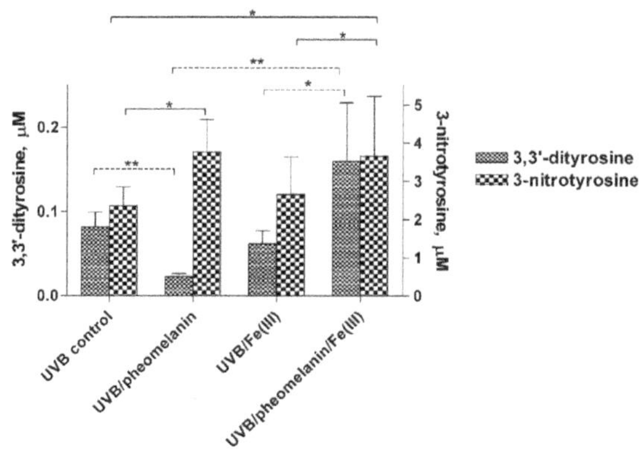

Figure 6. Photooxidation of tyrosine by the nitrite/pheomelanin/UVB system: effect of Fe(III). Pheomelanin 4.2 µg/mL is added to the reaction mixture containing 1 mM tyrosine, 10 mM K-nitrite, 0.1 mM DTPA in 0.2 M K-phosphate buffer at pH 5.5. The UVB control is pheomelanin free. Fe (III) is added as $FeCl_3$ at a concentration of 0.1 mM. The solution is exposed to UVB rays for 30 min. The reaction is stopped by placing the mixture in the dark and the supernatant, obtained after centrifugation, is analyzed by HPLC to determine the formation of 3-nitrotyrosine and 3,3′-dityrosine, as reported in Materials and Methods. ** $p < 0.01$, * $p < 0.05$.

2.4. *Pheomelanin Effect on Oxidative/Nitrative Modifications of* L*-Tyrosine Induced by UVB Radiation: Role of Singlet Oxygen*

The photooxidative reactions can be the result of radical type processes (type I) or of processes mediated by singlet oxygen (type II). Both mechanisms can contribute to the photooxidative reactions at the same time. In order to evaluate the role of singlet oxygen ($^1O_2$) in pheomelanin-sensitized nitration reaction of tyrosine at pH 5.5, the yields of 3-nitrotyrosine in H_2O and D_2O as solvent were compared. Replacement of H_2O by D_2O increases the lifetime of singlet oxygen by about 15 times [50] and, consequently, stimulates $^1O_2$-dependent reactions. As shown in Figure 7, the production of 3-nitrotyrosine is approximately 8.4 times greater in D_2O than in H_2O. This effect is indicative of the participation of singlet oxygen in the reaction. The formation of 3,3′-dityrosine is not affected by D_2O (Figures S1 and S2).

It has been also observed that the formation of 3-nitrotyrosine is significantly reduced in the presence of sodium azide (NaN_3), a known quencer of singlet oxygen (Figure 8). The inhibitory effect of azide confirms intermediacy of type II mechanism in the pheomelanin-sensitized formation of 3-nitrotyrosine.

2.5. *Pheomelanin Effect on Oxidative/Nitrative Modifications of* L*-Tyrosine Induced by Peroxyitrite*

Peroxynitrite induces both tyrosine oxidation to 3,3′-dityrosine and tyrosine nitration to 3-nitrotyrosine. Under our experimental conditions, peroxynitrite (100 µM, final concentration) added to a solution containing 100 µM of tyrosine generates 0.53 ± 0.02 µM of 3,3′-dityrosine and 6.75 ± 0.27 µM of 3-nitrotyrosine, respectively. As shown in Figure 8, pheomelanin, at a concentration of 4.2 µg/mL, is able to inhibit both the formation of 3,3′-dityrosine (~42%) and that of 3-nitrotyrosine (~47%). As reported [51,52], peroxynitrite reacts, in vivo, mainly with carbon dioxide, forming a peroxynitrite-CO_2 adduct which decomposes generating the nitrogen dioxide radicals ($^{\bullet}NO_2$) and carbonate radical anion ($CO_3^{\bullet -}$). In the presence of bicarbonate, tyrosine nitration mediated by peroxynitrite is generally increased due to high oxidative/nitrative properties of radicals generated by the decomposition of the peroxynitrite-CO_2 adduct. The results shown in Figure 8 indicate that pheomelanin is equally effective in protecting tyrosine from the nitrative and oxidative action of peroxynitrite also in the presence of 25 mM bicarbonate.

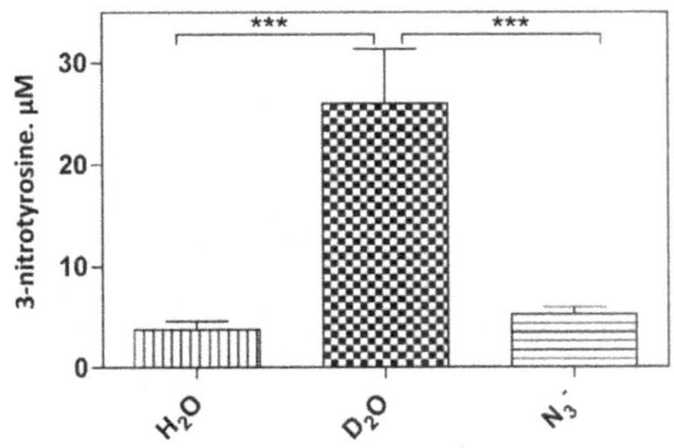

Figure 7. Photooxidation of tyrosine by the nitrite/pheomelanin/UVB system: effect of D_2O and NaN_3. Pheomelanin 4.2 µg/mL is added to the solution, containing 1 mM tyrosine, 10 mM K-nitrite in 0.2 M K-phosphate buffer at pH 5.5 and 0.1 mM DTPA. The solution is exposed to UVB rays for 30 min. The reaction is stopped by placing the mixture in the dark and the supernatant, obtained after centrifugation, is analyzed by HPLC to determine the formation of 3-nitrotyrosine, as reported in Materials and Methods. In D_2O, the pD (5.5) was taken as the measured pH + 0.4. NaN_3 is added to a final concentration of 1 mM. *** $p < 0.001$.

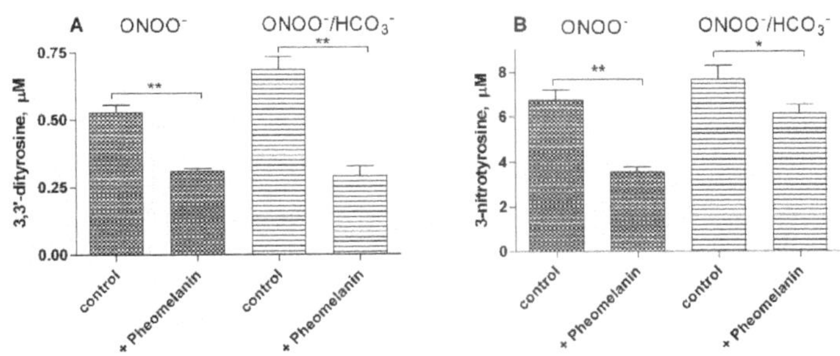

Figure 8. Pheomelanin effect on peroxynitrite-induced oxidation/nitration of tyrosine. To the reaction mix containing 100 µM tyrosine, 4.2 µg/mL pheomelanin, 0.1 mM DTPA in 0.2 M K-phosphate buffer, 100 µM peroxynitrite is added, Na-bicarbonate when present is at a concentration of 25 mM. After 5 min at room temperature, the reaction mixture is analyzed by HPLC to measure 3,3′-dityrosine (**A**) and 3-nitrotyrosine (**B**), as reported in Materials and Methods. ** $p < 0.01$, * $p < 0.05$

3. Discussion

The results of this study show that the photoproperties of pheomelanin can be modulated by various experimental conditions, ranging from the photoprotection to the triggering of potentially damaging photochemical reactions. These properties were studied by analyzing the effect of pheomelanin on UVB radiation-induced oxidation/nitration of tyrosine.

UVB irradiation leads, through tyrosyl radical intermediate, to the dimerization of tyrosine with the formation of 3,3′-dityrosine and in the presence of nitrite the photochemical reaction forms 3-nitrotyrosine as an additional product. The mechanism underlying the formation of 3-nitrotyrosine likely involves the combination of tyrosyl radical with nitrogen dioxide radical ($^{\bullet}NO_2$), which may be generated by photooxidation of nitrite [53,54]. In the presence of pheomelanin, tyrosine is dose-dependently protected from oxidation to 3,3′-dityrosine both at pH 5.5 and physiological pH (pH 7.4). Furthermore, pheomelanin can perform a protective function on the conversion of tyrosine to 3-nitrotyrosine at pH 7.4. It is known that UVB radiation induces the formation of oxyradicals capable of triggering oxidative reactions [55]. Therefore, the protective action of pheomelanin against the photooxidation of tyrosine could be related to its ability to act as a free radical scavenger.

The experiments conducted on the formation of 3,3′-dityrosine and 3-nitrotyrosine induced by peroxynitrite ($ONOO^-$) confirm this hypothesis. Peroxynitrite, which is generated in vivo from the reaction of nitric oxide ($^\bullet NO$) with the superoxide anion ($O_2^{\bullet -}$), is a very reactive species capable of nitrating and oxidizing tyrosine. This reactivity is mediated by the hydroxyl radical ($^\bullet OH$) and by the nitrogen dioxide radical ($^\bullet NO_2$) which are generated by the homolytic cleavage of peroxynitrite. In the presence of carbon dioxide (CO_2), a peroxynitrite-CO_2 adduct is formed which generates a further radical, the carbonate radical anion ($CO_3^{\bullet -}$). Pheomelanin showed protective properties both on the formation of 3,3′-dityrosine and on the conversion of tyrosine to 3-nitrotyrosine induced both by peroxynitrite and peroxynitrite-CO_2 adduct. These results indicate that pheomelanin can act as free radical scavenger and the observed protective action of the pigment on UVB-induced tyrosine modifications can be attributed to this property.

An interesting result that emerged from our investigations is that pheomelanin can have pro-oxidant properties under some experimental conditions. We observed that the nitration of tyrosine to 3-nitrotyrosine induced by UVB radiation in presence of nitrite at pH 5.5 is increased when carried out in the presence of pheomelanin. These results indicate that the properties of pheomelanin can be significantly influenced by the pH during UVB irradiation, switching from antioxidant (pH 7.4) to pro-oxidant (pH 5.5).

The photochemical experiments conducted with the addition of iron ion are also of particular interest. Pheomelanin has a remarkable ability to bind metals and this property leads often to a modification of the photoprotective capabilities of the pigment. In our experimental conditions by adding Fe(III), we observed a reduced ability to inhibit the oxidative reaction. These results indicate that the antioxidant properties of pheomelanin are sensitive to the effect of metal ions such as iron. It is plausible that the pigment bond with iron induces an increase in the production of highly oxidizing reactive species whose action can only be partially counteracted by the antioxidant activity of the pigment itself.

The production of reactive oxygen species (ROS) resulting from the interaction of oxygen with pheomelanin exposed to UV radiation has often been interpreted as the cause of its pro-oxidant properties. In these hypotheses, pheomelanin (Pheo) would act as a sensitizer and its ability to stimulate the formation of 3-nitrotyrosine (NO_2Tyr) in the pheomelanin/nitrite/UVB system could be rationalized with the following sequence of reactions:

$$\mathrm{Pheo} \rightarrow {}^1\mathrm{Pheo}^*$$

$${}^1\mathrm{Pheo}^* \rightarrow {}^3\mathrm{Pheo}^*$$

$${}^3\mathrm{Pheo}^* + O_2 \rightarrow \mathrm{Pheo}^{\bullet +} + O_2^{\bullet -}$$

$$2O_2^{\bullet -} + 2H_2O \rightarrow O_2 + H_2O_2 + 2OH^-$$

$$Fe^{3+} + O_2^{\bullet -} \rightarrow Fe^{2+} + O_2$$

$$Fe^{2+} + H_2O_2 \rightarrow Fe^{3+} + {}^\bullet OH + OH^-$$

$$\mathrm{Tyr} + {}^\bullet OH \rightarrow \mathrm{Tyr}^\bullet + OH^-$$

$$NO_2^- + {}^\bullet OH \rightarrow {}^\bullet NO_2 + OH^-$$

$$\mathrm{Tyr}^\bullet + {}^\bullet NO_2 \rightarrow NO_2\mathrm{Tyr}$$

In competition with the nitrogen dioxide radical ($^\bullet NO_2$), the tyrosyl radical can dimerize with the formation of 3,3′-dityrosine (Dityr):

$$\mathrm{Tyr}^\bullet + \mathrm{Tyr}^\bullet \rightarrow \mathrm{Dityr}$$

Another possible pathway of formation of the tyrosyl radical and of the nitrogen dioxide radical, responsible for the production of 3-nitrotyrosine, could occur through the direct interaction of the photoexcited pheomelanin in the excited triplet state ($^3Pheo^*$) with tyrosine and with nitrite (type I mechanism):

$$^3Pheo^* + Tyr \rightarrow Pheo^{\bullet -} + Tyr^{\bullet}$$

$$^3Pheo^* + NO_2{}^- \rightarrow Pheo^{\bullet -} + {}^{\bullet}NO_2$$

Photooxidation of pheomelanin-dependent tyrosine can also be mediated by singlet oxygen that is generated by energy transfer from photoexcited pheomelanin in the triplet state to ground state oxygen according to the following scheme (type II mechanism):

$$^3Pheo^* + O_2 \rightarrow {}^1O_2 + Pheo$$

$$^1O_2 + Tyr \rightarrow Tyr^{\bullet} + O_2{}^{\bullet -}$$

Tyrosyl radicals can either dimerize or react with a nitrite-derived species to form 3-nitrotyrosine. To our knowledge, the production of nitrating species by direct interaction of nitrite with singlet oxygen has not been reported. It is possible that indirect oxidation of nitrite by the radicals produced by type II mechanism can give rise to further oxidizing species which, as shown above, can contribute to the formation of 3-nitrotyrosine.

The investigations carried out to obtain information on the mechanism through which the nitrite/pheomelanin/UVB system induces the nitration of tyrosine at pH 5.5 indicate that, in our experimental conditions, the process can involve singlet oxygen, indeed, in the presence of D_2O, the production of 3-nitrotyrosine is considerably higher than that formed in H_2O. Accordingly, the inhibition exerted by sodium azide on the generation of 3-nitrotyrosine possibly obeys to the known quenching effect on singlet oxygen [21]. These investigations are in agreement with previous studies on nitrite-induced nitration of tyrosine in the presence of methylene blue as a photosensitizer [37]. On the other hand, previous studies carried out in our laboratory have shown that the nitration of tyrosine in presence of riboflavin as a photosensitive agent is mainly of type I [38]. It cannot be excluded that in our case, which uses pheomelanin as photosensitizer, the type I mechanism may participate in the photonitration reaction of tyrosine at slightly acid pH. UVB radiation exposure experiments under anaerobic conditions are currently underway to verify the role of the type I photochemical reaction.

Noteworthy, the photosensitizing effect exerted by pheomelanin is particularly efficient in a slightly acid environment and in the presence of metal ions, such as iron. These conditions, although not physiological, could acquire significance in some pathological situations such as during inflammatory processes or in the case of tissue ischemia. In both these states, pH values close to those used in our experiments are found in vivo (pH 5.8–6.1). Moreover, the forearm of a healthy man has an average surface pH around 5.4–5.9 [56]. Interestingly, pheomelanin synthesis is chemically promoted by weakly acid pH [24,57]. It has been reported that melanosomal pH regulates eumelanin/pheomelanin ratio in melanocytes with a shift towards a pheomelanic phenotype by lowering pH [58,59]. In our experimental conditions, supraphysiological concentrations of nitrite (10 mM) were used to highlight the nitration reaction of tyrosine. Nitrite is the main product of nitric oxide ($^{\bullet}NO$) catabolism, the production of which increases in inflammation [60]. Nitrite is also a constituent of sweat, where it can reach concentrations in the μM range following its formation on the surface of the skin by commensal bacteria. Furthermore, in normal conditions of exposure to the sun and heat, the surface layer of sweat undergoes a rapid increase in concentration caused by evaporation, so that the local concentration of nitrite can increase several times [61].

The result presented herein indicates that photoproperties of pheomelanin can be modulated by various experimental conditions. It is well-known that pheomelanin undergoes structural modifications by UV rays. In the course of the biosynthetic pathways, modification involves benzothiazine units which are gradually converted to benzothiazole [4]. The relative ratio of these two types of pheomelanin moieties appears important in determining whether pheomelanin acts as a pro-oxidant [5,62]. Recently, exploring the photoreactivity of pheomelanin by UVA radiation, the benzothiazole moiety has been shown to be more reactive than benzothiazine moiety [63]. Under our experimental conditions, UVB radiation and reactive nitrogen species could similarly influence pigment photoreactivity and induce structural modifications of pheomelanin worth to be further explored.

4. Materials and Methods

4.1. Chemicals

L-Cysteine, L-Dopa, L-Tyrosine, diethylenetriaminopentacetic acid (DTPA), sodium azide (NaN_3), mushroom tyrosinase, horseradish peroxidase was provided by Sigma Aldrich (St. Louis, MO, USA). Deuterium oxide (D_2O) was obtained from Aldrich (Milwaukee, WI, USA). The 3-nitrotyrosine was from Fluka (Buchs, Switzerland). All other reagents were used with the highest level of purity commercially available. The synthesis of 3,3′-dityrosine was enzymatically carried out from L-Tyrosine and hydrogen peroxide by horseradish peroxidase [64]. Peroxynitrite was prepared from K-nitrite and hydrogen peroxide under acid conditions as previously reported [65].

4.2. Synthesis of Pheomelanin

Pheomelanin was synthesized from L-cysteine and L-Dopa [66]. L-Dopa (25 µmoles) was dissolved in 20 mL of K-phosphate buffer (0.05 M pH 6.8) and incubated with mushroom tyrosinase (1.2 mg). After 30 s, L-cysteine (53 µmol) was added, and the reaction mixture was left overnight at 37 °C in agitation. To ensure total conversion to cysteinyl-dopa isomers, pheomelanin was prepared with cysteine/Dopa molar ratio of 2:1 [67]. The reaction was stopped by reducing pH to 2.2 with 6 N HCl, and the reaction mixture was subsequently centrifuged at 5000 rpm for 25 min. The residue is suspended in H_2O and subsequently lyophilized, thus obtaining approximately 3.4 mg of dark brown pheomelanin. The identification of pheomelanin as a synthetic product was performed by analyzing the absorption spectrum (λ = 800–200 nm) of a solution containing 2.5 µg/mL pheomelanin in 1 M K-phosphate buffer, pH 8.0. The absorbance of a 4 µg/mL pheomelanin solution was on average 0.092 at 400 nm [68]. Stock solutions of synthetic pheomelanin were in 1 M K-phosphate buffer, pH 8.0. The spectrum analysis was carried out using a UV-vis Cary 50 Scan spectrophotometer.

4.3. Nitration and Oxidation of L-Tyrosine Induced by UVB Radiation

The reaction mix containing 1 mM L-Tyrosine, 0.05 M K-phosphate buffer, pH 7.4, or 0.2 M K-phosphate, pH 5.5 and 0.1 mM DTPA, was incubated in the absence or by adding 10 mM K-nitrite to a Petri dish. To study the effect of pheomelanin on the oxidative/nitrative modifications of tyrosine induced by UVB rays, the final concentration of the pigment in the reaction mixture was 4.2 µg/mL. The photooxidation of tyrosine was initiated by exposing the reaction mixture to UVB radiation produced by two fluorescent lamps at room temperature. The irradiation was interrupted for 1 min every 5 min to mix the suspension and prevent overheating of the reaction mixture. After 30 min of irradiation, the samples were centrifuged for 5 min at 12,000 rpm, and the supernatant analyzed by HPLC to verify the formation of 3,3′-dityrosine and 3-nitrotyrosine.

4.4. Exposure to UVB Radiation

The exposure was carried out in an irradiation cabin built by the Bioltecnical Service, Nettuno, RM (Italy). Two Sankyo Denki G15T8E UVB fluorescent lamps (λ = 270–320 nm with a maximum peak at 313 nm) were mounted on the ceiling of a closed aluminum cabin equipped with a front door for loading the Petri dishes. The lamps emit a luminous flux, with energy administered in the unit of time equal to 2.5 J $m^{-2}s^{-1}$, perpendicular to the radiation plane placed at 23 cm distance. The total energy administered was 4500 J m^{-2}. The lamps have an efficiency equal to 1, i.e., all the absorbed power is transformed into

UV radiation; furthermore the radiation emitted, given the geometry of the cabin and the reflectivity of the walls, ends entirely on the irradiation plane.

4.5. Nitration and Oxidation of L-Tyrosine Induced by Peroxynitrite

The experiments with peroxynitrite were performed as described in [69]. The reaction mixture containing 4.2 μg/mL pheomelanin, 100 μM L-Tyrosine, 0.2 M K-phosphate buffer, pH 7.4 and 0.1 mM DTPA, was incubated in the absence or presence of 25 mM Na-bicarbonate. The reaction was started by the addition of peroxynitrite (final concentration of 100 μM). After 5 min at room temperature, the solution was centrifuged for 5 min at 12,000 rpm, and the supernatant analyzed by HPLC to verify the formation of 3-nitrotyrosine and 3,3′-dityrosine.

4.6. HPLC Analysis

3-nitrotyrosine and 3,3′-dithyrosine were analyzed by HPLC using a Waters chromatograph equipped with a model 600 pump and a model 600 gradient control module as reported [65,70]. Chromatographic separation was performed using a Nova-pak column. C18 (3.9 mm × 150 mm), 4 μm (Waters) and as mobile phase: (A) K-phosphate/H_3PO_4 buffer, 50 mM, pH 3.0; (B) acetonitrile-water (50:50, *v*/*v*) with a flow rate of 1 mL/min at room temperature and a linear gradient from A to 33% of B in 10 min. 3-nitrotyrosine was analyzed at 360 nm, using a Waters 996 photodiode spectrophotometric detector. 3,3′-dithyrosine was analyzed using a Waters 474 fluorescence detector, setting the wavelength at 260 nm for the excitation and at 410 nm for emission. Peaks were identified using external standards and sample concentrations were calculated using standard curves. The elution times of 3-nitrotyrosine and 3,3′-dithyrosine are 8.9 and 7.5 min, respectively. The limit of determination of 3-nitrotyrosine and 3,3′-dithyrosine is 20 pmoles and 1 pmol, respectively.

4.7. Data Analysis

The results are expressed as mean values ± SEM of at least three separate experiments performed in duplicate. The statistical analyses were performed using Student's *t*-test; $p < 0.05$ was deemed significant. The graphs and data analysis were performed using the GraphPad Prism 4 program.

5. Conclusions

UVB radiation induces the photooxidation/photonitration of tyrosine. Pheomelanin is able to perform a protective function both on the tyrosine oxidation to 3,3′-dityrosine and on the conversion of tyrosine to 3-nitrotyrosine when the exposure is conducted at physiological pH; conversely at pH 5.5, the presence of pheomelanin induces a 60% increase in the formation of 3-nitrotyrosine. The photosensitizing action of pheomelanin in the nitration reaction of tyrosine to 3-nitrotyrosine at pH 5.5, is further increased by about 8 times in D_2O, suggesting a role of $^1O_2$ in the reaction mechanism. The addition of Fe(III) during the irradiation of tyrosine in presence of nitrite provokes a decrease of the antioxidant activity of pheomelanin also against the formation of 3,3′-dityrosine, indicating that the photoproperties of pheomelanin may be affected by the presence of metal ions. Finally, pheomelanin showed protective properties on oxidation/nitration of tyrosine induced by peroxynitrite and by the decomposition of the peroxynitrite-CO_2 adduct An important implication of the results obtained is that the pheomelanin-dependent photonitration of tyrosine in presence of nitrite could exert toxic effects by inducing the nitration of tyrosine protein residues present in the skin with consequent functional alteration of proteins themselves.

Supplementary Materials: The following are available online at https://www.mdpi.com/article/10.3390/ijms23010267/s1.

Author Contributions: Conceptualization, A.F. and M.F.; methodology, M.F.; software, A.M.; resources, A.F.; validation, A.S.d., A.B.C. and M.F.; formal analysis, A.F. and M.F.; investigation, A.M.,

I.B. and S.M..; data curation, A.S.d. and M.F.; writing—original draft preparation, A.M., I.B. and S.M.; writing—review and editing, A.B.C., L.M. and M.F.; visualization, A.F. and M.F.; supervision, A.F., L.M. and M.F.; project administration, M.F.; funding acquisition, A.S.d. and M.F. All authors have read and agreed to the published version of the manuscript.

Funding: This research received no external funding.

Institutional Review Board Statement: Not applicable.

Informed Consent Statement: Not applicable.

Data Availability Statement: Not applicable.

Conflicts of Interest: The authors declare no conflict of interest. The funders had no role in the design of the study; in the collection, analyses, or interpretation of data; in the writing of the manuscript, or in the decision to publish the results.

References

1. Prota, G. *The Chemistry of Melanins and Melanogenesis. Progress in the Chemistry of Organic Natural Products*; Springer: Berlin/Heidelberg, Germany, 1995; Volume 64, pp. 93–148.
2. Simon, J.D.; Peles, D.; Wakamatsu, K.; Ito, S. Current challenges in understanding melanogenesis: Bridging chemistry, biological control, morphology, and function. *Pigment Cell Melanoma Res.* **2009**, *22*, 563–579. [CrossRef] [PubMed]
3. Napolitano, A.; De Lucia, M.; Panzella, L.; D'Ischia, M. The "Benzothiazine" Chromophore of Pheomelanins: A Reassessment. *Photochem. Photobiol.* **2008**, *84*, 593–599. [CrossRef] [PubMed]
4. Wakamatsu, K.; Ohtara, K.; Ito, S. Chemical analysis of late stages of pheomelanogenesis: Conversion of dihydrobenzothiazine to a benzothiazole structure. *Pigment Cell Melanoma Res.* **2009**, *22*, 474–486. [CrossRef]
5. Micillo, R.; Panzella, L.; Koike, K.; Monfrecola, G.; Napolitano, A.; D'Ischia, M. "Fifty Shades" of Black and Red or How Carboxyl Groups Fine Tune Eumelanin and Pheomelanin Properties. *Int. J. Mol. Sci.* **2016**, *17*, 746. [CrossRef]
6. Brenner, M.; Hearing, V.J. The Protective Role of Melanin Against UV Damage in Human Skin†. *Photochem. Photobiol.* **2008**, *84*, 539–549. [CrossRef] [PubMed]
7. Ito, S.; Wakamatsu, K. Human hair melanins: What we have learned and have not learned from mouse coat color pigmentation. *Pigment Cell Melanoma Res.* **2011**, *24*, 63–74. [CrossRef]
8. Rouzaud, F.; Kadekaro, A.L.; Abdel-Malek, Z.A.; Hearing, V.J. MC1R and the response of melanocytes to ultraviolet radiation. *Mutat. Res. Mol. Mech. Mutagenes.* **2005**, *571*, 133–152. [CrossRef]
9. Lopreiato, M.; Cocchiola, R.; Falcucci, S.; Leopizzi, M.; Cardone, M.; Di Maio, V.; Brocco, U.; D'Orazi, V.; Calvieri, S.; Scandurra, R.; et al. The Glucosamine-derivative NAPA Suppresses MAPK Activation and Restores Collagen Deposition in Human Diploid Fibroblasts Challenged with Environmental Levels of UVB. *Photochem. Photobiol.* **2020**, *96*, 74–82. [CrossRef]
10. Chedekel, M.R.; Smith, S.K.; Post, P.W.; Pokora, A.; Vessell, D.L. Photodestruction of pheomelanin: Role of oxygen. *Proc. Natl. Acad. Sci. USA* **1978**, *75*, 5395–5399. [CrossRef] [PubMed]
11. Panzella, L.; Szewczyk, G.; D'Ischia, M.; Napolitano, A.; Sarna, T. Zinc-induced Structural Effects Enhance Oxygen Consumption and Superoxide Generation in Synthetic Pheomelanins on UVA/Visible Light Irradiation†. *Photochem. Photobiol.* **2010**, *86*, 757–764. [CrossRef] [PubMed]
12. Takeuchi, S.; Zhang, W.; Wakamatsu, K.; Ito, S.; Hearing, V.J.; Kraemer, K.H.; Brash, D.E. Melanin acts as a potent UVB photosensitizer to cause an atypical mode of cell death in murine skin. *Proc. Natl. Acad. Sci. USA* **2004**, *101*, 15076–15081. [CrossRef]
13. Wenczl, E.; Van Der Schans, G.P.; Roza, L.; Kolb, R.M.; Timmerman, A.J.; Smit, N.P.M.; Pavel, S.; Schothorst, A.A. (Pheo)melanin photosensitizes UVA-induced DNA damage in cultured human melanocytes. *J. Investig. Dermatol.* **1998**, *111*, 678–682. [CrossRef]
14. Sarna, T.; Menon, I.A.; Sealy, R.C. Photoinduced oxygen consumption in melanin systems—II. Action spectra and quantum yields for pheomelanins. *Photochem. Photobiol.* **1984**, *39*, 805–809. [CrossRef]
15. Ezzahir, A. The influence of melanins on the photoperoxidation of lipids. *J. Photochem. Photobiol. B Biol.* **1989**, *3*, 341–349. [CrossRef]
16. Menon, I.A.; Persad, S.; Haberman, H.F.; Kurian, C.J. A comparative study of the physical and chemical properties of melanins isolated from human black and red hair. *J. Investig. Dermatol.* **1983**, *80*, 202–206. [CrossRef] [PubMed]
17. Menon, I.A.; Persad, S.; Ranadive, N.S.; Haberman, H.F. Role of superoxide and hydrogen peroxide in cell lysis during irradiation in vitro of Ehrlich ascitic carcinoma cells in the presence of melanin. *Can. J. Biochem. Cell Biol.* **1985**, *63*, 278–283. [CrossRef]
18. Ito, S.; Wakamatsu, K.; Sarna, T. Photodegradation of Eumelanin and Pheomelanin and Its Pathophysiological Implications. *Photochem. Photobiol.* **2018**, *94*, 409–420. [CrossRef] [PubMed]
19. Felix, C.C.; Hyde, J.S.; Sealy, R.C. Photoreactions of melanin: A new transient species and evidence for triplet state involvement. *Biochem. Biophys. Res. Commun.* **1979**, *88*, 456–461. [CrossRef]
20. Sarna, T.; Menon, I.A.; Sealy, R.C. Photosensitization of melanins: A comparative study. *Photochem. Photobiol.* **1985**, *42*, 529–532. [CrossRef] [PubMed]

21. Foote, C.S. Photosensitized Oxidation and Singlet Oxygen: Consequences in Biological Systems. In *Free Radicals in Biology*; Elsevier: Amsterdam, The Netherlands, 1976; pp. 85–133.
22. Panzella, L.; Leone, L.; Greco, G.; Vitiello, G.; D'Errico, G.; Napolitano, A.; d'Ischia, M. Red human hair pheomelanin is a potent pro-oxidant mediating UV-independent contributory mechanisms of melanomagenesis. *Pigment Cell Melanoma Res.* **2014**, *27*, 224–252. [CrossRef]
23. Ambrosio, A.L.; Boyle, J.A.; Aradi, A.E.; Christian, K.A.; Di Pietro, S.M. TPC2 controls pigmentation by regulating melanosome pH and size. *Proc. Natl. Acad. Sci. USA* **2016**, *113*, 5622–5627. [CrossRef] [PubMed]
24. Wakamatsu, K.; Nagao, A.; Watanabe, M.; Nakao, K.; Ito, S. Pheomelanogenesis is promoted at a weakly acidic pH. *Pigment Cell Melanoma Res.* **2017**, *30*, 372–377. [CrossRef]
25. Berlett, B.S.; Stadtman, E.R. Protein oxidation in aging, disease, and oxidative stress. *J. Biol. Chem.* **1997**, *272*, 20313–20316. [CrossRef]
26. Cai, S.; Saito, T.; Fujii, N. Simultaneous ultraviolet B-induced photo-oxidation of tryptophan/tyrosine and racemization of neighboring aspartyl residues in peptides. *Free Radic. Biol. Med.* **2013**, *65*, 1037–1046. [CrossRef] [PubMed]
27. Giulivi, C.; Traaseth, N.J.; Davies, K.J.A. Tyrosine oxidation products: Analysis and biological relevance. *Amino Acids* **2003**, *25*, 227–232. [CrossRef] [PubMed]
28. Giulivi, C.; Davies, K.J. Dityrosine and tyrosine oxidation products are endogenous markers for the selective proteolysis of oxidatively modified red blood cell hemoglobin by (the 19 S) proteasome. *J. Biol. Chem.* **1993**, *268*, 8752–8759. [CrossRef]
29. Ischiropoulos, H. Biological Tyrosine Nitration: A Pathophysiological Function of Nitric Oxide and Reactive Oxygen Species. *Arch. Biochem. Biophys.* **1998**, *356*, 1–11. [CrossRef]
30. Van der Vliet, A.; Eiserich, J.P.; Halliwell, B.; Cross, C.E. Formation of Reactive Nitrogen Species during Peroxidase-catalyzed Oxidation of Nitrite. *J. Biol. Chem.* **1997**, *272*, 7617–7625. [CrossRef] [PubMed]
31. Sampson, J.B.; Ye, Y.; Rosen, H.; Beckman, J.S. Myeloperoxidase and horseradish peroxidase catalyze tyrosine nitration in proteins from nitrite and hydrogen peroxide. *Arch. Biochem. Biophys.* **1998**, *356*, 207–213. [CrossRef]
32. Brennan, M.L.; Wu, W.; Fu, X.; Shen, Z.; Song, W.; Frost, H.; Vadseth, C.; Narine, L.; Lenkiewicz, E.; Borchers, M.T.; et al. A tale of two controversies. Defining both the role of peroxidases in nitrotyrosine formation in vivo using eosinophil peroxidase and myeloperoxidase-deficient mice, and the nature of peroxidase-generated reactive nitrogen species. *J. Biol. Chem.* **2002**, *277*, 17415–17427. [CrossRef]
33. Eiserich, J.P.; Hristova, M.; Cross, C.E.; Jones, A.D.; Freeman, B.A.; Halliwell, B.; Van Der Vliet, A. Formation of nitric oxide-derived inflammatory oxidants by myeloperoxidase in neutrophils. *Nature* **1998**, *391*, 393–397. [CrossRef] [PubMed]
34. Burner, U.; Furtmüller, P.G.; Kettle, A.J.; Koppenol, W.H.; Obinger, C. Mechanism of reaction of myeloperoxidase with nitrite. *J. Biol. Chem.* **2000**, *275*, 20597–20601. [CrossRef] [PubMed]
35. Gaut, J.P.; Byun, J.; Tran, H.D.; Lauber, W.M.; Carroll, J.A.; Hotchkiss, R.S.; Belaaouaj, A.; Heinecke, J.W. Myeloperoxidase produces nitrating oxidants in vivo. *J. Clin. Investig.* **2002**, *109*, 1311–1319. [CrossRef]
36. Capuozzo, E.; Pecci, L.; Giovannitti, F.; Baseggio Conrado, A.; Fontana, M. Oxidative and nitrative modifications of enkephalins by human neutrophils: Effect of nitroenkephalin on leukocyte functional responses. *Amino Acids* **2012**, *43*, 875–884. [CrossRef]
37. Pecci, L.; Montefoschi, G.; Antonucci, A.; Costa, M.; Fontana, M.; Cavallini, D. Formation of nitrotyrosine by methylene blue photosensitized oxidation of tyrosine in the presence of nitrite. *Biochem. Biophys. Res. Commun.* **2001**, *289*, 305–309. [CrossRef]
38. Fontana, M.; Blarzino, C.; Pecci, L. Formation of 3-nitrotyrosine by riboflavin photosensitized oxidation of tyrosine in the presence of nitrite. *Amino Acids* **2012**, *42*, 1857–1865. [CrossRef]
39. Silva, E.; Godoy, J. Riboflavin sensitized photooxidation of tyrosine. *Int. J. Vitam. Nutr. Res.* **1994**, *64*, 253–256.
40. Lu, C.Y.; Liu, Y.Y. Electron transfer oxidation of tryptophan and tyrosine by triplet states and oxidized radicals of flavin sensitizers: A laser flash photolysis study. *Biochim. Biophys. Acta Gen. Subj.* **2002**, *1571*, 71–76. [CrossRef]
41. Schopfer, F. NO-dependent protein nitration: A cell signaling event or an oxidative inflammatory response? *Trends Biochem. Sci.* **2003**, *28*, 646–654. [CrossRef]
42. Gow, A.J.; Farkouh, C.R.; Munson, D.A.; Posencheg, M.A.; Ischiropoulos, H. Biological significance of nitric oxide-mediated protein modifications. *Am. J. Physiol. Cell. Mol. Physiol.* **2004**, *287*, L262–L268. [CrossRef]
43. Atwood, C.S.; Perry, G.; Zeng, H.; Kato, Y.; Jones, W.D.; Ling, K.Q.; Huang, X.; Moir, R.D.; Wang, D.; Sayre, L.M.; et al. Copper Mediates Dityrosine Cross-Linking of Alzheimer's Amyloid-β. *Biochemistry* **2004**, *43*, 560–568. [CrossRef]
44. Monteiro, H.P. Signal transduction by protein tyrosine nitration: Competition or cooperation with tyrosine phosphorylation-dependent signaling events? *Free Radic. Biol. Med.* **2002**, *33*, 765–773. [CrossRef]
45. Park, J.Y.; Lee, Y.N. Solubility and decomposition kinetics of nitrous acid in aqueous solution. *J. Phys. Chem.* **1988**, *92*, 6294–6302. [CrossRef]
46. Schmitz, S.; Thomas, P.D.; Allen, T.M.; Poznansky, M.J.; Jimbow, K. Dual role of melanins and melanin precursors as photoprotective and phototoxic agents: Inhibition of ultraviolet radiation-induced lipid peroxidation. *Photochem. Photobiol.* **1995**, *61*, 650–655. [CrossRef]
47. Sealy, R.C.; Felix, C.C.; Hyde, J.S.; Swartz, H.M. Structure and Reactivity of Melanins: Influence of Free Radicals and Metal Ions. In *Free Radicals in Biology*; Elsevier: Amsterdam, The Netherlands, 1980; pp. 209–259.
48. Palumbo, A.; D'Ischia, M.; Misuraca, G.; Prota, G.; Schultz, T.M. Structural modifications in biosynthetic melanins induced by metal ions. *Biochim. Biophys. Acta Gen. Subj.* **1988**, *964*, 193–199. [CrossRef]

49. Zadlo, A.; Mokrzyński, K.; Ito, S.; Wakamatsu, K.; Sarna, T. The influence of iron on selected properties of synthetic pheomelanin. *Cell Biochem. Biophys.* **2020**, *78*, 181–189. [CrossRef] [PubMed]
50. Rodgers, M.A.J.; Snowden, P.T. Lifetime of oxygen (O2(1.DELTA.g)) in liquid water as determined by time-resolved infrared luminescence measurements. *J. Am. Chem. Soc.* **1982**, *104*, 5541–5543. [CrossRef]
51. Lymar, S.V.; Jiang, Q.; Hurst, J.K. Mechanism of carbon dioxide-catalyzed oxidation of tyrosine by peroxynitrite. *Biochemistry* **1996**, *35*, 7855–7861. [CrossRef] [PubMed]
52. Goldstein, S.; Czapski, G.; Lind, J.; Merényi, G. Tyrosine nitration by simultaneous generation of NO and O2/-under physiological conditions. How the radicals do the job. *J. Biol. Chem.* **2000**, *275*, 3031–3036. [CrossRef]
53. Mack, J.; Bolton, J.R. Photochemistry of nitrite and nitrate in aqueous solution: A review. *J. Photochem. Photobiol. A Chem.* **1999**, *128*, 1–13. [CrossRef]
54. Fischer, M.; Warneck, P. Photodecomposition of nitrite and undissociated nitrous acid in aqueous solution. *J. Phys. Chem.* **1996**, *100*, 18749–18756. [CrossRef]
55. Kadekaro, A.L.; Kavanagh, R.J.; Wakamatsu, K.; Ito, S.; Pipitone, M.A.; Abdel-Malek, Z.A. Cutaneous Photobiology. The Melanocyte vs. the Sun: Who Will Win the Final Round? *Pigment Cell Res.* **2003**, *16*, 434–447. [CrossRef] [PubMed]
56. Braun-Falco, O.; Korting, H.C. Der normale pH-wert der menschlichen haut. *Hautarzt* **1986**, *37*, 126–129.
57. Ito, S.; Suzuki, N.; Takebayashi, S.; Commo, S.; Wakamatsu, K. Neutral pH and copper ions promote eumelanogenesis after the dopachrome stage. *Pigment Cell Melanoma Res.* **2013**, *26*, 817–825. [CrossRef]
58. Ancans, J.; Tobin, D.J.; Hoogduijn, M.J.; Smit, N.P.; Wakamatsu, K.; Thody, A.J. Melanosomal pH Controls Rate of Melanogenesis, Eumelanin/Phaeomelanin Ratio and Melanosome Maturation in Melanocytes and Melanoma Cells. *Exp. Cell Res.* **2001**, *268*, 26–35. [CrossRef]
59. Fuller, B.B.; Spaulding, D.T.; Smith, D.R. Regulation of the Catalytic Activity of Preexisting Tyrosinase in Black and Caucasian Human Melanocyte Cell Cultures. *Exp. Cell Res.* **2001**, *262*, 197–208. [CrossRef]
60. Padgett, E.L.; Pruett, S.B. Evaluation of nitrite production by human monocyte-derived macrophages. *Biochem. Biophys. Res. Commun.* **1992**, *186*, 775–781. [CrossRef]
61. Weller, R.; Pattullo, S.; Smith, L.; Golden, M.; Ormerod, A.; Benjamin, N. Nitric oxide is generated on the skin surface by reduction of sweat nitrate. *J. Investig. Dermatol.* **1996**, *107*, 327–331. [CrossRef]
62. Napolitano, A.; Panzella, L.; Monfrecola, G.; D'Ischia, M. Pheomelanin-induced oxidative stress: Bright and dark chemistry bridging red hair phenotype and melanoma. *Pigment Cell Melanoma Res.* **2014**, *27*, 721–733. [CrossRef]
63. Tanaka, H.; Yamashita, Y.; Umezawa, K.; Hirobe, T.; Ito, S.; Wakamatsu, K. The Pro-Oxidant Activity of Pheomelanin is Significantly Enhanced by UVA Irradiation: Benzothiazole Moieties Are More Reactive than Benzothiazine Moieties. *Int. J. Mol. Sci.* **2018**, *19*, 2889. [CrossRef]
64. Malencik, D.A.; Sprouse, J.F.; Swanson, C.A.; Anderson, S.R. Dityrosine: Preparation, isolation, and analysis. *Anal. Biochem.* **1996**, *242*, 202–213. [CrossRef]
65. Fontana, M.; Pecci, L.; Duprè, S.; Cavallini, D. Antioxidant Properties of Sulfinates: Protective Effect of Hypotaurine on Peroxynitrite-Dependent Damage. *Neurochem. Res.* **2004**, *29*, 111–116. [CrossRef]
66. Krol, E.S.; Liebler, D.C. Photoprotective actions of natural and synthetic melanins. *Chem. Res. Toxicol.* **1998**, *11*, 1434–1440. [CrossRef]
67. Ito, S. Optimization of Conditions for Preparing Synthetic Pheomelanin. *Pigment Cell Res.* **1989**, *2*, 53–56. [CrossRef]
68. Mosca, L.; Foppoli, C.; Coccia, R.; Rosei, M.A. Pheomelanin Production by the Lipoxygenase-Catalyzed Oxidation of 5-S-Cysteinyldopa and 5-S-Cysteinyldopamine. *Pigment Cell Res.* **1996**, *9*, 117–125. [CrossRef] [PubMed]
69. Fontana, M.; Giovannitti, F.; Pecci, L. The protective effect of hypotaurine and cysteine sulphinic acid on peroxynitrite-mediated oxidative reactions. *Free Radic. Res.* **2008**, *42*, 320–330. [CrossRef]
70. Baseggio Conrado, A.; D'Angelantonio, M.; Torreggiani, A.; Pecci, L.; Fontana, M. Reactivity of hypotaurine and cysteine sulfinic acid toward carbonate radical anion and nitrogen dioxide as explored by the peroxidase activity of Cu, Zn superoxide dismutase and by pulse radiolysis. *Free Radic. Res.* **2014**, *48*, 1300–1310. [CrossRef] [PubMed]

International Journal of *Molecular Sciences*

MDPI

Article

Elastic Properties of Taurine Single Crystals Studied by Brillouin Spectroscopy

Dong Hoon Kang [1], Soo Han Oh [1], Jae-Hyeon Ko [1,*], Kwang-Sei Lee [2] and Seiji Kojima [3]

1 Nano Convergence Technology Center, School of Nano Convergence Technology, Hallym University, 1 Hallymdaehakgil, Chuncheon 24252, Korea; rkdehdgns99@naver.com (D.H.K.); soohanoh@naver.com (S.H.O.)
2 Center for Nano Manufacturing, Department of Nano Science & Engineering, Inje University, Gimhae 50834, Korea; kwangsei28@hanmail.net
3 Division of Materials Science, University of Tsukuba, Tsukuba 305-8573, Japan; kojima@ims.tsukuba.ac.jp
* Correspondence: hwangko@hallym.ac.kr; Tel.: +82-33-248-2056

Abstract: The inelastic interaction between the incident photons and acoustic phonons in the taurine single crystal was investigated by using Brillouin spectroscopy. Three acoustic phonons propagating along the crystallographic *b*-axis were investigated over a temperature range of −185 to 175 °C. The temperature dependences of the sound velocity, the acoustic absorption coefficient, and the elastic constants were determined for the first time. The elastic behaviors could be explained based on normal lattice anharmonicity. No evidence for the structural phase transition was observed, consistent with previous structural studies. The birefringence in the *ac*-plane indirectly estimated from the split longitudinal acoustic modes was consistent with one theoretical calculation by using the extrapolation of the measured dielectric functions in the infrared range.

Keywords: taurine; elastic constant; Brillouin scattering; acoustic mode

Citation: Kang, D.H.; Oh, S.H.; Ko, J.-H.; Lee, K.-S.; Kojima, S. Elastic Properties of Taurine Single Crystals Studied by Brillouin Spectroscopy. *Int. J. Mol. Sci.* **2021**, 22, 7116. https://doi.org/10.3390/ijms22137116

Academic Editor: Filipe Ferreira da Silva

Received: 2 June 2021
Accepted: 28 June 2021
Published: 1 July 2021

1. Introduction

Taurine (2-aminoethanesulfonic acid), $NH_3(CH_2)_2SO_3$, is a well-known and important organic compound that is found in many organs of animal bodies, the bile being the most representative one [1]. Taurine is one of the sulfur-containing amino acids, but it does not belong to the constituents of proteins and is thus regarded as a nonessential amino acid. However, taurine is known to have many biological functions, such as membrane stabilization [2], osmoregulation [3], and antioxidation [4], to name a few. Moreover, taurine has been studied for various applications, such as sensors [5], nanofiltration [6], etc.

On the other hand, fundamental aspects of taurine have been investigated by various methods because taurine is a model system of a sulfur-included organic compound. In particular, a single crystal form or a powder pellet has been used for the characterization of the exact structure and vibrational properties. Taurine can be formed into a colorless crystal and can thus be studied as one of the organic crystals. The structure of the taurine crystal was first determined by X-ray scattering to be monoclinic $P2_1/c$ [7,8]. There are trans and gauche forms in taurine molecules, but the taurine crystal is grown in the zwitterion configuration (NH_3^+–CH_2–CH_2–SO_3^-) where only the gauche configuration exists, as confirmed by x-ray crystallography [8–12]. According to these structural studies, the N—H···O hydrogen bonds form a three-dimensional network, and the electrostatic potential exhibits large negative and positive regions near SO_3^- and NH_3^+, respectively.

There are several vibrational studies on taurine material in its various forms, including solutions to solid states [13–22]. These studies, combined with normal mode analysis, revealed detailed vibrational modes of taurine molecules, power samples, and single crystals. Some of these vibrational studies include a surface-enhanced Raman study carried out on taurine absorbed on silver nanoparticles [19] and the combination of infrared and

Raman analyses for taurine single crystals [22]. In particular, temperature- or pressure-dependent spectroscopic investigations are a promising tool for studying any possible phase transitions in these organic materials.

Taurine crystal shows successive pressure-induced phase transitions at ~0.7 and ~5.2 GPa, as revealed by high-pressure Raman spectroscopy [15]. A high-temperature Raman study above room temperature did not reveal any high-temperature phase transition [18]. On the other hand, Lim et al. suggested a first-order phase transition at ~250 K (measured upon cooling) by using a temperature-dependent Raman study below room temperature [16]. They also presented an enthalpic anomaly in a similar temperature range where the phase transition temperature showed a large thermal hysteresis of about 21 K. On the other hand, Bajaj et al. reported a low-temperature infrared study of torsional vibrations of taurine [21]. They observed a splitting of the CSH torsional mode near 250 K consistent with the Raman study described above.

However, apart for these two spectroscopic studies, there has been no other report about the structural phase transition of taurine single crystals. Moreover, the crystal structure (space group) does not change below room temperature down to 120 K [23]. Other structural studies also confirmed that the space group at low temperatures below 250 K is $P2_1/c$, the same as that measured at room temperature [10,11]. In this context, the existence of a low-temperature phase transition near 250 K and its exact structure below this temperature are still controversial.

The purpose of the present study is aimed at tackling this issue by using Brillouin spectroscopy, which is one of the inelastic light-scattering methods. Brillouin light-scattering is based on the inelastic interaction between the incident photons and acoustic phonons in the crystal. From the measured inelastic photon signals, sound velocities and elastic stiffness coefficients (abbreviated as elastic constants) can be derived [24]. Elastic constants are directly associated with interatomic and intermolecular potentials and are thus affected by phase transitions [25]. Brillouin spectroscopy has been applied to various organic crystals including pharmaceutical materials, such as aspirin [26,27], ibuprofen [28], acetaminophen [29], ketoprofen [30], indomethacin [31], etc. Some of these studies revealed detailed temperature dependences of elastic constants and acoustic damping effects in these organic crystals. The present study is the first report on the temperature dependence of the elastic properties of taurine single crystals. We could not find evidence for a structural phase transition in the investigated temperature of −185 °C ~ 175 °C.

2. Materials and Methods

The powder of taurine (purity ≥ 99%) was purchased from Aldrich and was dissolved in doubly distilled water. Single crystals of taurine were grown from an aqueous solution by slow evaporation technique at about 38 °C, the form of the resulting crystals depending on the rate of evaporation or supersaturation of the solution, as mentioned in previous works [7,32–34]. Optically transparent crystals have a different morphology ranging from a prism shape to acicular shape, in agreement with references [7,32–34]. With a rapid evaporation or a high degree of supersaturated solution, needle-shaped crystals were formed, the length of the needle being parallel to the *a*-axis. On the other hand, with a slow evaporation or a low degree of supersaturated solution, the crystals adopted a tabular form, the greatest length being along the *a*-axis. The monoclinic *b*-axis is perpendicular to the largest face and the *a*-direction [32,33]. Thus, the natural plane perpendicular to the *b*-axis was chosen for the investigation.

Figure 1a,b shows the morphology of the taurine single crystal along with the crystallographic axes and the crystallographic unit cell, respectively. The directions of the incident and the scattered light were denoted in terms of the wave vectors $\mathbf{k_i}$ and $\mathbf{k_s}$, respectively. A conventional Brillouin scattering spectrometer (TFPI-1, JRS Co., Zürich, Switzerland) was used to perform the measurements. Sample excitation was achieved by using a diode-pumped solid-state laser with a wavelength of 532 nm. The sample was put in a cryostat stage (THMS 600, Linkam, Tadworth, England) for temperature variation. All measurements were carried out upon cooling. An optical microscope (BX41, Olympus, Tokyo, Japan) with an objective lens of ×20 magnification was employed for backscattering geometry along the monoclinic *b* axis, as shown in Figure 1. The polarization direction of the incident light was in the *ac* plane, and no analyzer was used for the scattered light. Before every measurement, the temperature was stabilized for at least 2 min.

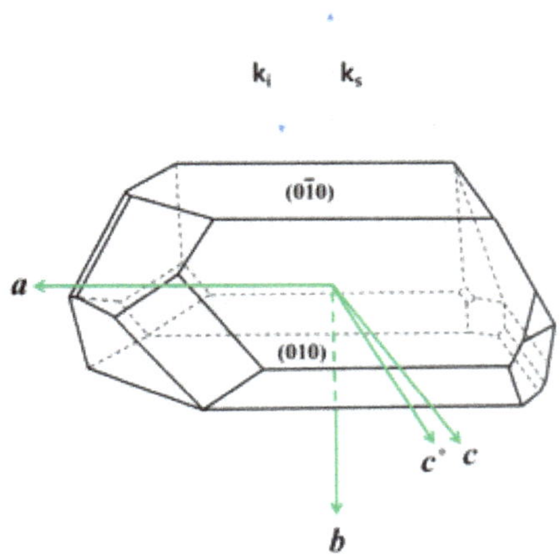

(a)

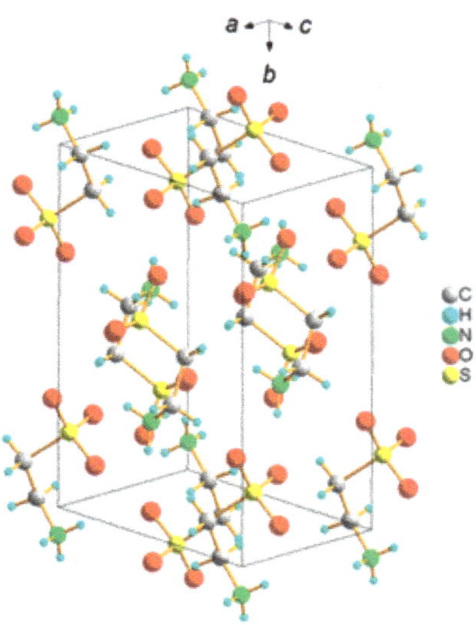

(b)

Figure 1. (**a**) The crystal morphology along with major crystallographic surfaces and axes. Two vectors $\mathbf{k_i}$ and $\mathbf{k_s}$ denote the wave vectors of the incident and the scattered lights, respectively. c^* direction is perpendicular to both *a* and *b* axes. (**b**) The crystallographic unit cell of the taurine single crystal.

3. Results and Discussion

Figure 2a shows the Brillouin spectra of the taurine single crystal at several temperatures. Each Brillouin spectrum consists of one longitudinal acoustic (LA) mode at ~24 GHz and two transverse acoustic (TA) modes near 15 and 10 GHz. However, a close inspection shows that the LA mode shows splitting at all temperatures. Figure 2b shows one example of the curve-fitting result for the spectrum measured at 30 °C. Splitting of the LA mode has been observed from several birefringent single crystals, such as aspirin and 1,2,4,5-tetrabromobenzene single crystals [26,35]. Thus, the most probable origin of the splitting of the LA mode is the birefringence of the taurine single crystal, which the incident polarization feels. Another possible origin for the splitting is the possible existence of monoclinic twins, where different twin structures might contribute to the splitting of the LA mode. The half widths do not show noticeable changes, but the mode frequencies seem to exhibit a slight hardening upon cooling. Figure 3 shows the intensity color plot of the measured Brillouin spectra as a function of the temperature. It includes the traces of the mode frequencies of two LA modes (LA1 and LA2, denoted with descending frequencies) and two TA modes (TA1 and TA2, denoted with descending frequencies), all being denoted by circle symbols. All acoustic modes seem to become hardened upon cooling.

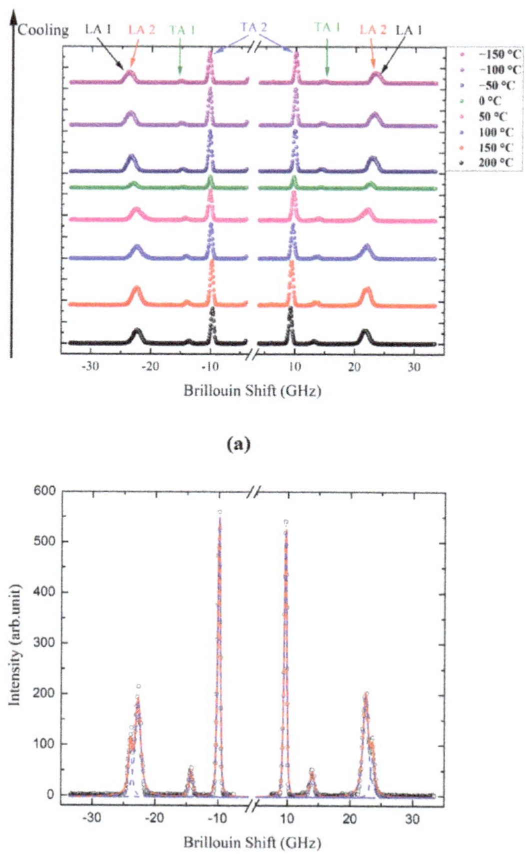

Figure 2. (**a**) Brillouin spectra of a taurine crystal at a few temperatures. (**b**) The Brillouin spectra of a taurine crystal measured at 30 °C and the fitting lines for each acoustic mode.

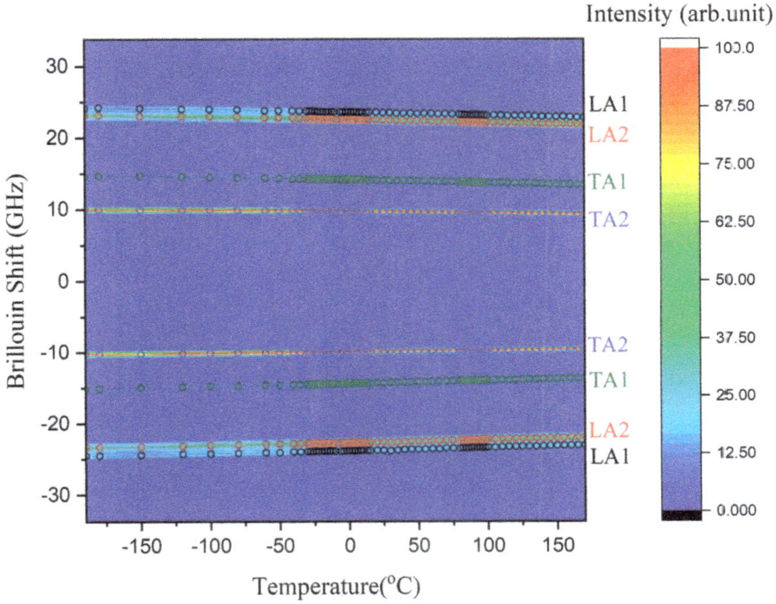

Figure 3. The intensity color plot of the measured Brillouin spectra as a function of the temperature. Mode frequencies are denoted in terms of circle symbols.

Each Brillouin spectrum was curve-fitted by using the Voigt function, which is a convolution of a Lorentzian function (an approximate form of the response function of the damped harmonic oscillator for the phonon modes) and a Gaussian function (an instrumental function of the interferometer which causes a broadening of the spectrum). The Brillouin frequency shift (ν_B) corresponding to the mode frequency and the full width at half maximum (FWHM, Γ_B) of each acoustic mode could be obtained as a function of the temperature. It clearly shows a splitting of the LA mode. There seems to be no central peak within the measured frequency range.

Figure 4 shows the temperature dependences of the Brillouin shifts of the acoustic modes. The split LA modes, as well as the two TA modes, show hardening upon cooling without any noticeable anomalies in the investigated temperature range. When there is a phase transition in condensed matters, one may observe various spectral changes in the light-scattering spectrum, such as the formation of strong central peaks in order-disorder phase transitions [36] or low-frequency soft optic modes [37], which may couple to the acoustic modes, resulting in substantial acoustic anomalies. These anomalies are in general associated with a downward softening or sudden, discontinuous changes in the mode frequencies concomitant with a significant increase in the half widths near or upon approaching the phase transition point [38]. The present result indicates that there is no experimental evidence for elastic anomalies associated with structural phase transitions in taurine single crystals at the present scattering geometry in the investigated temperature range of −185 °C to 175 °C. This result is in contrast to the result of the temperature-dependent Raman study where the authors suggested that a first-order phase transition occurs at −23 °C [16].

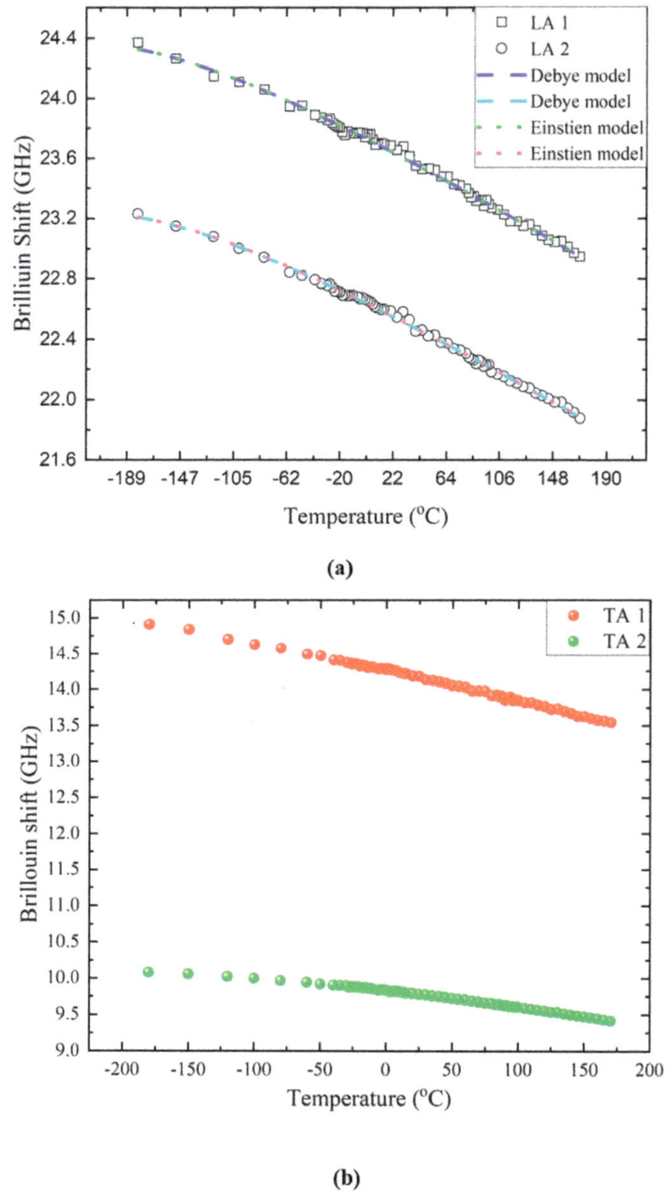

Figure 4. Temperature dependences of the Brillouin frequency shift of (**a**) the LA modes and (**b**) the TA modes. The dotted and dashed lines in (**a**) denote the best-fitted results by the Debye and the Einstein models. See the text for details.

The monotonic increase in the mode frequency upon cooling is conventionally observed from normal solids which do not show any structural phase transition. In this case, normal lattice anharmonicity plays a critical role in the change of acoustic properties. There are two theoretical models and corresponding functions to describe the temperature dependence of the acoustic mode frequency: the Varshni function based on the Einstein model [39] and the Lakkad function based on the Debye approach [40]. When the Einstein model is adopted, the temperature dependence of the Brillouin shift is expressed as:

$$\nu_B(T) = \nu_0^E \sqrt{1 - \frac{a}{\exp\left(\frac{\Theta_E}{T}\right) - 1}} \quad (1)$$

where ν_0^E, a, and Θ_E are the fitting parameters. The Θ_E is called the Einstein temperature. Lakkad's approach is based on the Debye model, and the Brillouin shift is described by the following equation:

$$\nu_B(T) = \nu_0^D \sqrt{1 - b\Theta_D F(T/\Theta_D)} \quad (2)$$

where

$$F(T/\Theta_D) = 3\left(\frac{T}{\Theta_D}\right)^4 \int_0^{\Theta_D/T} \frac{x^3}{e^x - 1} dx \quad (3)$$

is the Debye function and ν_0^D, b, and Θ_D are the fitting parameters. The Θ_D is called the Debye temperature. The ν_0^E and ν_0^D in each model are the mode frequencies extrapolated to 0 K. The ratio of Θ_E to Θ_D is known to be 0.75 [41].

The temperature dependences of the mode frequencies of the LA1 and LA2 modes were curve-fitted by using either Equation (1) or Equation (2). The dotted and dashed lines in Figure 4a denote the best-fitted results obtained by the Debye and the Einstein models described above. There is no noticeable difference between the two fitting results. Table 1 shows the obtained best-fitted parameters. It shows that the Einstein temperature is approximately 107 °C (~380 K) while the Debye temperature is near 237 °C (~510 K) in both modes. Moreover, the ratio of Θ_E to Θ_D is 0.74 in both cases, close to the theoretical prediction of 0.75 [41]. All these results indicate that the acoustic mode behavior in taurine single crystals is dominated by the usual lattice anharmonicity frequently observed in normal solids and that there is no direct experimental evidence for elastic anomalies associated with any structural phase transition in the investigated temperature range.

Table 1. The best-fitted results for the LA modes by using two models as described in the text.

	Einstein Model			Debye Model			Θ_E/Θ_D
	ν_0^E (GHz)	a	Θ_E (K)	ν_0^D (GHz)	b	Θ_D (K)	
LA 1	24.4	0.151	377	24.4	4.03×10^{-4}	508	0.74
LA 2	23.2	0.154	382	23.3	4.07×10^{-4}	514	0.74

The mode frequency and the FWHM measured in the backscattering geometry can be used to calculate the sound velocity (V) and the absorption coefficient (α) via the following equations:

$$V = \frac{\lambda \nu_B}{2n}, \quad (4)$$

$$\alpha = \frac{\pi \Gamma_B}{V}. \quad (5)$$

In these equations, λ and n denote the wavelength of the excitation laser light and the refractive index, respectively. For the calculation, the temperature dependence of the refractive index is necessary. However, only the average refractive index of 1.52 (space-averaged one) and the average refractive index of 1.545 on the *ac*-plane were reported experimentally [22]. Thus, the two LA mode frequencies were averaged to get the average mode frequency, which was used to calculate the average sound velocity and the absorption coefficient by using the average refractive index on the *ac*-plane (1.545).

Figure 5a–c shows the calculated sound velocities and the absorption coefficients of the LA and the two TA modes, respectively, as a function of the temperature. The longitudinal sound velocity increases from ~3860 m/s at 170 °C to ~4100 m/s at −180 °C upon cooling. The absorption coefficient of the LA waves is approximately 3×10^5 m^{-1} and does not show any substantial temperature dependence. The transverse sound velocities of the two TA modes show a monotonic increase upon cooling over the whole measured temperature range. The absorption coefficient is nearly constant in both cases. These results demonstrate that the investigated acoustic modes propagating along the [10] direction are not associated with a structural phase transition in the taurine single crystal over the temperature range of −185 °C to 175 °C.

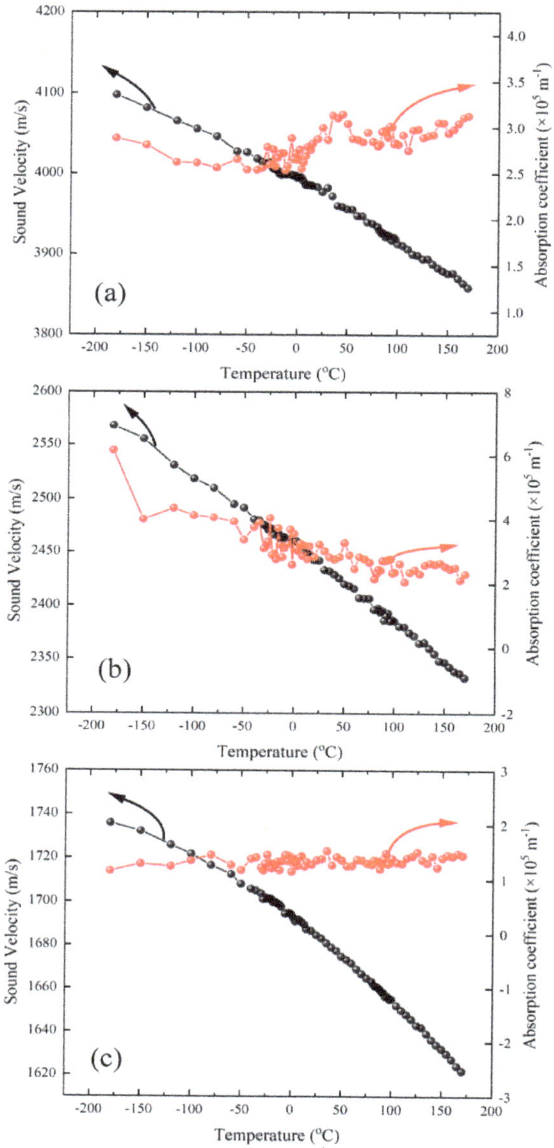

Figure 5. Temperature dependence of the sound velocity and the absorption coefficient of (**a**) the LA mode, (**b**) the TA1 mode, and (**c**) the TA2 mode.

Moreira et al. predicted the theoretical refractive indices n_1 = 1.601, n_2 = 1.597, and n_3 = 1.524 by using the extrapolation of the measured dielectric functions in the infrared range [22]. Since the sound velocity depends on the refractive index via Equation (4), it is worth checking the reliability of the calculated refractive indices of the taurine single crystal from the present study. The polarization of the incident light on the taurine crystal was in the *ac*-plane, and, thus, the two relevant refractive indices are n_1 and n_3. The sound velocities of the LA1 and LA2 modes were calculated by using these predicted refractive indices and Equation (4). The obtained results are shown in Supplementary Figure S1 in the Supplementary Materials. The two sound velocities are nearly the same within 0.21% on average, which indicates that the theoretical prediction for the refractive indices suggested by Moreira et al. [22] is reliable, at least from the viewpoint of the birefringence in the *ac*-plane.

The elastic constants of the three acoustic modes can be calculated from the obtained sound velocities via ρV^2, where ρ is the crystal density. The reported density of 1734 kg/m^3 [11] was used for the calculation. Table 2 shows the relationship between the acoustic mode and its associated elastic constant in the monoclinic phase [42]. The LA mode corresponds to C_{22}, while the two TA modes are related to two complex combinations of elastic constants. If the two elastic constants of the two TA modes are added, (C_{44} + C_{66}) can be derived. Figure 6a–c shows the temperature dependences of the three elastic constants shown in Table 2.

C_{22} is 27.5 GPa at room temperature, while the two elastic constants of the two TA modes are 10.3 and 4.9 GPa, resulting in (C_{44} + C_{66}) = 15.2 GPa. There is only one earlier report on the elastic constants of taurine crystals [32]. The C_{22} reported in this study is substantially different from this earlier value (~46.9 GPa), while (C_{44} + C_{66}) is similar ((C_{44} + C_{66}) in [32] is ~17.2 GPa). The origin of this large discrepancy for C_{22} is not clear at the moment, but it might be due to different experimental techniques having different probe frequencies and/or to different definitions of the crystal orientation. To check the reliability of the large value for C_{22} (~46.9 GPa) reported in [32], we made a taurine pellet sample by grinding the crystal into fine powders and pressing them into a circular plate. In the case of pellets and ceramics, the Brillouin light-scattering spectrum consists of continuously varying components whose scattering angle changes between 0 and 180° [43,44]. The Brillouin shift corresponding to the value of 46.9 GPa can be theoretically estimated to be about 31 GHz if we use the average refractive index of 1.574 [22]. This Brillouin shift corresponds to a longitudinal sound velocity of ~5200 m/s, which is unusually large for organic crystals like taurine. Figure 7 shows the comparison of the Brillouin spectrum of the taurine pellet with that of the single crystal measured at room temperature. Interestingly, the main peaks in the pellet spectrum are coincident with those of the single crystals measured at the present backscattering geometry. The broadening of the peaks in the pellet spectrum is due to the overlap of various acoustic modes with changing wave vectors. If the C_{22} value of ~46.9 GPa is correct, we should expect some significant contribution (or resonance peak) at ~31 GHz, where only the end of the tail of the main peak at ~23 GHz can be seen. Considering all these aspects, the previous report about the elastic constants of taurine [32] should be revisited.

Table 2. The correlation between the three acoustic modes and the elastic constants obtained from the present scattering geometry. The numerical values are measured at 20 °C.

Acoustic Modes	Elastic Constant	Elastic Constant (GPa)
L [010]	C_{22}	27.5
$T1$ [100]	$\frac{1}{2}(C_{44}+C_{66}+\sqrt{(C_{44}-C_{66})^2+4C_{46}^2})$	10.3
$T2$ [001]	$\frac{1}{2}(C_{44}+C_{66}-\sqrt{(C_{44}-C_{66})^2+4C_{46}^2})$	4.9
$T1+T2$	$C_{44}+C_{66}$	15.2

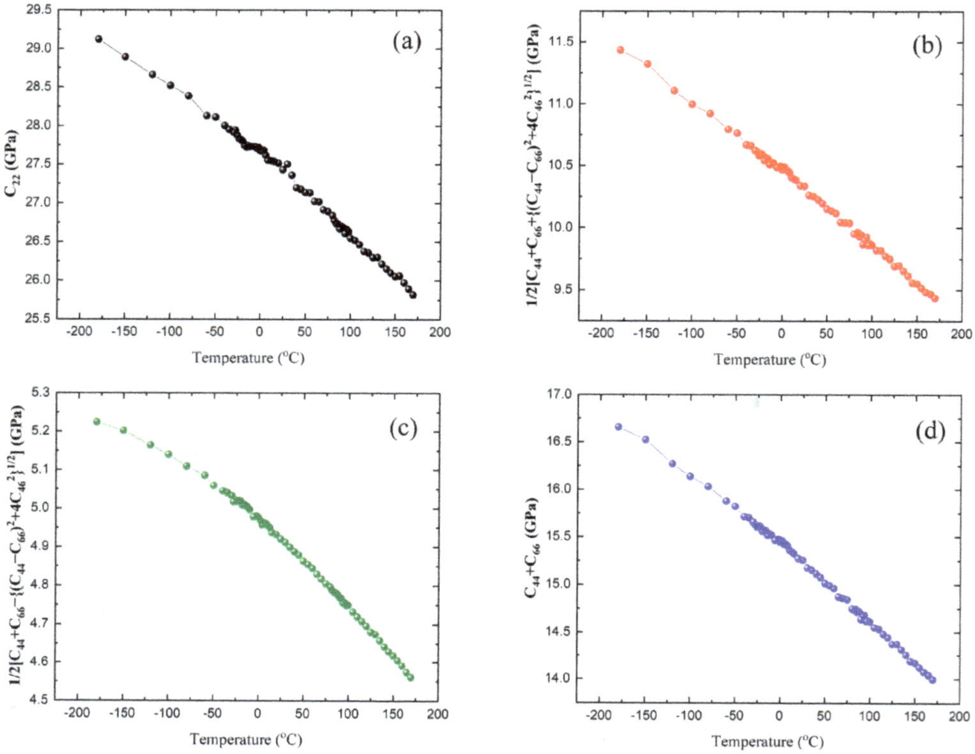

Figure 6. Temperature dependence of (**a**) C_{22}, (**b**) $(C_{44} + C_{66} + \sqrt{(C_{44} - C_{66})^2 + 4C_{46}^2}\ )/2$, (**c**) $(C_{44} + C_{66} - \sqrt{(C_{44} - C_{66})^2 + 4C_{46}^2}\ )/2$, and (**d**) $C_{44} + C_{66}$.

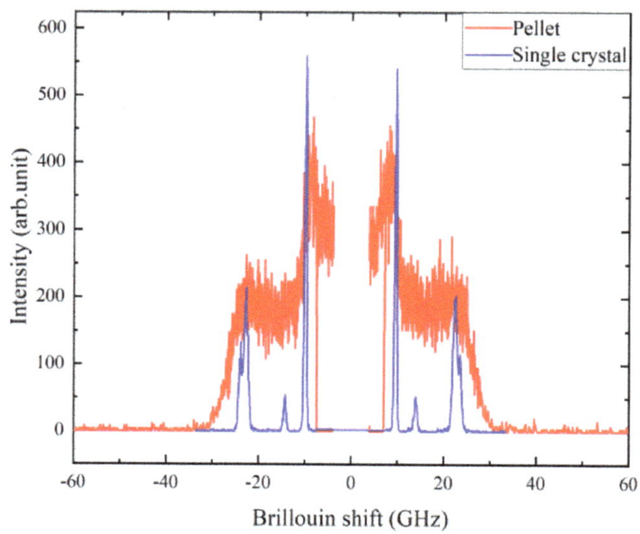

Figure 7. Comparison of the Brillouin spectrum of taurine pellet and single-crystal samples measured at room temperature.

The strong hydrogen bond network exists in the *ab* plane, with tightly-locked molecules being confined in this plane [23]. On the other hand, the molecules are loosely packed along the *c*-axis. Thus, although we have not investigated C_{33} in this study due to the limited available crystal size, we expect that C_{33} may be much smaller than C_{22}. Since elastic properties are important physical parameters required for various applications, such as pharmaceuticals or sensor applications, the present study and the succeeding studies will supply systematic elastic parameters which would accelerate the application of taurine molecules in various fields.

Lima et al. suggested that a structural phase transition existed near 250 K, where some changes in the Raman mode frequency occurred accompanied by enthalpic anomaly [16]. However, structural studies did not reveal any change in the space group of the taurine crystals. The acoustic anomalies and three sets of elastic constants showed monotonic increases upon cooling, which could be explained based on the normal anharmonic effect. Thus, the present study does not support the existence of low-temperature structural phase transitions in taurine single crystals. However, further studies based on other experimental techniques with different spatiotemporal probe scales are necessary in order to unambiguously settle this issue completely.

4. Conclusions

In summary, taurine single crystals were grown and investigated by Brillouin light scattering. One LA and two TA modes were observed, and their mode frequencies increased monotonically upon cooling from 175 °C to −180 °C. The temperature dependence of the LA mode could be satisfactorily explained by using both the Debye and the Einstein anharmonic models. The Einstein and Debye temperatures were 380 and 510 K, respectively, with a ratio of 0.74 consistent with theoretical predictions. The temperature dependences of the sound velocities, acoustic absorption coefficients, and associated elastic constants of taurine single crystals could be obtained for the first time. There was no noticeable anomaly in these acoustic properties, which does not support the existence of structural phase transitions suggested by Raman and calorimetric measurements in [16]. The birefringence of the split LA mode could be estimated and was consistent with the theoretical calculations of the three principal refractive indices [22].

Supplementary Materials: Supplementary materials can be found at https://www.mdpi.com/article/10.3390/ijms22137116/s1.

Author Contributions: J.-H.K., K.-S.L. and S.K. conceived and designed the experiments; K.-S.L. performed the crystal growth; D.H.K. and S.H.O. performed the Brillouin scattering experiments; D.H.K., S.H.O., J.-H.K., K.-S.L. and S.K. carried out a formal analysis of the experimental data; J.-H.K. prepared the original draft; J.-H.K., K.-S.L. and S.K. reviewed and did editing. All authors have read and agreed to the published version of the manuscript.

Funding: This work was supported by the Hallym University Research Fund, 2021 (HRF-202106-008).

Institutional Review Board Statement: Not applicable.

Informed Consent Statement: Not applicable.

Data Availability Statement: The data presented in this study is contained within the aricle.

Conflicts of Interest: The authors declare no conflict of interest.

References

1. Stapleton, P.P.; Charles, R.P.; Redmond, H.P.; Bouchier-Hayes, D.J. Taurine and human nutrition. *Clin. Nutr.* **1997**, *16*, 103. [CrossRef]
2. Timbrell, J.A.; Seabra, V.; Waterfield, C.J. The in vivo and in vitro protective properties of taurine. *Gen. Pharmacol.* **1995**, *26*, 453. [CrossRef]
3. Schaffer, S.; Takahashi, K.; Azuma, J. Role of osmoregulation in the actions of taurine. *Amino Acids* **2000**, *19*, 527. [CrossRef] [PubMed]

4. Aruoma, O.I.; Halliwell, B.; Hoey, B.M.; Butler, J. The antioxidant action of taurine, hypotaurine and their metabolic precursors. *Biochem. J.* **1998**, *256*, 251. [CrossRef]
5. Hasanzadeh, M.; Pournaghi-Azar, M.H.; Shadjou, N.; Jouyban, A. Electropolymerization of taurine on gold surface and its sensory application for determination of captopril in undiluted human serum. *Mater. Sci. Eng. C Mater. Biol. Appl.* **2014**, *38*, 197. [CrossRef]
6. Jin, J.B.; Liu, D.Q.; Zhang, D.D.; Yin, Y.H.; Zhao, X.Y.; Zhang, Y.F. Taurine as an additive for improving the fouling resistance of nanofiltration composite membranes. *J. Appl. Polym. Sci.* **2015**, *11*, 41620. [CrossRef]
7. Sutherland, H.H.; Young, D.W. The crystal and molecular structure of taurine. *Acta Cryst.* **1963**, *16*, 897. [CrossRef]
8. Okaya, Y. Refinement of the crystal structure of taurine, 2-aminoethylsulfonic acid. An example of computer-controlled experimentation. *Acta Cryst.* **1966**, *21*, 726. [CrossRef]
9. Briant, C.E.; Jones, D.W. Neutron refinement of the crystal structure of partially exchanged taurine. *J. Chem. Crystallogr.* **1997**, *27*, 481. [CrossRef]
10. Görbitz, C.H.; Pryds, K.; Ugland, S. Taurine. *Acta Cryst.* **2000**, *C56*, e23. [CrossRef]
11. Hibbs, D.E.; Austin-Woods, C.J.; Platts, J.A.; Overgaard, J.; Turner, P. Experimental and Theoretical Charge Density Study of the Neurotransmitter Taurine. *Chem. Eur. J.* **2003**, *9*, 1075.
12. Waller, M.P.; Howard, S.T.; Platts, J.A.; Piltz, R.O.; Willock, D.J.; Hibbs, D.E. Novel Properties from Experimental Charge Densities: An Application to the Zwitterionic Neurotransmitter Taurine. *Chem. Eur. J.* **2006**, *12*, 7603. [CrossRef]
13. Ohno, K.; Mandai, Y.; Matsuura, H. Vibrational spectra and molecular conformation of taurine and its related compounds. *J. Mol. Struct.* **1992**, *268*, 41. [CrossRef]
14. Frere, P.T.C.; Melo, F.E.A.; Filho, J.M. Polarized Raman and Infrared Spectra of Taurine Crystals. *J. Raman Spectrosc.* **1996**, *27*, 507. [CrossRef]
15. Lima, R.J.C.; Teixeira, A.M.R.; Freire, P.T.C.; Sasaki, J.M.; Ayala, A.P.; Melo, F.E.A.; Filho, J.M. High-pressure Raman study of taurine crystal. *J. Raman Spectrosc.* **2001**, *32*, 27. [CrossRef]
16. Lima, R.J.C.; Freire, P.T.C.; Sasaki, J.M.; Melo, F.E.A.; Filho, J.M.; Moreira, R.L. Temperature-dependent Raman study of taurine single crystal. *J. Raman Spectrosc.* **2001**, *32*, 751. [CrossRef]
17. Freire, P.T.C.; Melo, F.E.A.; Filho, J.M.; Lima, R.J.C.; Teixeira, A.M.R. The behavior of NH_3 torsional vibration of l-alanine, l-threonine and taurine crystals under high pressure: A Raman spectroscopic study. *Vib. Spectrosc.* **2007**, *45*, 99. [CrossRef]
18. Cavaignac, A.L.O.; Lima, R.J.C.; Filho, P.F.F.; Moreno, A.J.D.; Freire, P.T.C. High-temperature Raman study of L-alanine, L-threonine and taurine crystals related to thermal decomposition. *Phys. B* **2016**, *484*, 22. [CrossRef]
19. Maiti, M.; Thomas, S.; Debnath, A.; Kapoor, S. Raman and XPS study on the interaction of taurine with silver nanoparticles. *RSC Adv.* **2016**, *6*, 56406. [CrossRef]
20. Dai, Z.; Xu, X.; Gu, Y.; Li, X.; Wang, F.; Lian, Y.; Fan, K.; Cheng, X.; Chen, Z.; Sun, M.; et al. A terahertz study of taurine: Dispersion correction and mode couplings. *J. Chem. Phys.* **2017**, *146*, 124119. [CrossRef]
21. Bajaj, N.; Bhatt, H.; Vishwakarma, S.R.; Thomas, S.; Murli, C.; Deo, M.N. Low temperature IR spectroscopic study of torsional vibrations of taurine. *AIP Conf. Proc.* **2017**, *1942*, 030017.
22. Moreira, R.L.; Lobo, R.P.S.M.; Dias, A. Infrared dispersion analysis and Raman scattering spectra of taurine single crystals. *Spectrochim. Acta Part A* **2018**, *188*, 276. [CrossRef]
23. Beukes, J.A.; Mo, F.; van Beek, W. X-Ray induced radiation damage in taurine: A combined X-ray diffraction and Raman study. *Phys. Chem. Chem. Phys.* **2007**, *9*, 4709. [CrossRef] [PubMed]
24. Brillouin, L. Diffusion de la lumière et des rayons X par un corps transparent homogène. *Annal. Physique* **1922**, *17*, 88–122. [CrossRef]
25. Cummins, H.Z.; Levanyuk, A.P. *Light Scattering Near Phase Transitions*, 1st ed.; North-Holland Publishing Company: Amsterdam, The Netherlands, 1983; pp. 291–502.
26. Ko, J.-H.; Lee, K.-S.; Ike, Y.; Kojima, S. Elastic properties of aspirin in its crystalline and glassy phases studied by micro-Brillouin scattering. *Chem. Phys. Lett.* **2008**, *465*, 36. [CrossRef]
27. Ko, J.-H.; Kim, T.H.; Lee, K.-S.; Kojima, S. Acoustic properties of aspirin in its various phases and transformation stages studied by Brillouin scattering. *J. Non-Cryst. Solids* **2011**, *357*, 547. [CrossRef]
28. Ko, J.-H.; Kim, T.H.; Lee, K.-S.; Kojima, S. Brillouin scattering study on crystalline and glassy states of anti-inflammatory racemic S(+)–R(−) ibuprofen. *Chem. Phys. Lett.* **2011**, *515*, 221. [CrossRef]
29. Kwon, H.-J.; Kim, T.H.; Ko, J.-H.; Hwang, Y.-H. Relaxation phenomena in supercooled liquid and glassy acetaminophen studied by dielectric, photon correlation and Brillouin light scattering spectroscopies. *Chem. Phys. Lett.* **2013**, *556*, 117. [CrossRef]
30. Shibata, T.; Takayama, H.; Kim, T.H.; Kojima, S. Acoustic and thermal anomalies in a liquid–glass transition of racemic S(+)–R(−) ketoprofen. *Chem. Phys. Lett.* **2014**, *592*, 80. [CrossRef]
31. Shibata, T.; Kojima, S. Liquid, glass, and crystalline indomethacin studied by Brillouin scattering. *Jpn. J. Appl. Phys.* **2018**, *57*, 07LB03. [CrossRef]
32. Haussühl, S. Notizen: Elastische und thermoelastische Konstanten von $NH_2(CH_2)_2SO_3H$ (Taurin). *Z. Naturforsch. A* **1965**, *20*, 1235. [CrossRef]
33. Lind, G.; Kewley, R. Electron Spin Resonance of γ-Irradiated Taurine. *Can. J. Chem.* **1972**, *50*, 43. [CrossRef]

34. Takiyama, H. Supersaturation operation for quality control of crystalline particles in solution crystallization. *Adv. Powder Technol.* **2012**, *23*, 273. [CrossRef]
35. Ko, J.-H.; Lee, K.-S.; Sahoo, S.C.; Naumov, P. Isomorphous phase transition of 1,2,4,5-tetrabromobenzene jumping crystals studied by Brillouin light scattering. *Solid State Commun.* **2013**, *173*, 46. [CrossRef]
36. Ko, J.-H.; Kojima, S.; Koo, T.-Y.; Jung, J.H.; Won, C.J.; Hur, J. Elastic softening and central peaks in $BaTiO_3$ single crystals above the cubic-tetragonal phase-transition temperature. *Appl. Phys. Lett.* **2008**, *93*, 102905. [CrossRef]
37. Maczka, M.; Hanuza, J.; Ko, J.-H.; Kojima, S. High-resolution broadband Brillouin scattering study of antiferroelectric phase transition in $CsBi(MoO_4)_2$. *Phys. Rev. B* **2003**, *68*, 174101. [CrossRef]
38. Rehwald, W. The study of structural phase transitions by means of ultrasonic experiments. *Adv. Phys.* **1973**, *22*, 721. [CrossRef]
39. Varshni, Y.P. Temperature Dependence of the Elastic Constants. *Phys. Rev. B* **1970**, *2*, 3952. [CrossRef]
40. Lakkad, S.C. Temperature Dependence of the Elastic Constants. *J. Appl. Phys.* **1971**, *42*, 4277. [CrossRef]
41. Sonehara, T.; Tatsu, E.; Saikan, S. Temperature dependence of the Brillouin frequency shift in crystals. *J. Appl. Phys.* **2007**, *101*, 103507. [CrossRef]
42. Errandonea, G. Elastic and mechanical studies of the transition in LaP_5O_{14}: A continuous ferroelastic transition with a classical Landau-type behavior. *Phys. Rev. B* **1980**, *21*, 5221. [CrossRef]
43. Hernandez, J.; Li, G.; Cummins, H.Z.; Callender, R.H. Low-frequency light-scattering spectroscopy of powders. *J. Opt. Soc. Am. B* **1996**, *13*, 1130. [CrossRef]
44. Ko, J.-H.; Kim, D.H.; Kojima, S.; Kim, J.-H.; Choo, W.K. Brillouin Scattering Study on Polycrystalline Relaxor Ferroelectrics. *Jpn. J. Appl. Phys.* **2003**, *42*, 3076. [CrossRef]

International Journal of
Molecular Sciences

MDPI

Article

Electronic Circular Dichroism of the Cas9 Protein and gRNA:Cas9 Ribonucleoprotein Complex

Monika Halat [1], Magdalena Klimek-Chodacka [2], Jagoda Orleanska [1], Malgorzata Baranska [1,3,*] and Rafal Baranski [2,*]

[1] Faculty of Chemistry, Jagiellonian University, Gronostajowa 2, 30-387 Krakow, Poland; monika.halat@uj.edu.pl (M.H.); jagoda.orleanska@doctoral.uj.edu.pl (J.O.)
[2] Department of Plant Biology and Biotechnology, Faculty of Biotechnology and Horticulture, University of Agriculture in Krakow, AL. 29 Listopada 54, 31-425 Krakow, Poland; magdalena.klimek-chodacka@urk.edu.pl
[3] Jagiellonian Centre for Experimental Therapeutics (JCET), Bobrzyńskiego 14, 30-348 Krakow, Poland
* Correspondence: m.baranska@uj.edu.pl (M.B.); rafal.baranski@urk.edu.pl (R.B.)

Abstract: The *Streptococcus pyogenes* Cas9 protein (SpCas9), a component of CRISPR-based immune system in microbes, has become commonly utilized for genome editing. This nuclease forms a ribonucleoprotein (RNP) complex with guide RNA (gRNA) which induces Cas9 structural changes and triggers its cleavage activity. Here, electronic circular dichroism (ECD) spectroscopy was used to confirm the RNP formation and to determine its individual components. The ECD spectra had characteristic features differentiating Cas9 and gRNA, the former showed a negative/positive profile with maxima located at 221, 209 and 196 nm, while the latter revealed positive/negative/positive/negative pattern with bands observed at 266, 242, 222 and 209 nm, respectively. For the first time, the experimental ECD spectrum of the gRNA:Cas9 RNP complex is presented. It exhibits a bisignate positive/negative ECD couplet with maxima at 273 and 235 nm, and it differs significantly from individual spectrum of each RNP components. Additionally, the Cas9 protein and RNP complex retained biological activity after ECD measurements and they were able to bind and cleave DNA in vitro. Hence, we conclude that ECD spectroscopy can be considered as a quick and non-destructive method of monitoring conformational changes of the Cas9 protein as a result of Cas9 and gRNA interaction, and identification of the gRNA:Cas9 RNP complex.

Keywords: CRISPR/Cas9; DNA cleavage; ECD; spectroscopy; guide RNA; RNP complex; SpCas9

Citation: Halat, M.; Klimek-Chodacka, M.; Orleanska, J.; Baranska, M.; Baranski, R. Electronic Circular Dichroism of the Cas9 Protein and gRNA:Cas9 Ribonucleoprotein Complex. *Int. J. Mol. Sci.* **2021**, *22*, 2937. https://doi.org/10.3390/ijms22062937

Academic Editors: Filipe Ferreira da Silva and François Hache

Received: 15 February 2021
Accepted: 11 March 2021
Published: 13 March 2021

1. Introduction

In 2020, the Royal Swedish Academy of Sciences has awarded the Nobel Prize in Chemistry for the development of a method for genome editing, as the consequence of discovery and research on Clustered Regularly Interspaced Short Palindromic Repeats (CRISPR) and CRISPR-associated (Cas) proteins. The CRISPR/Cas system is a fast developing technology allowing for genome editing, i.e., for generation of small genome modifications at defined sites of the target DNA or, in other words, for a precise, site-directed mutagenesis [1]. A wide range of potential applications and the simplicity, precision and effectiveness of genome editing also justified the earlier announcement of CRISPR/Cas the Breakthrough of the Year by Science journal in 2015 [2]. Despite editing approach, various Cas variants have been used to create chimeric molecules to elucidate protein interactions, and to activate or repress gene expression [3]. The CRISPR/Cas-based methods have been revolutionizing biological sciences, nowadays. Clinical trials of CRISPR-mediated gene therapy have been initiated [4]. Advanced research has been carried out for the application of CRISPR/Cas in human disease treatments, including AIDS, hemophilia, autism spectrum disorder, and cancer [5]. Additionally, several molecular diagnostic tools were invented for virus detection, pathogen identification, cancer mutation analysis or patient genotyping [6]. In plant sciences, genome-edited soybean of high nutritional value has been adopted to agriculture recently while other crops are awaiting commercialization [7,8].

A native CRISPR/Cas system was discovered in microbes where it functions as an adaptive immune mechanism protecting against invading viruses [9]. The majority of prokaryotic genomes contain characteristic CRISPR locus with short (about 30 bp, on average) repeats interspaced by unique sequences of similar size [9–11]. These spacers are phage DNA fragments of variable sequences acquired by a microbe during past phage infections [12]. Each spacer and repeat pair are transcribed to CRISPR RNA (crRNA), which hybridizes with another RNA molecule, a small trans-activating crRNA (tracrRNA), to form a guide RNA (gRNA). Hence, gRNA is composed of the tracrRNA with a stem loop structure due to conserved nucleotide sequence and the crRNA molecule with 20 nt variable sequence at 5′ end that can hybridize to a complementary DNA strand. Cas protein, or usually a complex of various Cas proteins, forms a ribonucleoprotein (RNP) complex with gRNA, which guides the protein to the target phage DNA at the location with complementary sequence next to a short protospacer adjacent motif (PAM) of conserved, usually 3–6 nt sequence [13,14]. Then, Cas protein destroys phage DNA by its cleavage [15]. Further discoveries have paved the way for the utilization of this natural defense mechanism and have led to the development of CRISPR/Cas-based systems for DNA editing and research in eukaryotic organisms. In particular, it was found that about 100 nt long, single gRNA composed of crRNA and tracrRNA fragments linked with a tetraloop (Figure 1) can be used instead of a native hybrid gRNA [13]. In consequence, designing gRNA to target the desired DNA location as well as the delivery of gRNA to target cells was simplified and single gRNA has become routinely used.

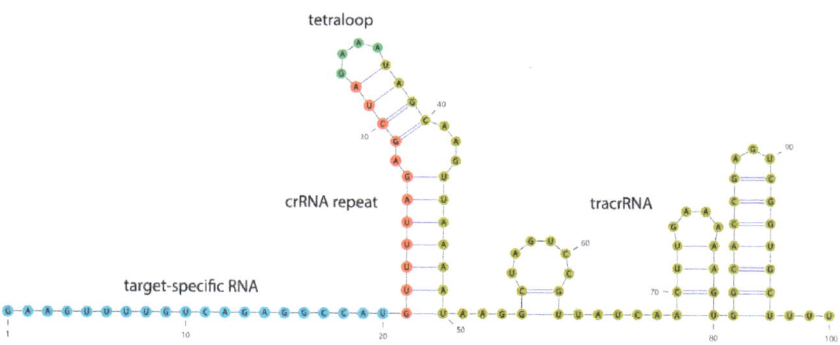

Figure 1. Single-guide RNA structure. In this work, the target-specific RNA sequence is complementary to the flavanone-3-hydroxylase (*F3H*) carrot (*Daucus carota* L.) gene (NCBI Acc. No. AF184270.1).

Microbes have developed diverse CRISPR/Cas systems differing in the number and function of Cas proteins which often must form functional complexes capable of recognizing, binding, and cleaving phage DNA. However, single proteins performing all these functions have been identified [11]. The best characterized, and now commonly used for genome engineering, is the *Streptococcus pyogenes* Cas9 (SpCas9, pCas9) protein. Cas9 is a class II, type II, nuclease consisting of seven domains: REC I, REC II, REC III, HNH, RuvC, PAM-interacting, and bridge helix (Figure 2) [16]. The REC I domain (blue colour) is responsible for a connection with gRNA. In turn, the REC III domain (cyan colour) is used to sense the formation of RNP complex. The PAM-interacting domain (red colour) recognizes the PAM sequence at the target DNA. The arginine-rich bridge helix (pink colour) plays a key role in binding target DNA and modulates cleavage activity of the Cas9 protein [17]. Bot, HNH (green colour) and RuvC (yellow colour) domains are responsible for cleavage; each of them cleaves a single DNA strand 3 nt upstream of PAM that leads to a double-strand break [18,19]. The role of REC II domain (gray colour) has not been fully elucidated, yet, but it seems not critical for DNA cleavage [16,20].

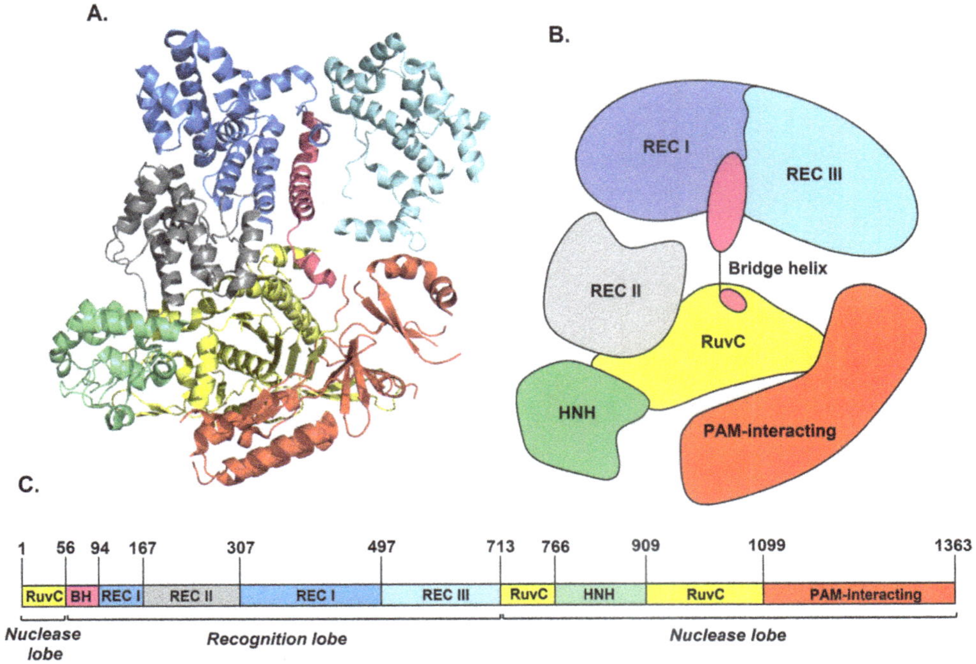

Figure 2. The structure of *Streptococcus pyogenes* Cas9 protein, which contains seven domains: Rec I, Rec II, REC III, Bridge Helix, RuvC, HNH, and PAM-interacting. Domains are presented in crystal (**A**), schematic (**B**), and map (**C**) form. Crystal image was rendered from RCSB PDB ID: 4CMP [19], using PyMOL software (Schrödinger Inc., New York, NY, USA).

The development of efficient CRISPR/Cas systems for genome editing or for other research purposes requires the selection of a proper Cas variant and designing correct gRNA. The ability of both components to form active RNP complex must be verified experimentally. Hence, a fast and non-destructive analytical method confirming RNP complex formation would be valuable. Here, we present an electronic circular dichroism (ECD) study on the Cas9 protein in its native form and bounded with gRNA. ECD spectroscopy is based on a differential absorption of left- and righthanded circularly polarized light in the UV-Vis range by chiral molecules. It is widely used to monitor structural changes of organic compounds or important macromolecules in solution [21,22] as nucleic acids [23,24] and proteins, especially to estimate the secondary structure of proteins including α-helix, β-sheet, or random coil [25–27]. Proteins give strong ECD signals in the far-UV region, dominated by the $n\pi^*$ (~220 nm) and $\pi\pi^*$ (~190 nm) transitions of amide groups, which are influenced by the geometries of the polypeptide backbones and reflective of the different types of secondary structures. In contrast to the other techniques like X-ray crystallography and NMR, ECD does not require much time and specific condition of preparing a sample [28]. Thus, a good quality ECD spectrum can be obtained for a quite low concentration of protein, in neutral water environment and with a short acquisition time (in 30 min or less). Hence, ECD measurement, possibly supported by theoretical calculations, is a powerful tool to examine the structure of proteins [28,29]. What is more, there are only a few reports engaging ECD spectroscopy, in a very little extent, to analyse CRISPR/Cas systems [30–33]. Two of them concern the Cas9 protein specifically [30,32]. So, we found that ECD can be considered as a quick and effective method to identify the gRNA:Cas9 RNP complex. To confirm this approach is non-destructive to the protein and its complex with RNA, we induced also in vitro DNA cleavage to show the biological activity of RNP after ECD measurements.

2. Results

2.1. Circular Dichroism of Cas9 Protein and gRNA

To determine secondary structure of Cas9 protein, ECD measurements were performed. Recorded ECD and UV-Vis spectra are presented in Figure 3 (blue line). Proteins give strong ECD signals in the range of 185–300 nm, which carry information about their secondary structure. For Cas9, ECD profile reveals two negative and one positive pattern located at 221, 209, and 196 nm, respectively. Intensities of negative bands are slightly different, as indicated by the calculated ratio $[\theta]_{221}/[\theta]_{209} = 1.12$. In addition, these bands are much more intense than the positive ones. In turn, the maximum of UV-Vis absorption for Cas9 protein is located at 191 nm (Figure 3).

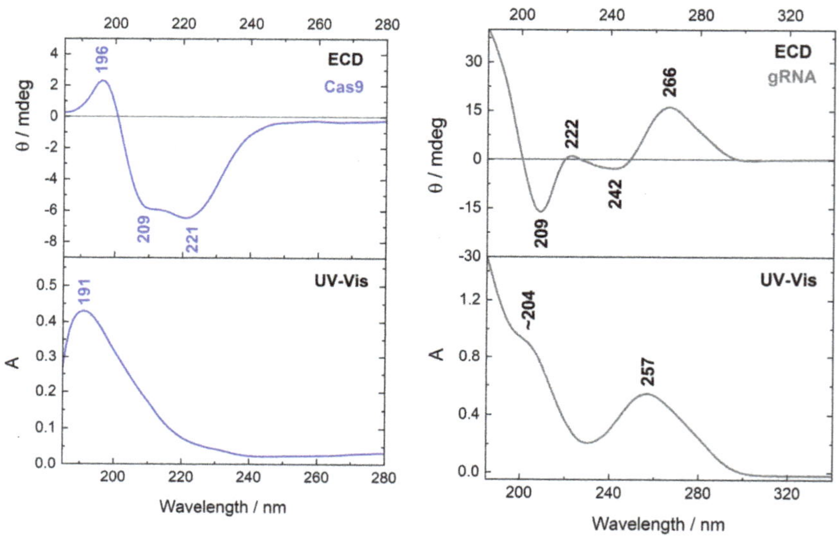

Figure 3. ECD and UV-Vis spectra of Cas9 protein (blue line, C = 9.35×10^{-7} M) and gRNA (gray line, C = 1.07×10^{-5} M).

The ECD spectrum registered for gRNA is also shown in Figure 3 (gray line). Going from longer to shorter wavelengths, the spectral profile exhibits positive/negative/positive/negative pattern of ECD bands situated at 266, 242, 222 and 209 nm, respectively. The UV-Vis spectrum of gRNA reveals maximum at 257 nm and a shoulder at around 204 nm.

2.2. Circular Dichroism of the RNP Complex

To find an effective and quick method for identification of the gRNA:Cas9 complex, ECD measurements were performed also for the RNP complex solution. Collected ECD and UV-Vis spectra are presented in Figure 4. The RNP ECD profile (red line) shows the positive/negative pattern of the bands located at 273 and 235 nm. The intensities of these bands are approximately one order in magnitude lower than the one in the spectrum of Cas9 protein (blue line). Moreover, the ECD spectrum of RNP complex differs significantly from the ECD spectrum of unassociated gRNA (light gray line). Both spectra have similar positive/negative pattern of two following bands going from longer to shorter wavelengths. However, those ECD bands have different positions and are shifted in relation to each other. The ECD spectrum of gRNA reveals also additional bands situated at 222 and 209 nm.

The UV-Vis spectrum of the gRNA:Cas9 complex shows tree maxima, observed at 258, 217 and 191 nm, two of which can be assigned to gRNA (258 nm) and protein (191 nm), and the band at 217 nm cannot be assigned to any of the components.

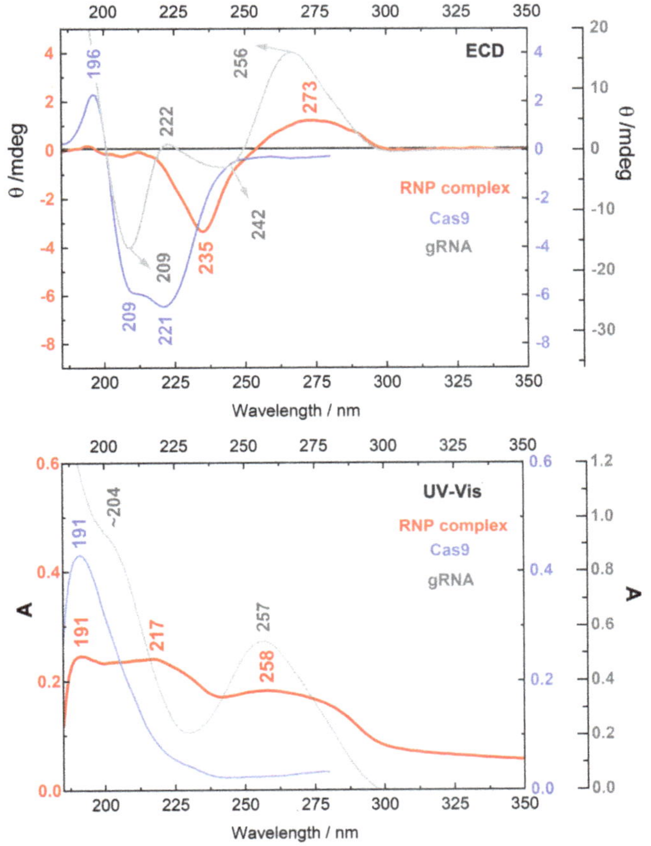

Figure 4. ECD and UV-Vis spectra of the Cas9:gRNA complex (red line and left scale; in the 1:1 molar ratio with individual concentrations $C_{CAS9} = 9.98 \times 10^{-6}$ M and $C_{gRNA} = 1.07 \times 10^{-5}$ M) in comparison to the spectra of the Cas9 protein (light blue line and the 1st right scale, $C = 9.35 \times 10^{-7}$ M) and gRNA (light gray line and the 2nd right scale $C = 1.07 \times 10^{-5}$ M). ECD: electronic circular dichroism.

2.3. Biological Activity of the Cas9 Protein and RNP Complex

The Cas9 biological activity after ECD measurements was verified by inducing DNA cleavage in vitro and the sizes of the obtained ROX-labelled DNA fragments were determined using capillary electrophoresis. Fluorescent signals from untreated sample revealed the presence of DNA, which size was exclusively 309 bp (Figure 5A). The same sample was incubated with the RNP complex, which was not used for ECD measurements (control). After incubation, fluorescent signals were obtained predominantly from DNA fragments of 162 bp in length (Figure 5B). Weaker signals from shorter (150–161 bp) fragments were also present while the 309 bp fragment signal was very weak. The presence of 162 bp or shorter fragments indicated that almost all DNA was cleaved by the control RNP complex.

The ECD measurements of Cas9 protein were performed and then such protein was used for RNP complex formation. Independently, a new RNP complex was prepared by incubating unused gRNA and Cas9, and then such RNP complex was subjected to ECD measurements. Both complex types, the RNP complex containing ECD-measured Cas9 and ECD-measured RNP complex, were used for DNA cleavage in vitro to check their biological activity. For both samples, electropherograms showed fluorescent signals indicating the presence of short DNA fragments of the same size as in the control, i.e., predominantly 162 bp fragments (Figure 5C,D).

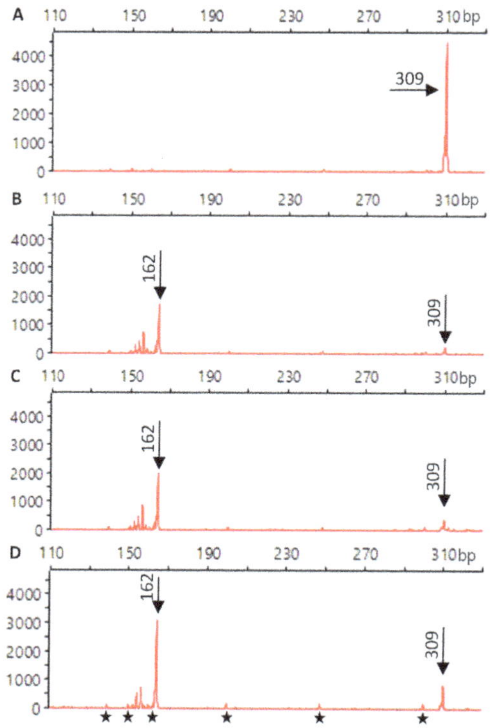

Figure 5. Electropherograms of ROX-labelled DNA fragments detected by a capillary electrophoresis. Untreated DNA of 309 bp in length (**A**), DNA incubated with the untreated RNP complex—control (**B**), DNA incubated with the RNP complex containing ECD-measured Cas9 (**C**), DNA incubated with the ECD-measured RNP complex (**D**). Stars—GeneScan™ 500 LIZ size marker.

3. Discussion

3.1. Cas9 and gRNA:Cas9 Complex Structures Revealed by ECD Spectroscopy

Based on ECD spectroscopy, a secondary structure of protein like α-helix, β-sheet or random coil can be identified [27,28]. Excitation of a number of closely connected chromophores (peptide bonds), being in a regular spatial disposition in protein, gives rise to characteristic spectral features below 240 nm. The experimental ECD spectrum of Cas9 protein (Figure 3, blue line) shows negative/positive pattern, like α-rich proteins, with two negative bands of nearly the same intensity, observed at 221 and 209 nm, accompanied by the positive one at 196 nm. The origin of these ECD bands is assigned to the $n\pi^*$, $\pi\pi^*||$ (parallel to the helix axis) and $\pi\pi^*\perp$ (perpendicular to the helix axis) electronic transitions of the peptide bond, respectively [27,34], where n is the nonbonding molecular orbital, π is the HOMO level, while π^* refers to the LUMO state [34]. In contrast to a typical α-helix ECD spectrum [35], the positive Cas9 band is slightly red-shifted (192→196 nm) and shows twice-lower intensity than negative ones, which in consequence shows an impact of the remaining secondary structures in the protein. According to the crystallographic data, the SpCas9 molecule adopts a crescent shape with approximate dimensions of 100 Å× 100 Å× 50 Å [19]. Its secondary structure is determined by α-helices with about a twice-less content of stranded antiparallel β-sheets that, in turn, are the most prevailing in the PAM-interacting and RuvC domains [36] (Figure 2A). Moreover, using an interactive interface to the STRIDE program [37] presented in detail by Frishman and Argos [38], the percentage content of individual secondary structure can be estimated based on knowledge of the atomic coordinates of proteins. Thus, according to calculations based on the STRIDE program, beyond the main contribution of α-helix (49.6%), the SpCas9 protein also contains:

4% of 3_{10}-helix, 9.4% of β-sheet, 1.6% of β-bridge and 17% of β-turn classified as I, I′, II, II′, VIa and VIb [39]. The remaining of 18.6% is a random coil structure. Hence, the decrease in intensity, as well as the position of positive band at 196 nm in the Cas9 ECD spectrum may be attributed to the presence of 3_{10}-helix, what was theoretically predicted by Manning and Woody [40]. The 3_{10}-helix is a relatively common structural element in globular proteins (3%), occurring often at the ends of α-helices [41,42]. The average conformational parameters of both helices are rather close to each other, but the 3_{10}-helix is gently tighter and more elongated, as well as is characterized by the different C=O···H−N intermolecular hydrogen bonding scheme [41,43]. Then, Toniolo et al. [44] reported the experimental ECD spectrum of 3_{10}-helix, confirming the positive band of very weak intensity near 195 nm. However, as opposed to the theoretical calculations, another negative strong maximum at 184 nm, not seen here (Figure 3, blue line), was also registered. The 3_{10}-helix seems to have a positive signal around 190 nm of an uncertain magnitude [45]. Therefore, it cannot be definitively determined whether the presence of the 3_{10}-helix in Cas9 protein contributed to the intensity reduction of the positive Cotton effect, and a higher contribution of β-sheet than one might expect cannot be ruled out.

Certainly, the β-sheet content has also an impact on the Cas9 ECD profile, what is observed in the spectra of α/β-rich proteins like ribonuclease A (21% α, 33% β) or subtilisin (30% α, 18% β) [45]. Signals of α-helix are stronger than β-sheet ones, so proteins with comparable amount of both secondary structures generally show the qualitive spectral features of the α-helix keeping the same negative/positive ECD profile. Although some differences in the intensity or localization of ECD bands are present, especially as a decrease in intensity of positive band around 195 nm in ribonuclease A spectrum [45]. Significant effect on protein ECD pattern may have also β-turns [45,46], enabling reversal of the polypeptide chain direction. These motifs constitute as the third most common secondary structural elements occurring in proteins. For example, the type I reveals α-helix-like ECD spectrum, while the type II shows a β-sheet-like one, what was confirmed experimentally [47]. The Cas 9 has β-turn motifs in its structure, what may be manifested in reduced intensity of the positive band at 196 nm, as well as in a gentle difference in intensity between negative ECD bands ([θ]221/[θ]209 = 1.12).

In contrast to ECD spectroscopy, UV-Vis spectra of different proteins secondary conformations do not reveal specific features, besides slight shift of the main electronic transition. The Cas9 UV-Vis spectrum (Figure 3, blue line) has only one maximum at 191 nm that is typical for pure α-helix structure and comes from the $\pi\pi^*$ excitation in the amide group [48].

Like proteins, the absorption spectra of nucleic acids are dominated by $\pi\pi^*$ electronic transitions, which are all polarized in the plane of the bases. Moreover, both wavelength maxima and transition intensities vary depending on the base sequence and structure adopted [48]. Nonbonding electrons of the amide-like oxygens can be also excited to π^* orbitals; however, $n\pi^*$ transitions are characterized by low intensity and are usually buried by the intense $\pi\pi^*$ ones [49]. Here, we observe maximum of UV-Vis gRNA band at 257 nm (Figure 3, gray line), which can be assigned to $\pi\pi^*$ excitation and corresponds to the literature data recorded for various polynucleotides [48]. In turn, the experimental ECD spectrum of gRNA (Figure 3, gray line) shows positive/negative/positive/negative pattern with bands observed at 266, 242, 222 and 209 nm, respectively, what is typical for the A-type helix geometry assigned to RNA duplexes [50,51]. Thereby, it confirms that gRNA has the repeat and anti-repeat regions adopting the A-form-like conformation [52]. In general, RNA is formed by canonical double-helical fragments, along with non-canonical secondary structure motifs like internal loops, hairpins, bulges, and the most widespread ones: single mismatches [53]. All these motifs play relevant functions in providing binding sites for proteins, small molecules, or nucleic acids, and have impact in folding the correct tertiary and quaternary structures. Theoretical calculation shows that ECD signal of polynucleotides mainly origins from the exciton coupling interaction between bases stacked in asymmetric structure of higher ordering like helices, than from the base-sugar interaction in monomeric nucleotides, as well as it arises for the most part from the $\pi\pi^*$ transitions, rather than $n\pi^*$ ones [54]. Thus, perturbation of a nucleic acid ECD profile, that appears due to binding of a protein, generally reflects changes in short-range base–base interactions, providing confirmation for transitions between secondary structures (A↔B, B↔Z) or alternations in base pairing, base stacking, or bending of a DNA or RNA duplex [55].

Despite a lot of ECD studies describing protein—nucleic acids interactions [55,56], here we present for the first time the experimental ECD spectrum of gRNA:Cas9 RNP complex (Figure 4, red line). This unique RNP hybrid reveals positive/negative ECD pattern with bands located at 273 and 235 nm, which differs significantly from individual spectral profiles of each macromolecule. The Cas9 upon gRNA binding changes dramatically its structure through wide intermolecular rearrangements. According to the literature, one of the most prominent changes takes place in the REC III domain, which moves ~65 Å toward the HNH domain due to gRNA connection [20], what is definitely reflected by the characteristic ECD spectral profile of RNP hybrid in comparison to the ECD spectrum of a free protein (Figure 4, light blue line). Moreover, using the STRIDE server [37] and the crystallographic data PDB ID 4ZT0 [57], the percentage content of individual secondary structure in active RNP complex can be provided: 41.5% of α-helix, 4.5% of 3_{10}-helix, 10.7% of β-sheet, 2% of β-bridge, 22.4% of β-turn and 18.8% of random coil structure. Based on presented data, it can be concluded that the Cas9 changes its secondary structure elements upon gRNA binding, mainly through loss of α-helicity (49.6→41.5%), and slight increase in the content of β-turns (17→22.4%). However, it is hard to determine those changes from the RNP ECD profile, which has no representative spectral features for still-prevalent α-helix structure. A major change in α-helical content undergoes also among a DNA binding proteins like transcription factors, which have been mostly studied among proteins by ECD below 250 nm [55]. Unlike the gRNA:Cas9 complex, transcription factors exhibit an enhancement of α-helix structure in the bound state around 10–15%. However, in some cases like recombination activation gene 1 protein (the RAG1) [58] and human translin protein [59], a strong decrease in the protein α-helicity after nucleic acids complexation was confirmed. On the other hand, the ECD spectrum of RNP complex is strongly red-shifted and do not reveal any characteristic bands below 220 nm, which can be assigned to any of the protein secondary structures. In addition, its ECD signals do not coincide with the gRNA spectrum indicating that the polynucleotide chain adopted also different geometry due to connection with the Cas9 protein. What is interesting, there are some similarities in ECD patterns between the gRNA:Cas9 complex and chromatin [60], especially in the range 240–300 nm that is practically the same in both cases. The ECD spectrum of the complex between EsColi ribonuclease and RNA also bears some resemblance to the RNP [61]. However, the RNP ECD profile is very individual and it is hard to compare with literature results obtained for other protein–nucleic acid hybrids. The UV-Vis spectrum of gRNA:Cas9 (Figure 4, red line) differs somehow from those of Cas9 and gRNA, but it includes absorption bands characteristic of each macromolecular component and does not allow to monitor changes in secondary structure geometry.

3.2. Cas9 Binds gRNA and Retains Cleavage Activity after ECD Measurements

We have previously demonstrated that carrot genome can be edited by delivering *S. pyogenes* Cas9 and gRNA expression vectors to living cells which led to the RNP complex formation in vivo [62]. The gRNA was designed to target the 20 nt sequence of the flavanone-3-hydroxylase (*F3H*) gene located upstream the GGG PAM sequence. Sequencing analysis revealed that the target DNA was cleaved usually 3 bp upstream of PAM, although small nucleotide deletions in close proximity were also identified in some samples.

In this work, we have verified that the measurements using ECD spectroscopy do not affect Cas9 biological activity by performing in vitro DNA cleavage experiments. Verification of RNP complex activity using in vitro cleavage has been reported earlier and such approach has several advantageous over experiments with living cells: it is fast, performed in controlled conditions, and the results are not affected by complex cell processes [63]. For that purpose, we assembled the above said gRNA molecules and obtained their complexes with SpCas9 to carry out in vitro cleavage of DNA utilizing a laboratory-amplified fragment of the *F3H* carrot gene as the target. The used gRNA was designed to guide Cas9 protein to the 309 bp DNA and to enable DNA cleavage at the selected target site to two, 162 bp and 147 bp, fragments. A fluorescent capillary electrophoresis was employed to confirm cleavage due to its high accuracy [64]. As the used DNA was ROX-labelled at only one end, the fluorescent signals registered during a capillary electrophoresis could originate from the unmodified DNA molecules and 162 bp cleaved products. The in vitro DNA cleavage and detection of predominantly 162 bp products confirmed the activity of control RNP complex, as expected.

Using the same approach, two other RNP complexes were also checked for their biological activity. In contrast to the control, either the whole RNP complex was subjected to ECD measurement before in vitro DNA cleavage experiments or only the Cas9 protein was measured prior to the RNP complex formation, and then this complex was measured using ECD. In all samples, the expected 162 bp cleaved fragments were identified. No essential difference in the activity of RNP complexes was found nor their activity differed from the control. Hence, in vitro DNA cleavage confirmed that: (1) gRNA was properly designed and synthesized, (2) Cas9 protein was able to form complex with gRNA, (3) gRNA properly directed Cas9 to the target DNA site, and (4) Cas9 had cleavage activity. Moreover, the same conclusions apply to the ECD-measured Cas9 protein used for RNP complex formation. The ECD measurements did not restrict Cas9 protein ability to interact with gRNA and DNA nor its enzymatic activity. Furthermore, ECD measurements of the already formed RNP complex also did not affect these properties. Hence, it can be concluded that after ECD measurements Cas9 retains its structure capable to form an active complex with gRNA to efficiently cleave DNA.

The obtained results provide new opportunities for analytical validation of correct gRNA designing. The ECD spectroscopy can be used to confirm the presence of RNP complexes in the sample without adverse effects on their biological activity. This is in particular valuable in case of unsuccessful DNA cleavage performed in vivo or in vitro. As indicated above, several requisites must be fulfilled for efficient DNA cleavage by RNP complex [13,16,20]. Therefore, the lack of cleaved products may be due, inter alia, to aberrant interaction of gRNA with the Cas protein. The detection of such events would be valuable in research aiming in verification of gRNA differing in their structure. In particular, the Cas activity depends on the gRNA length and tracrRNA fragment structure [65,66]. Also new Cas protein variants have been reported and they differ in structure which, in turn, determines the structure of interacting gRNA [3,13]. Hence, complementing experimental and bioinformatic approaches to studying RNP complexes [67], ECD spectroscopy can help reveal Cas–gRNA compatibility in a fast and simple fashion.

4. Materials and Methods

4.1. Cas9 Protein

Streptococcus pyogenes Cas9 (EnGen™ Spy Cas9 NLS) protein was purchased from New England Biolabs Inc. (Ipswich, MA, USA) as a buffer solution (300 mM NaCl; 10 mM Tris-HCl; 0.1 mM EDTA; 1 mM DTT; 50% glycerol) with a protein concentration of 3.22 mg/mL (100 μL; 2000 pmol).

4.2. gRNA Design and Transcription

The 20 nt target-specific RNA sequence of single-guide RNA (Figure 1) was designed to target the second exon in the flavanone-3-hydroxylase (*F3H*) carrot (*Daucus carota* L.) gene (NCBI Acc. No. AF184270.1) at 632–651 position (GAAGTTTTGTCAGAGGCCAT) using the CasOT software [68]. The complementary DNA oligonuclotide was synthesized commercially (Genomed, Warsaw, Poland). This fragment was assembled with oligonucleotides of the tracrRNA part of the gRNA using the GeneArt™ Precision gRNA Synthesis Kit (Invitrogen, Carlsbad, CA, USA), and the final product was transcribed to the gRNA molecules according to the kit manufacturer's recommendations. The correct assembly and nucleotide sequence of gRNA was confirmed by sequencing (Genomed, Warsaw, Poland).

4.3. Formation of the gRNA:Cas9 RNP Complex

The synthesized gRNA was mixed with Cas9 in 1:1 molar ratio and incubated at 37 °C for 20 min in order to form the RNP complex. In detail, 4.24 μL of the 7.58 mg/mL gRNA solution was added to 50 μL of the 3.22 mg/mL Cas9 solution. The obtained mixture was vortexed and then incubated. The RNP complex was used either for ECD measurement or for in vitro DNA cleavage.

4.4. UV-Vis/ECD Measurements

Samples of Cas9 protein, gRNA and RNP complex were measured by means of UV–Vis/ECD spectroscopy using JASCO J-815 spectrometer (JASCO Corporation, Tokyo, Japan). To avoid supersaturation of the detector, all samples were diluted with very clear nuclease-free water. The concentration of the individual solutions was: 9.35×10^{-7} M for the Cas9 protein, 1.07×10^{-5} M for the gRNA in both single and RNP form, as well as 9.98×10^{-6} M for the Cas9 nuclease in RNP complex. All UV–Vis/ECD spectra were recorded with the use of 0.1 cm path length quartz cuvette, as well as with the following settings: 185–350 nm spectral range, 100 nm/min scanning speed, 1 nm bandwidth, 0.2 nm step size, 0.5 s response time and 10 scans of accumulation. Finally, baseline and solvent corrections were done with the JASCO software (JASCO Corporation, Tokyo, Japan).

4.5. DNA Amplification

Two HPLC purified primers (F: 5′-GCAAGATTGGCGAGAGATAG-3′ and R: 5′-AGCAAGAGCGTAATTGTGCC-3′) designed for the amplification of 309 bp DNA fragment of the second exon in the *F3H* carrot gene (NCBI Acc. No. AF184270.1) during polymerase chain reaction (PCR) were ordered from Genomed company (Warsaw, Poland). The F primer was labelled with the ROX fluorescent dye at its 5′ end. The PCR reaction mixture set up in 20 μL volume contained 0.1 μM of each primer and a commercial 2× buffer composed of Taq polymerase and dNTPs (PCR Mix Plus, A&A Biotechnology, Gdynia, Poland). The reaction was carried out in the Eppendorf Mastercycler thermocycler (Eppendorf, Hamburg, Germany) with the following conditions: the initial denaturation at 94 °C for 4 min, 35 cycles of 45 s at 94 °C, 30 s at 60 °C, and 60 s at 72 °C, and the final extension at 72 °C for 5 min.

4.6. In Vitro DNA Cleavage

The RNP complex, in the amount of 9 μmol Cas9 and 9 μmol gRNA, was mixed with 0.9 μmol of the PCR amplified 309 bp DNA fragments, which were earlier purified on a column (Wizard® SV Gel and PCR Clean-Up, Promega, Madison, WI, USA). The final reaction mixture contained 10xNEBuffer 3.1 (100 mM NaCl, 50 mM Tris-HCl, 10 mM $MgCl_2$, 100 μg/mL BSA) adjusted to 30 μL volume in RNase free water. DNA cleavage was performed at 37 °C for 15 min followed by 10 min incubation at RT, after the addition of 1 μL of Proteinase K (Sigma-Aldrich, St. Louis, MO, USA). The DNA fragments were separated using a capillary electrophoresis (3730XL DNA Analyzer, Applied Biosystems, Foster City, CA, USA) combined with the detection of fluorescent dyes. DNA fragment sizes were determined using the GeneScan™ 500 LIZ size marker (Thermo Fisher Scientific, Waltham, MA, USA) and visualized using the PeakScanner v.2.0 software (Thermo Fisher Sci., Waltham, MA, USA).

5. Conclusions

In this work, characteristic spectra of the Cas9 protein, gRNA, and their complex were registered using ECD spectroscopy. In particular, the RNP complex showed a distinctive spectral pattern resulting from some structural rearrangements in the Cas9 protein. The formation of RNP complex as well as its activity is determined by the structure of Cas protein and interacting gRNA molecule. Hence, we show that ECD spectroscopy can be used for verification of the Cas9 ability to bind specific gRNA and for identification of the successfully formed RNP complex. Moreover, the activity of the RNP complex was further confirmed by conducting the in vitro DNA cleavage. The results show that either Cas9 protein or RNP complex retain their biological activities after the ECD measurements, and they are able to bind and cleave the target DNA. The experiments have shown the potential of the ECD in structural studies of various Cas proteins and their interactions with specific gRNAs, in an undoubtedly non-destructive manner. To facilitate the research and use of continuously and fast evolving CRISPR/Cas-based systems in biological sciences, reliable and simple techniques allowing characterization of the CRISPR/Cas individual components are needed, and ECD can be considered as such.

Author Contributions: Conceptualization, M.B. and R.B.; methodology, M.H., M.K.-C., M.B. and R.B.; formal analysis, M.H., M.K.-C. and J.O.; investigation, M.H., M.K.-C. and J.O.; resources, M.K.-C. and R.B.; data curation, M.H. and M.K.-C.; writing—original draft preparation, M.H., M.K.-C., J.O. and R.B.; writing—review and editing, M.H., M.B. and R.B.; visualization, M.H. and M.K.-C.; supervision, M.B. and R.B.; project administration, M.B. and R.B.; funding acquisition, M.H., M.K.-C. and R.B. All authors have read and agreed to the published version of the manuscript.

Funding: This research was partially funded by the National Science Center in Poland; grant Nos. 2016/21/B/NZ9/01054 (Opus, R.B.), and 2018/02/X/NZ9/02629 (Miniatura, M.K.-C.). The first author (M.H.) has received funding in the term of the PhD scholarship Etiuda 7 No. 2019/32/T/ST4/00230, also from the National Science Center.

Institutional Review Board Statement: Not applicable.

Informed Consent Statement: Not applicable.

Data Availability Statement: Publicly available datasets were analyzed in this study. This data can be found in the RCSB Protein Data Bank here: https://www.rcsb.org/structure/4CMP and https://www.rcsb.org/structure/4ZT0.

Acknowledgments: The project is co-financed by the Polish National Agency for Academic Exchange (No. PPN/BCZ/2019/1/00058/U/00001 and No. PPN/BCZ/2019/1/00046/U/00001).

Conflicts of Interest: The authors declare no conflict of interest.

References

1. Belhaj, K.; Chaparro-Garcia, A.; Kamoun, S.; Nekrasov, V. Plant genome editing made easy: Targeted mutagenesis in model and crop plants using the CRISPR/Cas system. *Plant Methods* **2013**, *9*, 39. [CrossRef] [PubMed]
2. Travis, J. Making the cut. *Science* **2015**, *350*, 1456–1457. [CrossRef] [PubMed]
3. Pickar-Oliver, A.; Gersbach, C.A. The next generation of CRISPR–Cas technologies and applications. *Nat. Rev. Mol. Cell Biol.* **2019**, *20*, 490–507. [CrossRef] [PubMed]
4. Uddin, F.; Rudin, C.M.; Sen, T. CRISPR gene therapy: Applications, limitations, and implications for the future. *Front. Oncol.* **2020**, *10*, 1387. [CrossRef]
5. Jacinto, F.V.; Link, W.; Ferreira, B.I. CRISPR/Cas9-mediated genome editing: From basic research to translational medicine. *J. Cell. Mol. Med.* **2020**, *24*, 3766–3778. [CrossRef]
6. Zhang, M.; Eshraghian, E.A.; Jammal, O.A.; Zhang, Z.; Zhu, X. CRISPR technology: The engine that drives cancer therapy. *Biomed. Pharm.* **2021**, *133*, 111007. [CrossRef]
7. Manghwar, H.; Lindsey, K.; Zhang, X.; Jin, S. CRISPR/Cas system: Recent advances and future prospects for genome editing. *Trends Plant Sci.* **2019**, *24*, 1102–1125. [CrossRef] [PubMed]
8. Ku, H.K.; Ha, S.H. Improving nutritional and functional quality by genome editing of crops: Status and perspectives. *Front. Plant Sci.* **2020**, *11*, 577313. [CrossRef]
9. Barrangou, R.; Fremaux, C.; Deveau, H.; Richards, M.; Boyaval, P.; Moineau, S.; Romero, D.A.; Horvath, F. CRISPR provides acquired resistance against viruses in Prokaryotes. *Science* **2007**, *315*, 1709–1712. [CrossRef]
10. Ishino, Y.; Shinagawa, H.; Makino, K.; Amemura, M.; Nakata, A. Nucleotide sequence of the iap gene, responsible for alkaline phosphatase isozyme conversion in Escherichia coli, and identification of the gene product. *J. Bacteriol.* **1987**, *169*, 5429–5433. [CrossRef]
11. Makarova, K.S.; Wolf, Y.I.; Alkhnbashi, O.S.; Costa, F.; Shah, S.A.; Saunders, S.J.; Barrangou, R.; Brouns, S.J.J.; Charpentier, E.; Haft, D.H.; et al. An updated evolutionary classification of CRISPR–Cas systems. *Nat. Rev. Microbiol.* **2015**, *13*, 722–736. [CrossRef] [PubMed]

12. Mojica, F.; Díez-Villaseñor, C.; García-Martínez, J.; Soria, E. Intervening sequences of regularly spaced prokaryotic repeats derive from foreign genetic elements. *J. Mol. Evol.* **2005**, *60*, 174–182. [CrossRef] [PubMed]
13. Jinek, M.; Chylinski, K.; Fonfara, I.; Hauer, M.; Doudna, J.A.; Charpentier, E. A programmable dual-RNA-guided DNA endonuclease in adaptive bacterial immunity. *Science* **2012**, *337*, 816–821. [CrossRef]
14. Anders, C.; Niewoehner, O.; Duerst, A.; Jinek, M. Structural basis of PAM-dependent target DNA recognition by the Cas9 endonuclease. *Nature* **2014**, *513*, 569–573. [CrossRef] [PubMed]
15. Garneau, J.E.; Dupuis, M.-È.; Villion, M.; Romero, D.A.; Barrangou, R.; Boyaval, P.; Fremaux, C.; Horvath, P.; Magadán, A.H.; Moineau, S. The CRISPR/Cas bacterial immune system cleaves bacteriophage and plasmid DNA. *Nature* **2010**, *468*, 67–71. [CrossRef] [PubMed]
16. Wang, G.; Li, J. Review, analysis, and optimization of the CRISPR *Streptococcus pyogenes* Cas9 system. *Med. Drug Discov.* **2021**, *9*, 100080. [CrossRef]
17. Babu, K.; Amrani, N.; Jiang, W.; Yogesha, S.D.; Nguyen, R.; Qin, P.Z.; Rajan, R. Bridge helix of Cas9 modulates target DNA cleavage and mismatch tolerance. *Biochemistry* **2019**, *58*, 1905–1917. [CrossRef]
18. Horvath, P.; Romero, D.A.; Coûté-Monvoisin, A.-C.; Richards, M.; Deveau, H.; Moineau, S.; Boyaval, P.; Fremaux, C.; Barrangou, R. Diversity, activity, and evolution of CRISPR Loci in streptococcus thermophilus. *J. Bacteriol.* **2008**, *190*, 1401–1412. [CrossRef]
19. Jinek, M.; Jiang, F.; Taylor, D.W.; Sternberg, S.H.; Kaya, E.; Ma, E.; Anders, C.; Hauer, M.; Zhou, K.; Lin, S.; et al. Structures of Cas9 endonucleases reveal RNA-mediated conformational activation. *Science* **2014**, *343*, 1247997. [CrossRef]
20. Jiang, F.; Doudna, J.A. CRISPR–Cas9 Structures and Mechanisms. *Annu. Rev. Biophys.* **2017**, *46*, 505–529. [CrossRef]
21. Berova, N.; Di Bari, L.; Pescitelli, G. Application of electronic circular dichroism in configurational and conformational analysis of organic compounds. *Chem. Soc. Rev.* **2007**, *36*, 914–931. [CrossRef]
22. Pescitelli, G.; Di Bari, L.; Berova, N. Conformational aspects in the studies of organic compounds by electronic circular dichroism. *Chem. Soc. Rev.* **2011**, *40*, 4603–4625. [CrossRef]
23. Brahms, J.; Mommaerts, W.F.H.M. A study of conformation of nucleic acids in solution by means of circular dichroism. *J. Mol. Biol.* **1964**, *10*, 73–88. [CrossRef]
24. Padula, D.; Jurinovich, S.; Di Bari, L.; Mennucci, B. Simulation of electronic circular dichroism of nucleic acids: From the structure to the spectrum. *Chem. A Eur. J.* **2016**, *22*, 17011–17019. [CrossRef]
25. Johnson, W.C. Secondary structure of proteins through circular dichroism spectroscopy. *Annu. Rev. Biophys. Biophys. Chem.* **1988**, *17*, 145–166. [CrossRef] [PubMed]
26. Whitmore, L.; Wallace, B.A. Protein secondary structure analyses from circular dichroism spectroscopy: Methods and reference databases. *Biopolymers* **2008**, *89*, 392–400. [CrossRef] [PubMed]
27. Rogers, D.M.; Jasim, S.B.; Dyer, N.T.; Auvray, F.; Réfrégiers, M.; Hirst, J.D. Electronic circular dichroism spectroscopy of proteins. *Chem* **2019**, *5*, 2751–2774. [CrossRef]
28. Kelly, S.M.; Jess, T.J.; Price, N.C. How to study proteins by circular dichroism. *Biochim. Biophys. Acta Proteins Proteom.* **2005**, *1751*, 119–139. [CrossRef]
29. Sreerama, N.; Woody, R.W. Computation and analysis of protein circular dichroism spectra. *Methods Enzymol.* **2004**, *383*, 318–351. [CrossRef]
30. Lim, Y.; Bak, S.; Sung, K.; Jeong, E.; Lee, S.H.; Kim, J.S.; Bae, S.; Kim, S. Structural roles of guide RNAs in the nuclease activity of Cas9 endonuclease. *Nat. Commun.* **2016**, *7*, 13350. [CrossRef]
31. Wang, M.; Sintim, H.O. Discriminating cyclic from linear nucleotides—CRISPR/Cas-related cyclic hexaadenosine monophosphate as a case study. *Anal. Biochem.* **2019**, *567*, 21–26. [CrossRef] [PubMed]
32. Thavalingam, A.; Cheng, Z.; Garcia, B.; Huang, X.; Megha, S.; Sun, W.; Wang, M.; Harrington, L.; Hwang, S.; Reyes, Y.H.; et al. Inhibition of CRISPR-Cas9 ribonucleoprotein complex assembly by anti-CRISPR AcrIIC2. *Nat. Commun.* **2019**, *10*. [CrossRef] [PubMed]
33. Pawluk, A.; Shah, M.; Mejdani, M.; Calmettes, C.; Moraes, T.F.; Davidson, A.R.; Maxwell, K.L. Disabling a Type I-E CRISPR-Cas Nuclease with a Bacteriophage-Encoded Anti-CRISPR Protein. *MBio* **2017**, *8*, e01751-17. [CrossRef] [PubMed]
34. Banerjee, R.; Sheet, T. Ratio of ellipticities between 192 and 208 nm (R1): An effective electronic circular dichroism parameter for characterization of the helical components of proteins and peptides. *Proteins Struct. Funct. Bioinform.* **2017**, *85*, 1975–1982. [CrossRef]
35. Hirst, J.D.; Colella, K.; Gilbert, A.T.B. Electronic circular dichroism of proteins from first-principles calculations. *J. Phys. Chem. B* **2003**, *107*, 11813–11819. [CrossRef]
36. Nishimasu, H.; Ran, F.A.; Hsu, P.D.; Konermann, S.; Shehata, S.I.; Dohmae, N.; Ishitani, R.; Zhang, F.; Nureki, O. Crystal structure of Cas9 in complex with guide RNA and target DNA. *Cell* **2014**, *156*, 935–949. [CrossRef]
37. Heinig, M.; Frishman, D. STRIDE: A web server for secondary structure assignment from known atomic coordinates of proteins. *Nucleic Acids Res.* **2004**, *32*, W500–W502. [CrossRef]
38. Frishman, D.; Argos, P. Knowledge-based protein secondary structure assignment. *Proteins Struct. Funct. Bioinform.* **1995**, *23*, 566–579. [CrossRef]
39. Zhang, Y.; Sagui, C. Secondary structure assignment for conformationally irregular peptides: Comparison between DSSP, STRIDE and KAKSI. *J. Mol. Graph. Model.* **2015**, *55*, 72–84. [CrossRef]

40. Manning, M.C.; Woody, R.W. Theoretical CD studies of polypeptide helices: Examination of important electronic and geometric factors. *Biopolymers* **1991**, *31*, 569–586. [CrossRef]
41. Enkhbayar, P.; Hikichi, K.; Osaki, M.; Kretsinger, R.H.; Matsushima, N. 310-helices in proteins are parahelices. *Proteins Struct. Funct. Bioinform.* **2006**, *64*, 691–699. [CrossRef] [PubMed]
42. Zhang, L.; Hermans, J. 3_{10} Helix Versus α-Helix: A molecular dynamics study of conformational preferences of aib and alanine. *J. Am. Chem. Soc.* **1994**, *116*, 11915–11921. [CrossRef]
43. Topol, I.A.; Burt, S.K.; Deretey, E.; Tang, T.H.; Perczel, A.; Rashin, A.; Csizmadia, I.G. α- and 3_{10}-Helix Interconversion: A quantum-chemical study on polyalanine systems in the gas phase and in aqueous solvent. *J. Am. Chem. Soc.* **2001**, *123*, 6054–6060. [CrossRef] [PubMed]
44. Toniolo, C.; Polese, A.; Formaggio, F.; Crisma, M.; Kamphuis, J. Circular dichroism spectrum of a peptide 3_{10}-helix. *J. Am. Chem. Soc.* **1996**, *118*, 2744–2745. [CrossRef]
45. Woody, R.W. Electronic circular dichroism of proteins. In *Comprehensive Chiroptical Spectroscopy*; Chapter 14; Berova, N., Polvarapu, P.L., Nakanishi, K., Woody, R.W., Eds.; Wiley: Hoboken, NJ, USA, 2012; Volume 2, pp. 475–497.
46. Migliore, M.; Bonvicini, A.; Tognetti, V.; Guilhaudis, L.; Baaden, M.; Oulyadi, H.; Joubert, L.; Ségalas-Milazzo, I. Characterization of β-turns by electronic circular dichroism spectroscopy: A coupled molecular dynamics and time-dependent density functional theory computational study. *Phys. Chem. Chem. Phys.* **2019**, *22*. [CrossRef]
47. Bandekar, J.; Evans, D.J.; Krimm, S.; Leach, S.J.; Lee, S.; Mcquie, J.R.; Minasian, E.; Nemethy, G.; Pottle, M.S.; Scheraga, H.A.; et al. Conformations of cyclo(L-alanyl-L-alanyl-ε-aminocaproyl) and of cyclo(L-alany1-D-alanyl-ε-aminocaproyl); cyclized dipeptide models for specific types of β-bends. *Int. J. Pept. Protein Res.* **1982**, *19*, 187–205. [CrossRef]
48. Rodger, A. UV absorbance spectroscopy of biological macromolecules. In *Encyclopedia of Biophysics*; Roberts, G.C.K., Ed.; Springer: Berlin/Heidelberg, Germany, 2013; pp. 2714–2718. [CrossRef]
49. Johnson, W.C. Determination of the Conformation of Nucleic Acids by Electronic CD. In *Circular Dichroism and the Conformational Analysis of Biomolecules*; Fasman, G.D., Ed.; Springer: Boston, MA, USA, 1996; pp. 433–468.
50. Baranowski, D.S.; Kotkowiak, W.; Kierzek, R.; Pasternak, A. Hybridization Properties of RNA Containing 8-Methoxyguanosine and 8-Benzyloxyguanosine. *PLoS ONE* **2015**, *10*, e0137674. [CrossRef]
51. Langkjær, N.; Pasternak, A.; Wengel, J. UNA (unlocked nucleic acid): A flexible RNA mimic that allows engineering of nucleic acid duplex stability. *Bioorg. Med. Chem.* **2009**, *17*, 5420–5425. [CrossRef]
52. Yamada, M.; Watanabe, Y.; Gootenberg, J.S.; Hirano, H.; Ran, F.A.; Nakane, T.; Ishitani, R.; Zhang, F.; Nishimasu, H.; Nureki, O. Crystal structure of the minimal Cas9 from campylobacter jejuni reveals the molecular diversity in the CRISPR-Cas9 systems. *Mol. Cell* **2017**, *65*, 1109–1121. [CrossRef]
53. Davis, A.R.; Kirkpatrick, C.C.; Znosko, B.M. Structural characterization of naturally occurring RNA single mismatches. *Nucleic Acids Res.* **2011**, *39*, 1081–1094. [CrossRef]
54. Johnson, W.C., Jr.; Tinoco, I., Jr. Circular dichroism of polynucleotides: A simple theory. *Biopolymers* **1969**, *7*, 727–749. [CrossRef]
55. Gray, D.M. Circular Dichroism of Protein–Nucleic Acid Interactions. In *Comprehensive Chiroptical Spectroscopy*; Chapter 19; Berova, N., Polvarapu, P.L., Nakanishi, K., Woody, R.W., Eds.; Wiley: Hoboken, NJ, USA, 2012; Volume 2, pp. 615–633.
56. Gray, D.M. Circular dichroism of protein-nucleic acid interactions. In *Circular Dichroism and the Conformational Analysis of Biomolecules*; Fasman, G.D., Ed.; Springer: Boston, MA, USA, 1996; pp. 469–500.
57. Jiang, F.; Zhou, K.; Ma, L.; Gressel, S.; Doudna, J.A. A Cas9–guide RNA complex preorganized for target DNA recognition. *Science* **2015**, *348*, 1477–1481. [CrossRef] [PubMed]
58. Ciubotaru, M.; Ptaszek, L.M.; Baker, G.A.; Baker, S.N.; Bright, F.V.; Schatz, D.G. RAG1-DNA Binding in V(D)J Recombination: Specificity and DNA-induced conformational changes revealed by fluorescence and CD spectroscopy. *J. Biol. Chem.* **2003**, *278*, 5584–5596. [CrossRef] [PubMed]
59. Kaluzhny, D.; Laufman, O.; Timofeev, E.; Borisova, O.; Manor, H.; Shchyolkina, A. Conformational Changes Induced in the Human Protein Translin and in the Single-stranded Oligodeoxynucleotides d(GT) 12 and d(TTAGGG) 5 Upon Binding of These Oligodeoxynucleotides by Translin. *J. Biomol. Struct. Dyn.* **2006**, *23*, 257–265. [CrossRef]
60. Hjelm, R.P.; Chih, R.; Huang, C. sheet of RNA and nonhistone proteins to the circular dichroism spectrum of chromatin. *Biochemistry* **1975**, *14*, 1682–1688. [CrossRef]
61. Oda, Y.; Iwai, S.; Ohtsuka, E.; Ishikawa, M.; Ikehara, M.; Nakamura, H. Binding of nucleic acids to E. coli RNase HI observed by NMR and CD spectroscopy. *Nucleic Acids Res.* **1993**, *21*, 4690–4695. [CrossRef]
62. Klimek-Chodacka, M.; Oleszkiewicz, T.; Lowder, L.G.; Qi, Y.; Baranski, R. Efficient CRISPR/Cas9-based genome editing in carrot cells. *Plant Cell Rep.* **2018**, *37*, 575–586. [CrossRef]
63. Mehravar, M.; Shirazi, A.; Mehrazar, M.M.; Nazari, M. In Vitro Pre-validation of Gene Editing by CRISPR/Cas9 Ribonucleoprotein. *Avicenna J. Med. Biotechnol.* **2019**, *11*, 259–263.
64. Ramlee, M.K.; Yan, T.; Cheung, A.M.S.; Chuah, C.T.H.; Li, S. High-throughput genotyping of CRISPR/Cas9-mediated mutants using fluorescent PCR-capillary gel electrophoresis. *Sci. Rep.* **2015**, *5*, 15587. [CrossRef]
65. Kocak, D.D.; Josephs, E.A.; Bhandarkar, V.; Adkar, S.S.; Kwon, J.B.; Gersbach, C.A. Increasing the specificity of CRISPR systems with engineered RNA secondary structures. *Nat. Biotechnol.* **2019**, *37*, 657–666. [CrossRef]

66. Creutzburg, S.C.A.; Wu, W.Y.; Mohanraju, P.; Swartjes, T.; Alkan, F.; Gorodkin, J.; Staals, R.H.J.; van der Oost, J. Good guide, bad guide: Spacer sequence-dependent cleavage efficiency of Cas12a. *Nucleic Acids Res.* **2020**, *48*, 3228–3243. [CrossRef] [PubMed]
67. Kim, N.; Kim, H.K.; Lee, S.; Seo, J.H.; Choi, J.W.; Park, J.; Min, S.; Yoon, S.; Cho, S.R.; Kim, H.H. Prediction of the sequence-specific cleavage activity of Cas9 variants. *Nat. Biotechnol.* **2020**, *38*, 1328–1336. [CrossRef] [PubMed]
68. Xiao, A.; Cheng, Z.; Kong, L.; Zhu, Z.; Lin, S.; Gao, G.; Zhang, B. CasOT: A genome-wide Cas9/gRNA off-target searching tool. *Bioinformatics* **2014**, *30*, 1180–1182. [CrossRef] [PubMed]

MDPI
St. Alban-Anlage 66
4052 Basel
Switzerland
Tel. +41 61 683 77 34
Fax +41 61 302 89 18
www.mdpi.com

International Journal of Molecular Sciences Editorial Office
E-mail: ijms@mdpi.com
www.mdpi.com/journal/ijms

www.ingramcontent.com/pod-product-compliance
Lightning Source LLC
Chambersburg PA
CBHW040743250926
51850CB00007BB/1392
* 9 7 8 3 0 3 6 5 6 5 7 9 8 *